ADVANCED FIRE ADMINISTRATION

Randy R. Bruegman

Fire Chief, Anaheim Fire Department
Anaheim, CA

Pearson

Boston   Columbus   Indianapolis   New York   San Francisco   Upper Saddle River   Amsterdam
Cape Town   Dubai   London   Madrid   Milan   Munich   Paris   Montreal   Toronto   Delhi
Mexico City   São Paulo   Sydney   Hong Kong   Seoul   Singapore   Taipei   Tokyo

Library of Congress Cataloging-in-Publication Data
Bruegman, Randy R.
    Advanced fire administration / Randy R. Bruegman.
      p. cm.
    ISBN-13: 978-0-13-502830-8
    ISBN-10: 0-13-502830-2
    1. Fire departments—United States—Management.   2. Fire departments—Management.
      I. Title.
    TH9503.B78 2011
    363.37068—dc22

                                                        2010054108

Publisher: Julie Levin Alexander
Publisher's Assistant: Regina Bruno
Executive Editor: Marlene McHugh Pratt
Senior Acquisitions Editor: Stephen Smith
Associate Editor: Monica Moosang
Developmental Editor: Jill Rembetski, iD8 Publishing Services
Editorial Assistant: Samantha Sheehan
Director of Marketing: David Gesell
Senior Marketing Manager: Katrin Beacom
Marketing Specialist: Michael Sirinides
Marketing Assistant: Crystal Gonzalez
Managing Production Editor: Patrick Walsh
Production Liaison: Julie Boddorf
Production Editor: Karen Fortgang, bookworks publishing services
Senior Media Editor: Amy Peltier
Media Project Manager: Lorena Cerisano
Manufacturing Manager: Alan Fischer
Creative Director: Jayne Conte
Interior Designer: Wanda España
Cover Designer: Bruce Kenselaar
Cover Photos: Joel L. Aranaz, Fresno Fire Department; Rick Black, Center for Public Safety Excellence; and Tony Escobedo,
Fresno Fire Department
Composition: Aptara®, Inc.
Printing and Binding: R.R. Donnelley/Willard
Cover Printer: Coral Graphics

10 9 8 7 6 5 4 3 2 1

www.pearsonhighered.com

ISBN 13: 978-0-13-502830-8
ISBN 10:   0-13-502830-2

DEDICATION

A special thanks to my wife, Susan, for her continued support and understanding of these special projects that I undertake! To my executive assistant, Maria Campos; without her diligent work and ability to decipher my handwriting, this book would have never been completed. To three chiefs who have been mentors to me: Chief Ray Picard, Chief Ron Coleman, and the late Charlie Rule. Thanks for the vision.

Be safe.

Chief Bruegman

Chief Randy R. Bruegman began his career as a volunteer firefighter in Nebraska. He was hired as a firefighter in Fort Collins, Colorado, where he served in a variety of positions including engineer, inspector, lieutenant, captain, and battalion chief. He has served as the fire chief for the City of Campbell, California; the Village of Hoffman Estates, Illinois; Clackamas County Fire District No. 1 (Oregon); the City of Fresno, California; and currently is the fire chief for the City of Anaheim, California.

He is a noted author and lecturer on such topics as Leadership and Managing Change in the Fire Service and a contributing author of fire service literature including the following: *Fire Attack: Strategy and Tactics of Initial Company Response*; *Making a Difference: The Fire Officer's Role*; *Surviving Haz-Mat, Haz-Mat for First Responders*; and *The Volunteer Firefighter, A Breed Apart*. He has authored four books for Brady: *Exceeding Customer Expectations*, *The Chief Officer: A Symbol Is a Promise*, *Fire Administration I*, and the new *Advanced Fire Administration*.

He served on the Accreditation Task Force for the International Association of Fire Chiefs for eight years, six as vice chairman. Chief Bruegman served as chairman of the Commission of Fire Accreditation International (CFAI) for three years, 1997–2000. He is a Chief Fire Officer Designate, holds member status with the Institute of Fire Engineers, and is a member of NFPA. He serves on the editorial advisory board for *Fire Chief* magazine and as a member of the NFPA 1710 Technical Committee. He was elected as president of the International Association of Fire Chiefs in August of 2002 and has served as the president of the board of directors of the Center for Public Safety Excellence since 2003. Chief Bruegman was selected as *Fire Chief* magazine's Career Fire Chief of the Year in 2009. He has an associate's degree in fire science, a bachelor's degree in business, and a master's degree in management.

CONTENTS

Chapter 7 Fiscal Management in Difficult Times 214

Chapter 9 Firefighter Health and Safety: Making a Difference 304

Chapter 10 National and International Trends in the Fire Service 337

As I began to think of the context for an advanced fire administration textbook, I realized the complex challenges for leaders of the fire service today are significant.

If one takes a retrospective look at where the fire service has evolved from in the 20th century, it is truly quite remarkable how far the profession has progressed; yet, at the same time it still faces many of the same issues our colleagues did at the turn of the last century. At the turn of the 20th century, our counterparts were confronted with rapidly growing cities. Our society was becoming an industrial world power with westward growth to levels not seen previously. An influx of immigrants arrived from all over the world wanting to take advantage and be part of what America was to become. Our cities exploded with wealth and new challenges including high-rise buildings, manufacturing plants, congestion, and the creation of a new urban landscape.

The stage was set for the continued westward expansion of this country's growth, which was to create a whole new set of issues respective to urban–wildland interface. The immigration that began in the early 20th century has continued, as America is now a mosaic of different cultures from throughout the world. This mosaic is reflected today in the workforce, which demands that the leaders of today and those of tomorrow incorporate different approaches to how they lead and manage their organizations.

Over the last hundred years, the world has seen innovations that surpass all that has come in the preceding thousands of years that humans have lived on Earth. In the fire service, technology began to improve the capabilities of the profession with the introduction of crude self-contained breathing apparatus, aerial ladders, chemical hose wagons, new fog-pattern hose nozzles, and of course Henry Parmelee's new fusible link automatic fire sprinkler. During the past 20 years, the innovation and technology that have been introduced into the fire service are nothing short of amazing. One has to wonder what the next 25 to 50 years will bring, and how it will impact the way the fire service conducts its business and delivers services in the future.

When I entered the service, starting as a volunteer firefighter with absolutely no training other than attending a social meeting to meet other volunteers, I was issued a helmet and turnout coat, given a pager, and told that when the pager went off to report to the firehouse, where I would be told what to do. There was very little training or education other than what was received when we responded to an emergency. I know from speaking to many of my colleagues at the time and in later years that this was the norm in many organizations.

The evolution of the fire service can be seen not only in how firefighters are brought in and prepared to enter the service but also in every aspect of how to lead and manage fire organizations today. What was done 30 years ago, whether from a training perspective or about how organizations were managed, would in many cases be found unacceptable today. The level of professionalism brought to the fire service over the course of the last 30 to 40 years, the leadership that has risen up in the fire service, the focus on performance measurement, and community risk and deployment are so much more sophisticated than they were even just a few years ago.

One has to begin to wonder what the fire chief in 2025 will be faced with. What are the leading challenges that organizations today must address to prepare their communities, personnel, and the organization as a whole to meet the demands of a changing society? What will the improvements in technology and the effect of a more robust preventive application have on the service? Think about how far the fire service has come in the use of technology in just the last 25 years.

I remember when I became part of a HazMat team in 1980. There were no commercial plug-and-patch kits available to purchase. In fact, we had to design and make our own. The gas monitors were very simplistic, difficult to use, and their reliability was questionable. The available suits were largely developed for the military, and although we adapted applications for the fire service, they were cumbersome at best. Today, this is not the case as many of the HazMat teams are traveling laboratories with tools and equipment that were not even thought of 25 years ago.

In every segment of the fire profession, one can see the impact technological improvements have had on how business is conducted: whether it is in hazardous materials (HazMat), urban search and rescue (USAR), geographical information systems (GIS), information technology, radio systems, and the list goes on. Compare each to where the fire service was just 25 years ago, and one has to ask, what will it look like in the future? It is really exciting to think about. As much as the fire service has changed over the course of the last two decades, I do not think we can imagine how much technology will impact and change the fire service over the course of the next 25 years.

Another significant shift we have seen through the course of the last two decades has been the expectations of our customer, the taxpayer. As competition for local governmental funds has turned out to be more aggressive, it has become critical for the profession to articulate the services being delivered in respect to measurable performance that the community can understand. Not only is the fire service there as an insurance policy in the community to combat fires, mitigate disasters, and respond to medical emergencies, but in many cases it becomes a part of the economic development engine for the community. Good fire protection is an essential ingredient to attracting quality business to a community.

The service's thought process has changed from a focus on the number of fires being extinguished and the amount of property saved to looking at how the service integrates to a more global view with that of other departments and the community as a whole. How does the fire service impact such things as economic development; the psyche of the resident in feeling safe; and overall community risk? The fire service has changed in regard to community outreach and stakeholder investments not only within our own organization but externally as well. The engagement of the community in the development of what our strategic vision is to be in the future is a departure from what our practice was in the recent past.

As fire officials begin to contemplate what the 21st century may look like for the fire service, it is a picture that cannot be painted today, as it will be ever-changing. Yet, it points to the importance of the ability for leadership to understand that managing change within the fire service will become a fundamental element to our success in the future. The changes of community expectation, technology, workforce, higher level of community expectation, and different view on community risk management, will all have a significant impact on what the fire service profession will evolve into in the future.

One of the biggest challenges leaders will face in managing change in the 21st century will be how the fire service will continue to refine resource deployment and services as community risk changes. It appears the fire service has been successful in obtaining sprinklers to be required in residential structures and most business occupancies. Over the course of the next hundred years, this will have a dramatic impact upon the level of fire activity and the type of fires that are being experienced today. With this change will come a different deployment system than that currently being used. Not only may it change the resource distribution patterns in the future, but it may change the type of equipment currently being purchased and the tactics utilized by the profession. Although it is impossible to predict what the future deployment patterns may be, it is safe to say that as communities begin to fully incorporate the technology such as fixed building fire protection systems into their management of community risk, the fire service will evolve from being largely reactive to emergencies to being proactive in prevention. This in itself will change how the profession deploys and delivers its services.

This does not mean there will be fewer firefighters in the future. It does mean our firefighters in 2050 will probably experience a very different job than the firefighters of today. The reality is this same thing is true for the firefighter of 1950 and the firefighter of today. If you take a look at the 1950 fire department, it was primarily doing one thing—responding to fires. Very little effort was spent with community involvement and public education. At that time, medical emergencies were usually handled by the local funeral home or in some cases a private ambulance provider. The fire service was not even thinking of hazardous materials response, USAR, or other specialties currently being engaged in today, nor were there standards, laws, and best practices in place to guide the delivery of services and the management of organizations.

What firefighters were doing and how they trained and operated 50 years ago was vastly different from today. One could surmise the same will hold true for those of us in the service today and those of the service who will be writing books 50 years from now looking back at what was happening at the turn of the 21st century. Much will continue to change and evolve over the course of the next century. The only difference is the rapidness of this change will be much greater in the future than it has been in the past.

As we begin to contemplate leading the fire service in the 21st century, several points are critical to the success of leaders who will help make this transition successful. The fire administrator of the future will have to be one who engages a community approach and is open to options in fire and emergency service delivery. Developing the framework for articulating those various options to community and policy leaders will become increasingly important. There must be a comprehensive community risk management process in place so the fire service can measure the impacts technology and community education are having on the risk of the community. Leaders will have to become much more adept at organizational planning with five-year snapshots of where the organization needs to be and what it will take to continue to move forward. At the same time, meeting the changes in the dynamic of the communities being served will be essential. Leaders have to be focused on leading change within their organizations and developing a culture that is accepting of the change process. Managing in this changing environment will require future fire service leaders to be focused on a capital management process to ensure the infrastructure is current and ready to meet the demands of the community.

For fire service leaders, the future will focus on system performance measurement: benchmarking, exploring quality improvement opportunities, and measuring the department's performance against that of others. The fire service must be on a continual quest for best practices, and fire service leaders must understand often these are not within their own organizations. A focus on the cost benefit of local tax dollars that are spent on fire and emergency services, with an underlying theme, is the community receiving the best results for the money expended. *Return on investment* is a term common in the private sector and one relevant in government today, especially in light of the current economic climate. Future leaders will need to be focused on the trends in the fire service not only within the United States but also internationally. Progressive leaders will determine how to utilize such information within their own organizations to improve services and enhance the safety of the public and firefighting force.

Leaders have to begin to think how they are going to create their own opportunities for the future within their communities. Although they are very similar in the type of services being delivered throughout the world, our communities, jurisdictions, and the governance structure firefighters work under are often significantly different. Developing a framework to have these critical discussions locally will continue to be one of the challenges fire service leaders must face in the future.

This text will focus on those areas having the most impact on our profession, on our departments, and on each of us as leaders. The text is divided into 11 chapters:

Chapter 1, Our Past Is a Window to Our Future
Chapter 2, A Community Approach to Fire Protection and Emergency Services
Chapter 3, The Leadership of Change
Chapter 4, Planning for Organizational Achievement
Chapter 5, An Integrated Approach to Community Risk Management
Chapter 6, Managing in a Changing Environment: Measuring System Performance
Chapter 7, Fiscal Management in Difficult Times
Chapter 8, Effective Personnel Management
Chapter 9, Firefighter Health and Safety: Making a Difference
Chapter 10, National and International Trends in the Fire Service
Chapter 11, The Future: The Legacy of Leadership

In authoring a text, one cannot help but to share some of one's own experiences and opinions. To provide a balance of perspective and thought, I solicited the help of many of the nation's fire and city manager experts to share their insights as well. These include the following:

Debra H. Amesqua, Chief, City of Madison Fire Department, Wisconsin
Max Baker, County Administrator, Los Alamos, New Mexico
Jim Broman, Fire Chief, Lacey, Washington
Alan Brunacini, Fire Chief, Retired, Phoenix, Arizona
Steve Carter, City Manager, Champaign, Illinois
Jim Crawford, Vision 20/20, Vancouver, Washington
I. David Daniels, Fire Chief, City of Woodinville, Washington
James Gigsby, Assistant City Manager, Roanoke, Virginia
Billy Goldfedder, Battalion Chief, Loveland, Ohio
Randy Knight, City Manager, City of Winter Park, Florida

Bob Lamkey Director of Public Safety, Sedgwick County, Kansas
David Limardi, City Manager, City of Highland Park, Illinois
Bruce Moeller, City Manager, Sunrise, Florida
Brian Nakamura, City Manager, Hemet, California
Alan W. Perdue, Director of Emergency Services, Guilford County, North Carolina
Jeffrey A. Pomeranz, City Manager, West Des Moines, Iowa
James (Jay) P. Reardon, Fire Chief (Ret.), President and CEO, MABAS-Illinois
Christopher P. Riley, Fire Chief, City of Pueblo, Colorado
Jon Ruiz, City Manager, Eugene, Oregon
Ronald J. Siarnicki, Executive Director, National Fallen Firefighters Foundation
Janet Wilmoth, Editor, *Fire Chief* magazine
Shana Yelverton, City Manager, City of Southlake, Texas

For each chapter the fire service leaders have offered their insights and perspectives, which can be found on the MyFireKit Web site. A special stand-alone section has been created on MyFireKit to highlight the insights of the city/county managers.

I would also like to recognize the work done by the National Fire Protection Association Fire Analysis and Research Division, including that of John Hall, Jr., Michael J. Karter, Jr., and Marty Ahrens. Pieces of their work on national fire data and reports for the U.S. Fire Administration can be found within this text. Also I would like to recognize Phil Schaenman of TriData for his team's work on the Global Concepts in Residential Fire Safety Study.

My thanks to each for taking the time to provide their thoughts in this text.

As this text looks at how far the fire service has come in the last hundred years and contemplates how far it will move forward in the next hundred years, it truly is an exciting time to be in the fire service. The last hundred years have seen a marked evolution within the fire service in the application of the fire service as a profession. The next hundred years will experience a revolution in the way service is delivered, how resources are managed, the need for expanded community engagement, and very possibly the roles the fire service will undertake in local government. It is and will be an exciting time, one in which we will not realize our full success unless we have the ability to lead the fire service into the 21st century.

It is my hope you will find this text useful in your pursuit of leading a quality organization, in your opportunities to promote, and in shaping the future of the fire service.

Chief Bruegman

PEARSON
myfirekit

As an added bonus, *Advanced Fire Administration* features a **myfirekit**, which provides a one-stop shop for online chapter support materials and resources. You can prepare for class and exams with multiple-choice and matching questions, weblinks, study aids, and more! To access **myfirekit** for this text, please visit **www.bradybooks.com** and click on **mybradykit**.

ACKNOWLEDGMENTS

I would like to thank and acknowledge the following reviewers:

John P. Alexander
Adjunct Instructor, Connecticut Fire Academy
Captain, Hazardville Fire Department
Enfield, Connecticut

Andrew Byrnes
Faculty, Emergency Services Department
Utah Valley University
Provo, Utah

Dane Carley
Captain
Fargo Fire Department
Fargo, North Dakota

Michael Falese
Assistant Fire Chief
Bartlett Fire Protection District
Bartlett, Illinois

J. Robert Griffin
Adjunct Instructor
A-B Tech
Asheville, North Carolina

William M. Kramer, Ph.D.
Director of Fire Science Education
University of Cincinnati
Cincinnati, Ohio

Ronald R. Lowe
Captain
Palm Beach County Fire Rescue
Training and Safety Division
Palm Beach, Florida

Mark Martin
Fire Chief
Perry Township Fire Department
Stark County, Ohio

Joseph Mercieri
Fire Chief/Adjunct Instructor
Numerous New Hampshire fire stations
Connecticut Fire Academy
Vermont Fire Training School
Littleton Fire Rescue
Littleton, New Hampshire

John Moschella
Adjunct Professor, Fire Science and Administration
Anna Maria College
Paxton, Massachusetts

Donnie P. West, Jr., MS, EMT-P, EFO, CFO
Assistant Fire Chief/Fire Marshal
Center Point Fire Department
Center Point, Alabama

The following grid outlines the Fire and Emergency Service Administration course requirements and where specific content can be located within this text:

Course Requirements	1	2	3	4	5	6	7	8	9	10	11
Define and discuss the elements of effective departmental organization.	X	X		X	X	X	X	X	X	X	X
Classify what training and skills are needed to establish departmental organization.		X	X				X	X	X	X	X
Analyze the value of a community-related approach to risk reduction.		X		X	X	X			X	X	X
Outline the priorities of a budget planning document while anticipating the diverse needs of the community.			X			X		X			
Assess the importance of positively influencing community leaders by demonstrating effective leadership.		X	X	X		X	X			X	X
Analyze the concept of change and the need to be aware of future trends in fire management.	X	X	X	X	X	X	X	X	X	X	X
Report on the importance of communications technology, fire service networks, and the Internet when conducting problem-solving analysis and managing trends.	X		X	X		X		X	X	X	
Develop a clear understanding of the national assessment models and their respective approaches to certification.			X	X	X	X				X	

CHAPTER **1**

Our Past Is a Window to Our Future

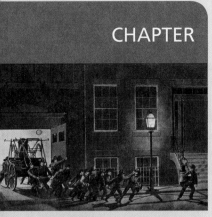

(Source: Currier & Ives, National Archives)

KEY TERMS

conflagration, *p. 3*

cost estimates of human loss, *p. 11*

cost of the fire service, *p. 6*

direct economic losses, *p. 8*

indirect economic losses, *p. 10*

metropolitan fire problem, *p. 15*

net insurance cost, *p. 10*

rural fire problem, *p. 16*

Stockdale Paradox, *p. 29*

synergy, *p. 27*

total cost of fire, *p. 6*

OBJECTIVES

After completing this chapter, you should be able to:

- Describe how history has impacted today's fire service.
- Describe the total cost of fire in the United States.
- Explain the difference of the fire problem between rural and metropolitan areas in the United States.
- Explain the fire rates in residential occupancies.
- Explain how the changes occurring in the profession today may impact the future.

PEARSON

> For additional review and practice tests, visit www.bradybooks.com and click on MyBradyKit to access book-specific resources for this text!

Introduction

Then: The shouts of "fire, fire" would ring out and able men would run from their homes and businesses, buckets in hand, to join the fight. The church bells would sound and all within earshot would move to the column of smoke rising in the distance.

At the first sign of fire, the sound of wooden rattles used by the fire warden would echo through the night and volunteer firefighters would rush to their neighborhood firehouses and pull their equipment to the scene.

The bells in the firehouse would toll 3-5-3, denoting the alarm box location of the fire and sending horse-drawn equipment and firefighters to the scene of the emergency.

The Klaxon would be loud enough to be heard throughout the neighborhood, and firefighters would climb onto their new motorized fire pumper, responding with speed and efficiency.

Now: The lights in the station come on as the alarm sounds and an automated voice states: "Engine 1, Engine 3, Truck 4, Battalion 1, residential structure 123 Main Street, time out 1536." With a computer keystroke, information is transmitted in a millisecond to multiple stations and onboard mobile computers, providing incident information, maps showing the best response route, structure floor plans, and a list of the on-site hazards.

The sounding of the alarm has evolved over time, mirroring the evolution of the fire service. This rich history created the fabric from which today's profession has been cut.

History Impacts the Fire Service

Throughout history, fire has been an agent of progress, a means of protection and survival, and a devastating force of destruction. Although the fire service has made great progress in gaining the knowledge and developing the technology needed to harness fire and use it to its advantage, too frequently firefighters still face out-of-control fire as it assumes its destructive role.[1]

The fire service's evolution is evident in every aspect of how organizations are led and managed today. The elements of professionalism brought to the fire service over the last 30 to 40 years, including hiring practices, improvement in the quality of leadership, focus on performance measurement, development of community risk and deployment modeling, and use of technology, are more sophisticated than they were even a few years ago. Yet fire continues to have not only a tremendous impact on the health and welfare of U.S. citizens but also a significant economic impact on society.

Looking back in history, Caesar Augustus established one of the first known organized firefighting forces in Rome around 100 B.C. Augustus created a force of approximately 600 men belonging to the *familia publica*, servants of commonwealth, called the Corps of Vigiles. These men were stationed near the city gates for the specific purpose of fighting fire.[2] Fast-forward to the founding of America, and the challenges of fire suppression can be seen in the early settlements as the country began to expand west, grow in population, and become industrialized.

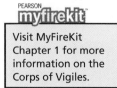

Visit MyFireKit Chapter 1 for more information on the Corps of Vigiles.

During the last three centuries, America has experienced more dangerous and tragic fires than any other nation in history.[3] Each tragedy has helped shape today's fire service profession. The first permanent settlement in Jamestown, Virginia, was destroyed by fire on January 7, 1608. As a result of this first American **conflagration,** most of the colonists' lodgings and provisions were lost. Hence, during that winter many died from exposure and hunger.[4]

As more European settlers arrived and founded colonies, a common and repeated experience was fire loss. Many of these fire events were catastrophic because of unsafe living conditions, no effective legislation to minimize the hazards of fire, and no readily available water supply for firefighting. Settlers used the most expedient means to create temporary, livable shelters to protect themselves against the harsh winter elements; hence, safety was not a priority. Within these buildings, the fireplaces were made of bricks and mud or clay. The chimneys, situated to vent smoke and heat, were made of "daub and wattle," brush and stalks, or wood planks, covered by a heavy mixture of mud or clay. The open cooking utensils of earthenware or metal in constant use were sometimes filled with highly combustible stock, such as beer, whiskey, brandy, wax, and gunpowder. In addition, the small dwellings, barns, and storage buildings in these first settlements were built close together, usually within the protection of a nearby fort as a defense against Indian attacks. Such close groupings enabled most fires to spread quickly from building to building.[5]

The use of wood as a predominant construction material in cities during this time undoubtedly contributed to establishing a poor fire record. Lack of enforcement of the fire prevention building laws also factored in the great fire loss.[6]

Then certain individuals used their authority and influence to campaign for fire prevention. Governor Peter Stuyvesant of New Amsterdam (later to become New York City) was truly one of those pioneers. Recognizing the hazards of combustible chimneys, in 1648, Governor Stuyvesant succeeded in having laws passed prohibiting the construction of chimneys made of wood or plaster. These were the first of many such laws that American colonists passed in an effort to prevent fire disasters.[7] In 1658, Stuyvesant appointed eight young men to roam the streets of New Amsterdam at night and watch for fires. These men, clad in long capes, carried wooden rattles, sounded the alarm, and are considered to be the first step in organized firefighting in America.[8] Essentially, this group functioned as the first municipal alarm system. Another contributing factor to such catastrophic fire loss was the lack of available water. Groundwater systems were not in use, creating the challenge of transporting water to the fire through various means to effect extinguishment.

Faced with the problem of a fast-growing colony, the general assembly established the Volunteer Fire Department of New York in December 1737. After the Revolutionary War, the department was reorganized and renamed the Fire Department of the City of New York. This volunteer fire department continued to protect the lives and property of the city's citizens until after the end of the Civil War in 1865, when it was replaced by a paid fire department.[9]

Benjamin Franklin, one of the Founding Fathers of America, worked as an early champion of fire prevention, being the most notable among famous Americans who helped to shape the fire service. His writings in the *Pennsylvania Gazette* increased the public's safety awareness and formed opinions of the importance of fire prevention. Franklin coined one of his most familiar epigrams, "An ounce of prevention is worth a pound of cure," in a letter warning Philadelphia citizens about the hazards of carrying burning firebrands or coals in a full shovel from one room to another.

conflagration
■ A large destructive fire.

He recommended the use of a closed warming pan. Franklin also campaigned for clean chimneys and chimney sweeps. In addition to his contributions to fire prevention, Franklin founded the volunteer Union Fire Company in 1736.[10]

The 19th century was an exciting time for the fire service as the United States expanded westward and thousands of immigrants swelled its population. As industry grew in cities to the west, several significant fires occurred, which influenced the course of the American fire service into the 20th and 21st centuries.[11]

On October 8, 1871, the *Chicago Tribune*'s front page read, "Firefighters Prepare for Fall and Winter Fires." Later that day fire broke out in the vicinity of a barn owned by Mr. and Mrs. Patrick O'Leary at 137 Dekoven Street (now the site of the Chicago Fire Training Academy). Tradition blames the start of the conflagration on Mrs. O'Leary's cow. The cow allegedly kicked over a kerosene lantern, igniting the fire. This report may be questioned, but the extent of the losses is well known. The fire burned for 27 hours, destroyed 17,500 buildings, killed nearly 300 people, and left approximately 100,000 homeless.[12]

A small lumbering community in Peshtigo, Wisconsin, also suffered one of the most significant fires in U.S. history in terms of lives lost. On October 8, 1871, the same day as the Great Chicago Fire, a forest fire in the area surrounding Peshtigo developed into a firestorm that swept through the town, destroying every building except for one house under construction. Nearly 800 people died.[13]

Another major conflagration was the Great Boston Fire of 1872, which destroyed over 776 buildings in the business district, covering approximately one square mile. The fire claimed 13 lives, including two firefighters, with losses estimated at $75 million. As a result of the great conflagrations of the 1800s, the National Board of Fire Underwriters (now the Insurance Services Office [ISO] Fire Insurance Grading Schedule) was developed, which is still in use today.[14]

The year 1873 marked the culmination of nearly a decade of significant trial for the American fire service. Many of the large cities had recently changed from using volunteer firefighters to employing full-paid professionals. In addition, the sociological and psychological changes resulting from the end of the Civil War had an impact on many fire departments as returning volunteers were war weary and many had fought on different sides of the war. Several fire chiefs had experienced conflagrations in their own cities, and many others anticipated facing similar ordeals in their own communities. Consequently, these chiefs corresponded with one another, discussing methods of meeting their mutual challenges. Boston's Chief John S. Damrell, as the most persistent of these correspondents, suggested creating a national association or a convention of fire chiefs to exchange ideas and discuss the possibilities of new equipment to facilitate the workings of fire departments. On August 20, 1873, the following notice was sent to all known fire departments:

> The occurrence of large fires in the principal cities of the United States has demonstrated the fact that upon the chief officers of our fire departments, there rests a fearful responsibility, both in the protection of life and property. That this responsibility shall be intelligently assumed, and that the very best means may be employed both in the prevention and extinguishment of fires, you are earnestly invited to attend a national convention in the United States to be held in the city of Baltimore on Monday, October 20, for consultation and advisement.[15]

Chief Damrell (see Figure 1.1) stated at the 25th annual conference of the International Association of Fire Engineers (1897): "The primary objective of the

FIGURE 1.1 Chief Damrell
(Source: A Centennial History, IAFC)

convention was to jointly consider the conditions of organized fire departments and to discover, if possible, the causes leading to the fearful conflagrations of Portland, Chicago, and Boston."[16]

PEARSON
myfirekit
Visit MyFireKit Chapter 1 for a list of Key Events in History.

THE LESSONS OF EXPERIENCE

Many aspects of the current fire profession reflect lessons learned from past events. The list of significant fires that occurred solely in the United States and Canada and resulted in significant loss of life is extensive. Each fire produced a historic change, whether in organizing fire-control forces, changing fireground operations to create a safer environment for firefighting personnel, improving the viability of a community, instituting new technology in the detection or suppression of fire, or developing new fire and building codes.

Civic response to fires in colonial times brought into being the first legislation, the first volunteer firefighting action with buckets and hooks, the beginning of fire patrols and mutual-aid societies, the establishment of the first public fire departments, and the institution of paid fire departments with authority to inspect properties and extend fire prevention.

Improvements were made in the methods used to apply water to the fireground—from handheld buckets, to handheld pumps, to horse-drawn pumps, to steam-powered pumps, to self-propelled apparatus, to centrifugal pumps, to above-ground application by aircraft. Improvements to water systems also were developed—from wells, to cisterns, to wooden piping, to metal and concrete asbestos underground water mains, to the complex and highly controlled pumping and hydrant systems serving modern cities and towns.

Changes occurred in the type of building materials used—from grass and thatched roofs, to mud or clay covering, to brick and cement, to reinforced concrete, to the current fire-resistant structures having modern fire detection, alarm, and extinguishing systems. Changes also occurred in the use of power for energy—from wood, to coal, to steam, to dammed rivers and lakes, to wind power, to oil, to electricity, to nuclear power, to hydrothermal power, to solar power.

New extinguishing agents and methods were developed—from dirt to water, to water additives, to other liquids and chemicals, to gases, to portable extinguishers, to wheeled extinguishers, to fixed extinguishing systems, to aircraft dropping extinguishing agents on wildland fires. New ways to analyze the causes of fires were developed—educating the public and specific groups who could take responsible actions, granting statutory authority to municipal fire departments, and establishing fire protection associations; and new national standards and building codes on construction practices and fire protection were written.

Surely, from what has been learned from fighting the significant fires of the last 150 years, controlling fire is within the collective grasp of the profession. Considering the many problems of protecting against fire and the tragedies and devastation wrought by fire in past years, it is imperative to remember and act on the lessons learned.[17] Firefighting has arrived at a historic crossroads. The question arises: What does the future hold for this profession?

The Cost of Fire in the United States

Leaders of today's fire service must have a clear understanding of the critical issues surrounding the nation's fire problem. One such significant issue is calculation of fire loss. As a profession, the fire service has done itself and its communities a disservice by not capturing the **total cost of fire,** choosing instead to report only direct property losses from fire. If the general public and decision makers were to realize the full impact of fire on society, the fire profession would be in a better position to raise their awareness of the economic magnitude of the fire problem, to obtain needed resources, to adopt stricter safety and building codes, and to exercise a more significant role at the national level regarding fire-related issues.

Over the last three decades, many attempts have been made to estimate the total annual cost of fire in the United States, which is far greater than solely the value of the property destroyed by fire. It includes the **cost of the fire service;** the cost of fire protection built into buildings and equipment; the cost of fire insurance overhead; the annual maintenance of fire protection systems; the many indirect costs (such as business interruptions, medical expenses, and temporary lodging); the cost to society from the injuries and deaths caused by fire; the cost of maintaining government and private fire-related organizations; and a myriad of other related costs that translate into a significant economic impact.

The total cost of fire ranges from $130 billion to $250 billion annually, depending on the definition of losses and costs and the estimation methodology used. Some believe that disasters such as fires stimulate the economy and that the economic multiplier effects of recovery activities, such as rebuilding and redevelopment, may offset some of the costs. Despite potential offsets, it is important to estimate these costs, specifically the cost of fire, as a measure of losses incurred and of expenditures caused by these losses, money that society might prefer to see spent elsewhere. Although the focus of this effort of determining costs has been on national statistics, the application of total cost modeling would also be beneficial at the local level.

Even though the costs associated with fire are often underestimated and overlooked, they are significant, accounting for up to 2.5 percent of the gross domestic product. This alone should encourage a national strategic plan incorporating all levels of government to determine the actions necessary to reduce the costs considerably. It also is useful to compare the fire problem with other problems

total cost of fire
■ The total cost of fire includes the cost of fire services; the cost of fire protection built into buildings and equipment; the cost of fire insurance overhead; the annual maintenance of fire protection systems; the many indirect costs (such as business interruptions, medical expenses, and temporary lodging); the cost to society of the injuries and deaths caused by fire; the cost of maintaining government and private fire-related organizations; and a myriad of other related costs that translate into a significant economic impact.

cost of the fire service
■ Cost to the public and government to maintain a "ready army" of firefighters, equipment, and stations.

facing the nation to apply some rationale in the allocation of resources. Finding a common denominator for these comparisons is difficult. The two most common ways that fire service decision makers track changes over time are by (1) cost of direct fire loss and (2) number of casualties. The question becomes whether this analysis is sufficient to gain a better understanding of the real-world effects of fire and to stimulate prevention and mitigation efforts. Unfortunately, using only two of the economic indicators in respect to the total cost of fire is not sufficient.

Estimating and tracking trends in the magnitude of the main components of the total cost of fire is a vital step in assisting with fire protection policy trade-offs. Moreover, the apparent and hidden costs of fire protection need to be compared with the losses averted and incurred. Eventually, a quantitative understanding of how investments in protection affect total costs needs to be established so that policy makers at all levels of government have a clear understanding of the impacts of the money invested in fire protection. To do so, an understanding of the total cost of fire is an important factor.

ESTIMATES OF THE COST OF FIRE

A team of fire protection engineering students from Worcester Polytechnic Institute (WPI) in 1978 undertook one of the first attempts in the last 30 years to estimate the total cost of fire in the United States. This estimate was based on initial thoughts and concepts regarding how to quantify the total cost of fire.[18]

Economist William Meade made a further effort to estimate the total cost of fire in 1991 for the National Institute of Standards and Technology (NIST).[19] Drawing upon the WPI study and relying on in-depth discussions with experts in a variety of fields, including many from the fire profession, Meade expanded his research into the wide range of areas in which fire protection is built into society. The inclusion of fire stops, fire doors, sprinkler systems, or alarm systems provides examples of built-in protection factors. Each adds cost to the building during construction and for annual maintenance. Using a broader definition of costs, including an estimate of the value of volunteer firefighters but resisting placing a value on the human loss equivalent, Meade estimated the total cost of fire in the United States at the time to be between $92 billion and $139 billion. This study sparked interest in not only the magnitude of the cost of fire but also how he determined and used these costs.

John Hall of the National Fire Protection Association (NFPA) made a series of estimates of the total cost of fire built on the WPI and Meade estimates, further developing the methodology.[20] Notably, his improvements segregate cost estimates into those that are more solid, based on verifiable data sources and inputs, and those that are less well-defined, based on broad understandings of the cost.

The cost of core elements of fire protection, those more solid costs, has grown larger. In 1980, Hall conservatively calculated the cost of these elements to be $28.3 billion. By 2002, the estimate tripled to $84.9 billion. Almost half of this increase can be credited to inflation, with the remainder largely attributable to increases in fire service costs. In terms of 2002 dollars, Hall estimated the overall net increase in these core costs to be $23.1 billion, or 37 percent. The remaining costs included a $36.7 billion component, based on Meade's 1991 estimate, for expenses associated with fire protection built into equipment, fire maintenance, and other areas; $39.0 billion for human loss; and a range of $47.0 billion to $90.0 billion as an estimate of the value of volunteer firefighters' time.

In a related 1994 study, Phil Schaenman of TriData Corporation (Arlington, Virginia) expanded the U.S. work on the total cost of fire by applying it to the Canadian fire experience.[21] The Canadian National Research Council had performed original research on estimating the incremental cost of fire protection in structures. This research was incorporated into the analysis of the total cost of fire in Canada, solidifying the basis for some of the Canadian estimates. As a point of comparison, these estimates, converted and inflated to 2002 U.S. dollars, adjusted for the U.S. population, and using U.S. estimates for direct and human losses, yield a range of $119 billion to $159 billion.

RELATED FIRE COST STUDIES

A variety of studies have estimated different aspects of the cost of fire or, more specifically, the losses resulting from fire. For example, as part of the Fire Safe Cigarette Act, the Consumer Product Safety Commission developed estimates on the societal costs of cigarette-ignited fires, valued at the time (1992) at approximately $4 billion.[22] Much of this particular study focused on the economic costs resulting from burns and anoxia and was factored as part of the initial analysis.

To understand the impact of its fire programs, in 2004 the NIST issued a research report conducted by the TriData Corporation on the cost of firefighter injuries.[23] Based on methods applied from economic studies, the estimated cost of addressing firefighter injuries and the efforts to prevent them ranged from $2.8 billion to $7.8 billion. This later research incorporated newly published injury cost methodologies from the National Highway Traffic Safety Administration. When this cost methodology is applied to all fire casualties, the resulting estimate of the cost of human loss is $30 billion.

In 2003, the NFPA, as part of its annual report on fire loss in the United States, produced statistically derived estimates on the **direct economic losses** from fire. These estimates of the direct cost of fire in terms of property and human loss are widely used to determine estimates for the total cost of fire.

direct economic losses
■ The monetary loss experienced by what was burned or damaged by fires.

TOTAL COST COMPONENTS

Although the different approaches to computing the cost of fire may group subsets of the costs differently, there are generally six main cost components: direct economic losses, cost of the fire service, cost associated with equipment and buildings, net fire insurance, indirect economic losses, and estimates of human loss. When people refer to the cost of fire, the most common statistic quoted is direct economic loss from fire—what was burned or damaged by fires. As has been discussed previously, direct losses constitute only a small fraction, approximately 5 to 6 percent, of the total cost of fire (see Table 1.1). Other categories of losses and costs must be taken into account to fully estimate the total cost of fire.

Direct Economic Losses

When estimating direct economic losses, the following questions often arise: What costs are reflected for property loss? Are these insurance estimates of actual loss or replacement cost? Are they fire department loss estimates? How are uninsured losses accounted for? What is the extent of the unreported losses? It is safe

TABLE 1.1 | Total Cost of Fire Components

COST COMPONENT	CONTRIBUTION TO TOTAL COST (%)
Direct economic losses	5–6
Cost of the fire service	30–45
Cost associated with equipment and buildings	25–35
Net fire insurance	5
Indirect economic losses	5–15
Estimates of human loss	10–15

Source: Patricia Frazier, *Total Cost of Fire in the United States*, Fire Protection Engineering Archives, retrieved September 13, 2010, from http://www.fpemag.com/archives/article.asp?issue_id=18&i=53

to say there is no standardized format, methodology, or legal mandate to do so, which raises the question, how much more or less would economic loss be if calculations were standardized?

Cost of the Fire Service

The cost of fires often is the primary focus, whereas the cost to the public and government to maintain a "ready army" of firefighters, equipment, and stations is forgotten (see Figure 1.2). The cost of the fire service includes the cost of local paid and volunteer departments (although the latter includes an estimate of the equivalent cost of volunteer time, which is not a direct cost), forest fire management, and capital outlays for equipment. It may not include infrastructure improvements

FIGURE 1.2 Fire Station *(Courtesy of Steve Derenia)*

necessary to accommodate firefighting (for example, increased water main capacity or road improvements to accommodate the fire equipment) and generally does not consist of the cost of federal and private fire brigades. This cost component can vary, depending on the inclusion of the volunteer time by as much as 30 to 45 percent of the total cost.[24]

Cost Associated with Buildings and Equipment

The cost of business operations affected by fire considerations is the cost associated with equipment and buildings. It also includes the training of employees in fire safety, cost of special transportation for flammables, use of special containers for flammables, and work time lost evacuating buildings from false alarms. The cost of fixed fire protection systems, the maintenance of these systems, and the maintenance of the structure to the appropriate code also need to be considered. The cost associated with equipment and buildings is 25 to 35 percent of the total cost.

Net Fire Insurance

Net insurance cost, or insurance overhead, is the cost paid by the public for insurance, less what is returned to the public in payments for insured losses, which are accounted for as part of direct losses. The issues include how to separate fire-related insurance from other hazards and peril insurance, and how to accurately estimate the overhead and profit paid for fire-related insurance. Insurance costs are generally less than 5 percent of the overall cost of fire.

Indirect Economic Losses

Indirect economic losses from fire include business interruptions (see Figure 1.3), costs of temporary lodging, taxes, loss of market share, legal expenses, and many other categories related to the effects of having a fire.

net insurance cost
■ Paid by the public for insurance, less what is returned to the public in payments for insured losses, which are accounted for as part of direct losses.

indirect economic losses
■ Business interruptions, costs of temporary lodging, taxes, loss of market share, legal expenses, and many other categories related to the effects of having a fire.

FIGURE 1.3 Burned Out Business *(Courtesy of CPSE, Rick Black)*

This is one of the most difficult categories to estimate with any degree of accuracy, partly because of the numerous and disparate categories that comprise this loss group and partly because disruptions in one category may be offset by transfers to others. Depending on what is included and how the costs are derived, this category can range from 5 to 15 percent of the total cost of fire. Many of these costs require in-depth study to determine, which most agencies have not yet undertaken as part of their data collection and analysis.

Estimates of Human Loss

The cost of deaths and injuries to society also must be considered. Part of these costs is fairly easy to determine, such as the cost of medical treatment (see Figure 1.4), funeral expenses, and time lost from work. Other costs, such as **cost estimates of human loss,** are more difficult to ascertain and include the value of a life and pain and suffering.

To estimate these aspects of loss, cost studies in other fields—such as medical, health care, worker disability, statistics on lost time worked—often are examined. The total of these attributed costs is about 10 to 15 percent of the estimated total cost of fire.

THE CHALLENGE OF REDUCING COSTS

The fire protection engineering community is faced with the remarkable challenge of reducing costs. For example, using Hall's 2003 estimates, in 2002 between $144 billion and $187 billion was spent (total cost less losses and insurance) to avert an untold number of fires and their resulting losses. A substantial portion of this estimated cost ($61 billion) was spent on built-in fire protection such as sprinklers. Even with this amount spent on prevention, combined economic and

cost estimates of human loss
■ Human losses as a result of fire were estimated at $42.5 billion in 2007 using formulas developed by the U.S. Consumer Product Safety Commission (CPSC). Economists at the CPSC have an ongoing program of studies of injury costs. Periodically, it reviews the literature, including its own studies, and selects dollar values for use in policy analysis of fire safety and other hazard analysis. It must be acknowledged that no amount of money can compensate for the loss of a loved one.

human losses of $52 billion still occurred. This fact raises several confounding questions, not only about the resources spent to prevent fires and protect against them but also about how to maximize the benefits of these resources.

The foremost questions are: How much more would it cost to reduce the current losses? Would the increased cost to prevent losses be worth more or less than the losses? Using Hall's 2003 estimates, if 40 percent more could be spent on built-in protection, which then resulted in a 50 percent reduction in losses, a near zero sum would be achieved, spending as much as was saved—a situation that might not necessarily make monetary sense but could achieve the valued societal goals of saving lives and property.

A corollary question is: With increasing improvements in fire safety products and construction materials, should more be spent now to achieve lower losses in the future? This question is critically important because the non-loss components (for example, built-in protection) are the driving elements in determining the total cost of fire. Whereas investments such as residential sprinklers increase the short-term cost, the long-term cost savings may be substantial, not only in reducing fire loss but also in protecting civilians and firefighting personnel.

Spending $150 billion to $200 billion might avert losses in addition to containing the losses currently experienced. How many incidents would be averted? How could this number be determined? What would the losses be from these averted fires? It could be that the value of averted losses would more than compensate for what was spent in fire protection services.

In examining the costs of fire, it is vital to recognize that the role of the fire service has changed remarkably in the last 20 years. The fire service no longer comprises "just" firefighters; it now includes "first responders" who provide an increasingly wide array of services. Although fire protection remains its primary job, the fire service has expanded its role in not only delivering emergency medical services (EMS) but also responding to and mitigating hazardous materials incidents and, most recently, increasing its responsibilities in homeland security. The costs discussed thus far do not account for these increased responsibilities. Requiring built-in protection to reduce the incidence of fire will help to increase the agency's efficiency (and hence reduce costs); first responders will have the added benefit of being able to perform other equally important functions, such as home fire inspections/education. Innovative designs and cost-efficient solutions along with the increased use of installation of sprinklers and fire-graded materials in residential structures surely will impact future costs significantly.

Although the total cost of fire is a major national problem in terms of its economic impact, it often is not viewed as such by elected officials and the general public. The same may also hold true at the local level. Recognizing the importance of considering each major cost element will facilitate creating a more comprehensive fire protection policy in the future. For example, the size of the fire service affects losses, the extent of built-in protection and engineering affects the cost of fire services and the losses incurred, and factors such as the number and size of losses affect net insurance costs. Changes in incremental costs of all the major components of the total cost of fire covered previously should be analyzed, with the results being given more consideration in setting priorities for the use of resources.

Current estimates of the total cost of fire include a large component for the fire service. With the burgeoning roles of the fire service, the cost of the fire service protection and other first-responder roles may need to be separated from its other services, or it becomes necessary to assess the overall costs of providing first-responder services where fire protection is just one facet of the services delivered. Further refinement may be needed in this area.[25]

The U.S. Fire Experience

In assessing the total cost of fire, those countries experiencing the best fire statistics, at least regarding loss, have put more effort into fire prevention education and the incorporation of building safety technology. This is apparent in Japan and Europe, where a societal awareness of fire prevention exists. This still is not the case in the United States; other countries instill a higher degree of personal responsibility and an acceptance of more restrictive codes relating to their building environments. Built-in fire protection systems, along with comprehensive programs on fire prevention education addressing attitudes about fire, have proven to be more effective in saving lives than has trying to extinguish the fires once they have started.

Although it is difficult to change societal attitudes and instill the importance of individual responsibility, the fire service must understand these are key factors in reducing overall fire loss. Therefore, it is vital to design educational efforts to influence these attitudes and to promote personal responsibility related to fire and safety. Two distinct challenges for the fire service include (1) overcoming the mind-set that tolerates a cavalier attitude toward fire safety by the general public and (2) overcoming the suppression-only mentality found in many departments. In developing future strategies, the fire service must revisit and focus on the following:

- Categorizing civilian fire deaths and injuries occurring in residences, targeting specific age-groups by need
- Funding fire prevention activities at a level to make them effective and realistic
- Redefining the local fire prevention needs based on the needs of the individual community
- Changing attitudes in the civilian population regarding the acceptance of fires in the United States
- Changing the culture of the fire service suppression forces to accept and fully engage in fire prevention efforts
- Increasing the cooperation and focus of all government and industry agencies regarding fire prevention
- Continuing to improve engineering practices
- Using and requiring fixed fire protection systems in all structures
- Understanding the cultural diversity of individuals in communities and finding more effective ways to communicate fire and life safety issues to them
- Using performance-based codes to promote a comprehensive system for evaluating and designing system performance measures

FIGURE 1.5 Residential Structure Fire
(Courtesy of Tony Escobedo)

OVERVIEW OF 2008 U.S. FIRE EXPERIENCE

In 2008, U.S. fire departments responded to an estimated 1,451,500 fires. These fires resulted in 3,320 civilian fire fatalities, 16,705 civilian fire injuries, and a direct property loss of approximately $15.5 billion[26] (see Figure 1.5).

A civilian fire death occurred every 158 minutes and a civilian fire injury every 31 minutes in 2008. Home fires caused 2,755, or 83 percent, of the civilian fire deaths. Fires accounted for approximately 6 percent of the total emergency calls; 9 percent of the calls were false alarms; and 62 percent of the calls were for aid such as EMS.[27]

In 2000, the Federal Emergency Management Agency (FEMA) issued a report titled *America at Risk,* which indicated the frequency and severity of fires in America do not result from a lack of knowledge of the causes, means of prevention, or methods of suppression. A fire problem exists because the nation has failed to apply and fund known loss reduction strategies adequately. As a consequence, the United States has one of the highest fire losses in terms of both frequency and total losses of any modern technological society.

Reviewing a profile of fire in the United States from 1999 to 2008 shows millions of fires, thousands of deaths and injuries, and billions of dollars lost, thus emphasizing the scope of the U.S. fire problem. During this time period, an average of 1.6 million fires resulted in an estimated dollar loss of over $11.6 billion each year. Over 3,500 Americans lost their lives and another 19,000 were injured annually as the result of fire. These averages do not reflect the events of September 11, 2001 (see Table 1.2).

Clearly there is a dual interest in reducing U.S. fire losses, including a reduction of civilian and firefighter casualties and implementation of ways to achieve more effective fire safety. There is need for product innovations and additional educational programs, which can improve fire safety and reduce the impact on human losses, thereby reducing total fire cost. Need also exists for improved methods and models for calculating fire performance and costs, so the advantages of different choices can be considered and judged more comprehensively.[28]

TABLE 1.2 — U.S. Fire Department Responses

NUMBER OF FIRES, DEATHS, INJURIES, AND DOLLAR LOSS IN THE UNITED STATES, 1999–2008

YEAR	FIRES	DEATHS	INJURIES	DIRECT DOLLAR LOSS IN MILLIONS
1999	1,823,000	3,570	21,875	$10,024
2000	1,708,000	4,045	22,350	11,207
2001*	1,734,500	3,745	20,300	10,583
2001**	—	2,451	800	33,440
2002	1,687,500	3,380	18,425	10,337
2003	1,584,500	3,925	18,125	12,307
2004	1,550,500	3,900	17,875	9,795
2005	1,602,000	3,675	17,925	10,672
2006	1,642,500	3,245	16,400	11,307
2007	1,557,500	3,430	17,675	14,639
2008	1,451,500	3,320	16,705	15,478

* Excludes the events of September 11, 2001.

**These estimates reflect the number of deaths, injuries, and dollar loss directly related to the events of September 11, 2001.

Source: U.S. Fire Administration, FEMA, 2009.

It is evident that a significant opportunity for improvement still exists for the U.S. fire service.

Fire in America

Although the aggregate statistics provide a broad overview of the U.S. fire experience, to understand America's fire problem fully, the focus must be on discovering what drives these numbers. A good place to begin is an overview of the research conducted on the behavioral characteristics, societal factors, and related impacts of rural versus **metropolitan fire problems.** The fire may burn fundamentally the same way in every community, but the fire problem and the related risk factors can be substantially different. The same modeling or methodology cannot be applied universally with the same expected outcome. Just as communities are unique so, too, are their risk factors for fire.

metropolitan fire problem
■ The fire problem found in incorporated or unincorporated areas with a population of over 200,000 people and/or a population density of over 3,000 people per square mile.

FIRES IN RURAL AMERICA

In spring 2004, the U.S. Fire Administration (USFA) partnered with the NFPA in a cooperative project entitled *Mitigation of the Rural Fire Problem*, which examined what could be done to reduce the high death rate from fires in rural U.S. communities. Rural communities, defined by the U.S. Census Bureau as communities with a population of less than 2,500, had a fire death rate twice the national average.

rural fire problem

■ The fire problem unique to rural America, typically incorporated or unincorporated areas with a total population less than 10,000, or with a population density of fewer than 1,000 people per square mile.

The objectives of this project were to (1) conduct research on behaviors and other factors contributing to the **rural fire problem**; (2) identify mitigation programs, technologies, and strategies to address these problems; and (3) propose actions that the USFA Public Education Division could take to better implement programs in rural communities. Research sources included a review of published literature, original statistical analysis, and information from national technical experts who had worked with the NFPA.

Characteristics of Rural America

Rural America has several distinct characteristics. One is its population, typically in incorporated or unincorporated areas with a total population less than 10,000, or with a population density of fewer than 1,000 people per square mile. A second distinct characteristic is the frequent separation of communities and residents from one another. This separation causes longer travel times in emergency situations, which translates into loss of not only economies of scale but also the concentration of tax base. The potential market for any business requiring travel—either for delivery of the product to the home or for the resident to acquire the product at a store—is smaller, which affects operating costs and revenues. Print media are thus affected, which impacts the quantity and ease of communication within and to a rural community. This directly affects how fire protection is provided and the amount of revenue needed to do so.

A third relevant characteristic of rural America is the greater likelihood of poverty. For example, in 2003, the percentage of the population below the poverty level was 12.1 percent inside of metropolitan areas and 14.2 percent outside.[29] A smaller income means fewer resources. This often translates into a greater need for safety—in the form of safer (often newer) products and devices designed to provide safety (such as smoke alarms)—and a reduced ability to fill this need without outside help.

In the United States, communities with the highest risk are the smallest rural communities and the largest metropolitan areas. Although rural communities and large cities do not share the characteristics of population and the problems of distance and separation, both have a higher likelihood of poverty and other socioeconomic factors that increase fire risk.[30]

Characteristics of the Rural Fire Problem

The distribution of types of fire incident is roughly the same in rural and non-rural areas, including the proportion of reported outdoor fires that do not involve either a structure or a vehicle. However, the cause profiles in rural areas are different for both outdoor fires and residential structure fires. In 2007, 45 percent of the rural outside fires were caused by open flame, 16 percent by arson, and 9 percent by natural causes. In contrast, arson caused 44 percent of the non-rural outside fires. Rural residential fires more likely were caused by heating equipment, occurred in properties without smoke alarms, and had flame damage extending to the entire structure (see Table 1.3).

Table 1.3 is a testimony of how ineffective education and enforcement have been in the area of residential fires in rural districts. Also note that flame damage extended to the entire structure in 29 percent of rural residential structure fires but in only 17 percent of such incidents in non-rural areas. This may indicate the extended travel times in rural areas and/or lack of responding resources.[31]

TABLE 1.3 — Causes of Residential Fires

RURAL RESIDENTIAL FIRES		FATAL RURAL RESIDENTIAL FIRES		NON-RURAL FATAL RESIDENTIAL FIRES	
CAUSE	PERCENTAGE	CAUSE	PERCENTAGE	CAUSE	PERCENTAGE
Heating	36	Heating	36	Smoking	28
Cooking	13	Smoking	23	Arson	17
Electrical	12	Electrical	17	Heating	12

Source: U.S. Fire Administration, *Mitigation of the Rural Fire Problem*, December 2007, retrieved September 1, 2010, from http://www.usfa.dhs.gov/downloads/pdf/publications/MitigationRuralFireProblem.pdf

Rural Differences by Region

Rural communities and their fire problems differ considerably among U.S. regions. From 2000 to 2004, the rural fire death rates per million population were as follows:

- 29.0 for the South
- 28.2 for the West
- 27.0 for the Northeast
- 22.8 for the north-central region

Historically, the South, sometimes referred to as the Southeast, has had the highest fire death rate and the highest rural fire death rate. However, the gap between the South and the other regions has shrunk recently, and neither of these rates has been the highest every year in the same region. The South also is the most populous region (more than one-third of total U.S. population) and contains nearly half the total U.S. rural population; therefore, one may assume that the rural fire problem in the United States is due to the fire experienced in the South. However, this can be misleading because, in most years, rural communities have the highest fire incident and fire death rates in every one of the four U.S. regions.

Because the South has half of the nation's rural population as well as a higher rural fire incident rate and a higher rural fire death rate than other regions, it dominates total national rural fire statistics. For example, the heightened share of rural fire deaths involving heating equipment is as much a southern phenomenon as it is a rural phenomenon. Of the four regions, the South has the most consistently mild and shorter heating season. Therefore, poorer households in the South are the ones finding it most feasible to use space heating exclusively, with documentation repeatedly showing more fire is caused by space heating compared with central heating.

Often, fire risk is correlated with poverty, and rural areas tend to be poorer than non-rural areas throughout the country. The gap in poverty rates between rural and non-rural areas is largest in the South and is associated with African Americans more than in other regions. The rural South has not only most of the nation's rural African Americans but also most of the nation's poor rural African Americans. Overall, rural populations tend to have a lower percentage of African Americans than do non-rural populations, but this is not the case in the South.

Historically, the West has had the lowest overall fire death rates yet not always the lowest rural fire death rates. This region contains some distinctive and important subgroups of the rural population, which require specific attention because they may not be representative of the typical rural population of these four regions. The West includes Native American communities, migrant worker communities, and Mexican border communities (sometimes called *colonias*). As with the African American population of the rural South, so it is with the Native American and Mexican American populations of the West; each region's poor rural populations have a distinctive character.

Rural America also is mistakenly characterized primarily as agricultural. Nonfarm rural dwellers outnumber rural farmers by approximately 18 to 1; however, this imbalance varies by region. Much of America's agricultural activity and farms is still located in the north-central region.

In addition, poor housing quality is more prevalent in the rural South. The South has the highest proportion of housing units in manufactured homes—12 percent versus 3 to 7 percent in the other regions. Historically, this has been a factor in the elevated fire death rate in the South because manufactured homes formerly had a higher fire death rate than conventional "stick-built" homes or apartments. However, most manufactured homes currently in use were built after 1976 when the U.S. Department of Housing and Urban Development (HUD) introduced construction requirements, making manufactured homes a lower-risk environment. This factor may partly explain why fire death rates in the South are no longer consistently much higher than those rates in other regions of the United States.[32]

The Rural Fire Service

Just as rural communities are distinctive in their makeup, so is the rural fire service. The principal distinguishing characteristics of the rural fire service follow:

1. Most (99.5 percent) of the rural fire departments are primarily volunteer.
2. Rural fire departments are more likely to have insufficient companies and personnel to meet national guidelines for effective response because travel distances and travel times to fires and other emergencies tend to be longer because of the low population density of such communities.
3. Rural fire departments are less likely to have adequate equipment.
4. Rural firefighters are less likely to have adequate training.
5. Rural fire departments are less likely to conduct educational programs on fire prevention and code enforcement.

Rural Fire Safety Programs

Rural communities have another special challenge: how to distribute the programs they offer effectively. Programs that operate through the adoption of local codes and regulations need to address gaps in code enforcement, which tend to be greater in rural communities. Additionally, programs delivered via mass media whose target population includes rural communities need to be realistic about usage and access rates of those communities. Mass media may be less established in rural communities, making it a less effective tool for fire safety programs. Under these conditions, door-to-door distribution may prove to be an effective option. Programs delivered in person consider the lower density of the rural populations by going to the customers versus having them come to you. With these greater distances in rural communities, attention should be paid as to whether and how the program is able to reach all or most of its target audience.[33]

THE METROPOLITAN FIRE PROBLEM

A 2007 NFPA report focused on the need to reach high-risk populations more effectively in cities with populations of 250,000 or more.[34] Large city fire departments have particular challenges that involve conducting fire prevention programs with limited resources, staffing smoke alarm installation programs, and working in high-crime areas where bars on windows may block exits. Additional challenges include leveraging relationships with large city-wide institutions serving preschoolers, school-age children, or older adults; reaching multicultural communities using their own languages; and getting community members to focus on fire safety when other issues may seem more compelling.

Characteristics of Metropolitan Areas

As part of this initiative, the NFPA study focused on cities large enough to be eligible for membership in the Metropolitan Fire Chiefs Association, which is limited to fire departments having a minimum staff of 400 fully paid career firefighters. This loosely corresponds to those cities having a population of 250,000 or more. Of the nation's 64 metropolitan areas, the 14 with the highest population density included 10 of the 23 oldest cities; a third of these cities' housing units were built before 1940, or more than 50 years before the 1990 census from which this study took the data. In the research considering metropolitan areas, there exists a distinct difference between the skyscraper-filled cities of the Northeast and the newer cities in the rest of the country. The latter more accurately represent the greater number of cities defined as a metropolitan city.

Metropolitan areas can be divided into three groups. One group, mostly in the northern tier and particularly in the Northeast, consists of older cities with small geographic areas built skyward because developers could not build out. This has created a high-density environment favoring apartments but without significant crowding in the individual housing units. A second group, mostly in the southern tier, has rapidly added people and housing, creating communities that may or may not have high community density but often have considerable crowding in the housing units. Finally, the largest group of metropolitan areas does not reflect either of these patterns.

Characteristics of metropolitan population centers can impact the demand placed on the local fire service. These general features include a higher percentage of the population living below the poverty level, higher rates of violence and property crimes, and a higher percentage of people speaking a primary language other than English at home. Cities with the highest percentages of housing units containing more than one person per room also were the most affected by immigration, especially emigration from Central and South America. The demand for service for this population group is higher due to their living in less well-maintained buildings, which experience a higher incidence of fire; the use of first responders as the primary vehicle to access health care; higher arson rates to cover up other crimes; and cultural difference in respect to fire safety.

The second fire service needs assessment conducted by the NFPA and the U.S. Fire Administration in 2005[35] found metropolitan areas are more likely to

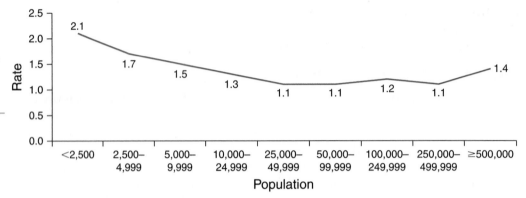

FIGURE 1.6 Home Structure Fires per Thousand Population by Size of Community, 2002–2006 *(Source: NFPA Urban Fire Safety Project, November 2008)*

have more robust fire safety and fire prevention programs than are smaller communities.

The percentage of departments protecting communities in which fire code inspections are conducted was 100 percent for metropolitan areas (or comparably sized communities) compared with 75 percent for all departments combined. Fire safety initiatives may depend as much on the resources and capabilities of the school system as on those of the fire department. The resources available to metro city schools do not appear to be significantly less than those available to U.S. schools in general.[36]

Fire Rates in Metropolitan Areas

The largest metropolitan areas have home fire incident rates somewhat higher than those of smaller cities but lower than those of rural communities. Studies of metropolitan area fire rates by census tract have established that many—perhaps most—cities have an "inner city" area in which the fire rates are higher than in the rest of the city. The 2007 Urban Fire Safety project, a metropolitan fire safety report, noted fire rates per 1,000 population in smaller communities may be higher than in metropolitan areas; however, an important caveat to this statistic is fire rates found in core inner cities may be higher than the average rates in rural communities because of the unique characteristics previously discussed (see Figure 1.6).

Most fire deaths occur in home fires. Fire incident rates relative to population are lower in one- and two-family dwellings than in apartments, whereas fire death rates relative to population are higher in one- and two-family dwellings.

Metropolitan areas have a lower share of owner-occupied housing units. One may speculate that renters are less consistent in exhibiting fire-safe behaviors than owners. However, linear regression statistical studies of factors correlated with fire incident rates show a weak relationship for owner/renter status. Age of the buildings shows an even weaker relationship, whereas vacancy rates for the community show a stronger relationship.[37]

Fire Causes in Metropolitan Areas

The most dramatic difference in the profile of fire cause between metropolitan areas and other communities lies in the number of intentionally set fires. Larger metropolitan areas have roughly four times the number of intentionally set fires than do small towns with populations of 2,500 to 4,999.

Home heating fires generally are not a problem in metropolitan areas but may be an issue in any neighborhood where residents use space heaters extensively, typically for cost-related reasons. Therefore, the incidences of home heating fires would be worth examining on a city-by-city basis as part of setting local priorities for fire safety programs.

Electrical distribution equipment is the only major cause of fire for which statistical studies show a strong correlation with the age of the building or system. This suggests electrical distribution equipment may be worthy of priority attention in older cities but less so in metropolitan areas in general.[38]

Challenges of Large Urban Areas

The 2007 study probed for the challenges fire departments face when reaching out to the community, especially among high-risk groups, in metropolitan areas. Representatives of the fire departments and organizations funded by smoke alarm installation grants from the Centers for Disease Control and Prevention (CDC) discussed many of the same issues.

The challenges raised by this research and found in the literature review for this study include dealing with high-risk populations, children, child education, older adults, people with disabilities, race, immigrant population, homeless population, and crime.[39]

To understand the fire problem in America and the factors that impact community risk, it is vital to gain familiarity with the distinct factors that contribute to risk of fire and associated losses in the specific rural and metropolitan area environments. The fire service works to be diligent in all facets of response and prevention, which include education, enforcement, and engineering.

PEARSON
myfirekit

Visit MyFireKit Chapter 1, The Fire Problem Among High-Risk Populations.

However, the true impact the fire profession makes on the fire problem and associated cost will be limited in certain situations. When buildings and their systems are engineered incorrectly in the first place, it creates unnecessary hazards for the life of the building. Also, budget constraints in many departments often force leaders to choose between funding prevention and education programs, and keeping firefighters for response; many organizations choose the latter. As statistics indicate, without a systems approach, education, enforcement, and engineering will not solve America's fire problem. Suppression without education and prevention (enforcement) is limited. Reactive education and prevention without adequate fire suppression capability, although proactive, limits the firefighting options once an incident occurs. Unfortunately, with the fiscal constraints faced by the fire service and the escalating cost to provide services, no easy solution exists.

One area that has seen dramatic evolution over the last 30 years is the fire engineering/code adoption arena. The cumbersome and often political nature of the code consensus process can dramatically lengthen the time between when a problem is acknowledged and when strategies are written and included in the code. The fire profession has witnessed significant changes in building and fire regulations over the course of the last century.

America's Fire Problem

As the statistics indicate, when it comes to America's fire problem, it is in the home (see Table 1.4).

TABLE 1.4	Estimates of 2008 Fires, Civilian Injuries, and Property Loss in the United States		
	ESTIMATED	**RANGE**[1]	**PERCENT CHANGE FROM 2007**
Number of fires	1,451,500	1,424,000	−6.8*
Number of civilian deaths	3,320	3,000 to 3,640	−3.2
Number of civilian injuries	16,705	15,755 to 17,655	−5.5
Property loss[2]	$15,478,000,000[3]	$15,188,000,000 to $15,768,000,000	+5.7*

Note: The estimates are based on data reported to the NFPA by fire departments, that responded to the 2008 National Fire Experience Survey.

[1]These are 95 percent confidence intervals.

[2]This includes overall direct property loss to contents, structures, vehicles, machinery, vegetation, and anything else involved in a fire. It does not include indirect losses. No adjustment was made for inflation in the year-to-year comparison.

[3]This figure includes the California wildfires in 2008 with an estimated property loss of $1,400,000,000. Loss by specific property type was not available.

*Change was statistically significant at the .01 level.

Source: Michael J. Karter, Jr., *Fire Loss in the United States 2008*, NFPA Fire Analysis and Research Division (Quincy, MA: National Fire Protection Association, August 2009).

The following estimates are based on data reported to the NFPA by fire departments responding to the 2008 National Fire Experience Survey and include the direct property loss to contents, structures, vehicles, machinery, vegetation, and anything else involved in a fire. They do not include indirect losses, such as business interruption or temporary shelter costs. No adjustment was made for inflation in the year-to-year comparison (see Tables 1.5 and 1.6).

The fire service has made significant improvements in reducing civilian fire deaths in the home (see Figure 1.7). The 2008 civilian fire death toll of 3,320 was 3 percent lower than the 3,430 reported in 2005 and 55 percent lower than the 7,395 reported in 1977. However, much more needs to be accomplished to reduce or eliminate this risk.

With home fire deaths still accounting for 2,755 fire deaths, or approximately 84 percent of all civilian deaths, fire safety initiatives must focus on those major causes of fire in the home. Attending to these sources of fire will be a key element in reducing the overall fire death toll (see Figure 1.8).

Figure 1.9 shows how the risks of death and injury in home structure fires by age-group vary, depending on the highest risk groups for both death and injury.

FIRE PREVENTION STRATEGIES

The five major fire prevention strategies follow:

1. Far-reaching public education on fire safety is needed, specifically how to prevent fires and how to avoid serious injury or death should fire occur. Information on the common causes of fatal home fires should continue to be

| TABLE 1.5 | Estimates of 2008 Structure Fires and Property Loss by Property Use |

	STRUCTURE FIRES		PROPERTY LOSS	
PROPERTY USE	ESTIMATE	PERCENT CHANGE FROM 2007	ESTIMATE	PERCENT CHANGE FROM 2007
California wildfires 2008			$1,400,000,000	
Public assembly	14,000	−3.5	518,000,000	+4.0
Educational	6,000	−7.7	66,000,000	−34.0*
Institutional	6,500	−7.1	22,000,000	−46.3*
Residential (Total)	403,000	−2.7	8,550,000,000	+13.3*
One- and two-family homes[1]	291,000	−3.2	6,982,000,000	+10.7*
Apartments	95,500	−3.1	1,351,000,000	+16.1*
Other residential[2]	16,500	+10.0	307,000,000	+95.5*
Stores and offices	20,500	−4.7	684,000,000	+5.0
Industry, utility, defense[3]	10,000	−13.0*	684,000,000	+5.0
Storage in structures	30,000	−3.2	661,000,000[4]	+79.9*
Special structures	25,000	+2.0	459,000,000	+26.8
Total	515,000	−2.9	$12,361,000,000[5]	+16.2*

[1]This includes manufactured homes.

[2]Includes hotels and motels, college dormitories, boarding houses, etc.

[3]Incidents handled only by private-fire brigades or fixed-suppression systems are not included in the figures shown here.

[4]This total reflects three industrial property incidents, which results in $775 million in property damage.

[5]This total does not include the California wildfires of 2008.

*Change was statistically significant at the .01 level.

Source: Michael J. Karter, Jr., *Fire Loss in the United States 2008*, NFPA Fire Analysis and Research Division (Quincy, MA: National Fire Protection Association, August 2009).

used in the design of fire safety education materials, which must underscore messages of personal responsibility and accountability.

2. People must use and maintain smoke detectors and develop and practice escape plans. Although the fire service has been successful in distributing smoke detectors, a high percentage of those installed are found not to be functioning at the time of a fire.

3. Installation of sprinklers must be mandated on all new construction.

4. Additional research must be conducted to make home products (both building and consumer) more fire safe.

5. The special fire safety needs of high-risk groups (for example, the young, older adults, those with physical and mental challenges, and the poor) need to be addressed.[40]

TABLE 1.6 Estimates of 2008 Civilian Fire Deaths and Injuries by Property Use

PROPERTY USE	CIVILIAN DEATHS			CIVILIAN INJURIES		
	ESTIMATE	PERCENT CHANGE FROM 2007	PERCENT OF ALL CIVILIAN DEATHS	ESTIMATE	PERCENT CHANGE FROM 2007	PERCENT OF ALL CIVILIAN INJURIES
Residential (Total)	2,780	–4.0	83.7	13,560	–3.1	81.2
One- and two-family homes[1]	2,365	+0.5	71.2	9,185	–4.8	55.0
Apartments	390	–24.7	11.8	3,975	+0.6	23.8
Other residential[2]	25	–16.7	0.7	400	0	2.4
Nonresidential structures[3]	120	+14.3	3.6	1,400	+3.7	8.4
Highway vehicles	350	–4.1	10.5	850	–43.3*	5.1
Other vehicles[4]	15	–25.0	0.5	215	+22.9	1.3
All other[5]	55	+22.2	1.7	680	+4.6	4.0
Total	3,320	–3.2		16,705	–5.5	

Note: Estimates are based on data reported to the NFPA by fire departments that responded to the 2008 National Fire Experience Survey. Note that most changes were not statistically significant; considerable year-to-year fluctuation is to be expected for many of these totals because of their small size.

[1]Includes manufactured homes.

[2]Includes hotels and motels, college dormitories, boarding houses, etc.

[3]Includes public assembly, educational, institutional, store and office, industry, utility, storage, and special structure properties.

[4]Includes trains, boats, ships, farm vehicles, and construction vehicles.

[5]Includes outside properties with value, as well as brush, rubbish, and other outside locations.

*Change was statistically significant at the .01 level.

Source: Michael J. Karter, Jr., *Fire Loss in the United States 2008*, NFPA Fire Analysis and Research Division (Quincy, MA: National Fire Protection Association, August 2009).

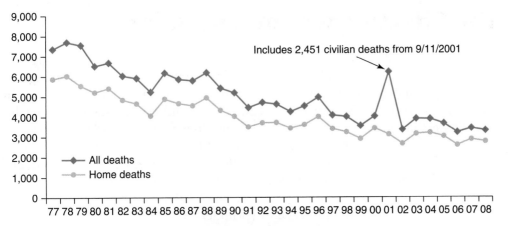

FIGURE 1.7 Civilian Fire Deaths in the Home in the United States, 1977–2008 *(Source: NFPA, Fire Analysis and Research, Trends and Patterns of U.S. Fire Loss, September 2009)*

Includes 2,451 civilian deaths from 9/11/2001

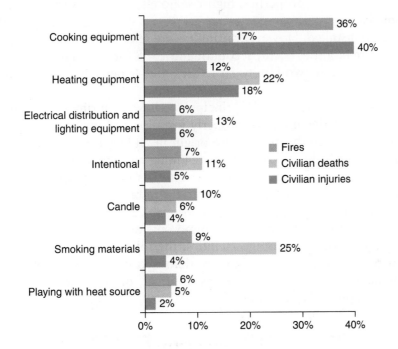

FIGURE 1.8 Major Causes of Home Structure Fires, 2003–2006 *(Source: NFPA, Fire Analysis and Research, Trends and Patterns of U.S. Fire Loss, September 2009)*

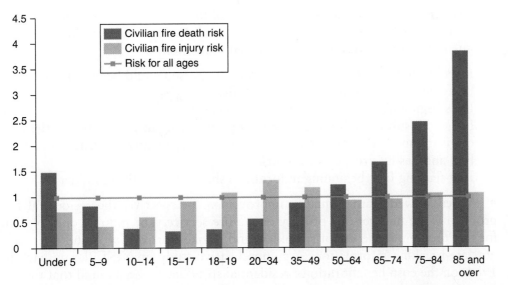

FIGURE 1.9 Risk of U.S. Civilian Fire Deaths and Injuries in Home Structure Fires, 2003–2007 *(Source: NFPA, Fire Analysis and Research, Trends and Patterns of U.S. Fire Loss, September 2009)*

The Effectiveness of Sprinklers

Automatic sprinklers are a highly effective element of a building's overall fire protection system. When sprinklers are present, the chances of dying in a fire are reduced by one-half to three-fourths, and the average property loss per fire is cut by one-half to two-thirds. Moreover, this comparison understates the potential value of sprinklers because it lumps together all sprinkler types, regardless of coverage or operational status, and is limited to reported fires. If unreported fires could be included in this evaluation and if the concept of a complete, well-maintained, and properly designed and installed system could be defined, the effectiveness of sprinklers would be even more impressive. As of 1998—the latest year with reliable multiyear data on the presence of sprinklers in buildings—most of the fires reported in health care properties, high-rise hotels, high-rise office buildings, and selected mercantile and manufacturing properties occurred where sprinkler systems had been installed. However, sprinklers are seldom found in homes, which is where most fire fatalities occur, accounting in 1998 for less than 1 percent of fires reported in one- and two-family structures and 8 percent of fires reported in apartment structures.[41]

In the last decade, the NFPA's principal statistic on sprinkler effectiveness has drawn attention to the ability of properly installed and maintained sprinklers to prevent deaths outside the area of fire origin in all but a few unusual situations.

In fact, the NFPA has no record of fire killing more than two people in a completely sprinklered building where the system was properly operating, except in an explosion or flash fire or where industrial fire brigade members or employees were killed during fire suppression operations. Because explosions, flash fires, and industrial fire brigades are rarely found outside of mercantile and industrial properties and associated storage facilities, the following statement is also true: NFPA has no record of fire killing more than two people in a completely sprinklered public assembly, educational, institutional, or residential building where the system was properly operating.[42]

WHO DIES WHEN SPRINKLERS ARE PRESENT?

A NIST report concisely describes the kinds of fire scenarios that account for the deaths that would still occur even if operational detector and sprinkler systems, designed to current codes, were in universal use in homes:

- Fires beginning so close to a victim that the person could be described as being intimate with the ignition of the fire
- Fires beginning in combustibles in a concealed space
- Fires with substantial smoldering periods occurring in a room where the victim is immobile (for example, bedridden or incapacitated by drugs or alcohol) and has no prospects for quick rescue
- Fast-flaming fires beginning in locations shielded from the sprinkler[43]

When viewed from a quality improvement perspective, the residential fire problem is the number-one priority of the fire service. The number-one solution for this problem is the use of sprinklers and smoke alarms. Unfortunately, this is not late-breaking information. A comprehensive study by NIST, dating back to 1984, on the cost–benefit ratio of residential sprinkler systems found that the use of sprinklers and smoke alarms together decrease the risk of dying in a home fire by 82 percent.

Numerous studies support the use of sprinkler systems to effectively minimize fire loss, reduce injuries, and save lives. Although the evidence is clear, most home builders continue to strongly resist mandates to include sprinkler technology into houses. In time, as this technology is incorporated into building and fire codes, and as the fire service becomes more adept at and astute in using performance-based design, the day may come when all structures will be built with these life-saving and cost-saving devices.

Mapping the Future

When considering the future of the fire service, several questions come to mind: What will the fire chief have to face in 2025? What are the primary challenges organizations must address to prepare their communities, personnel, and organizations to meet the demands of a changing society? How will the improvements in technologies and prevention techniques affect the fire service?

One key factor in addressing these concerns that is often discussed and yet overlooked is leadership. The fire service has benefited from several key leaders who have made a profound impact on the profession over the last 30 years. Many destined for leadership will have similar opportunities, yet will fail to act; and unfortunately, many will fail to recognize the opportunities before them.

The leaders who have had a positive influence on the fire service possess the unique ability to perceive a long-term vision for the profession and to engage the right people to facilitate both short- and long-term change. They are in effect leadership brokers, helping to connect the right people, to place effective people in key positions, and to create **synergy**. The concept of synergy is that every individual working on a team achieves more than he or she would as an individual. These influential leaders have another common trait: they have mapped their careers with clear objectives, which they continually strive to achieve. They recognize that a good map is required to navigate into the future, to be successful, and to leave a legacy. This map includes the routes to enhance a leader's professional development, to develop the department's strategic vision, and to have a positive impact and influence within the community and with coworkers and those within the network of fire professionals. Unfortunately, many in leadership roles start and end their journey without a map. The U.S. fire problem will be resolved only when a sufficient number of quality leaders are in place to create a tipping point. This requires a clear and effective road map for the profession as a whole and for each person in the fire service.

synergy
■ The Greek root of this word means "working together." The concept of synergy is that every individual working on a team achieves more than he or she would as an individual. The situation in which the whole is greater than its parts. In organizational terms, the fact that departments that interact cooperatively can be more productive than if they operate in isolation.

Customer/Community Expectations

One significant shift seen by the fire service throughout the last two decades has been the expectations of its customer, the taxpayer. Because of the increasing competition for local governmental funds, it is critical for the fire service to articulate to the community the services it delivers in terms of measurable performance. Not only does the fire service act as an insurance policy in the community to

FIGURE 1.10
Residential Sprinkler
Heads

combat fires, avert disasters, and respond to medical emergencies; it also is an integral part of the community's economic development engine. Good fire protection is an essential ingredient to attracting quality businesses to a community.

The thought process of the fire service has changed from a focus on the number of fires being extinguished and the amount of property being saved to a more global view of how the fire service integrates with other departments and the community. How does the fire service impact such things as economic development, the psyche of residents in feeling safe, and overall community risk? The fire service has changed with regard to community outreach and stakeholder investments not only within the organization but also externally. This will help to form the fire service's strategic vision of the future. Everything will continue to change from what was practiced 20 to 30 years ago.

One of the biggest challenges the fire service may face in this century, as its leaders lead and manage change, will be how it continues to refine resource deployment and services as community risk changes. At some point the fire service will be successful in mandating the installation of sprinklers in all new construction. Over the next one hundred years, this will change the level of fire activity and the type of fires. This modification also will entail a different deployment system than is currently in use. It also may change future staffing patterns, as well as the type of equipment being used. Although any specific future deployment pattern cannot be predicted, it is safe to say that as the fire service begins to incorporate the technology and fixed fire protection systems into community risk management fully, it will evolve from being largely reactive to being proactive (see Figure 1.10).

FUTURE FOCUS

While contemplating how to lead the fire service in the 21st century, consider that the success of the transition depends on several critical characteristics of its leaders: forward thinking, having strong communication skills, ability to develop collaborative work teams with those in and outside of the fire service, excellent organizational skills, adaptable, capable of leading change, a person of integrity with the courage to make difficult decisions, and a visionary.

Future fire administrators will have to adopt a community approach and must be open to different options in the delivery of fire and emergency services. Developing the framework for articulating various options to the community and policy leaders will become increasingly important. Putting a comprehensive community risk management process in place is essential to measure the impact of technology and community education on community risk of fire.

Leaders will also have to become more adept at organizational planning, developing five-year snapshots of where the organizations need to be and what is required for continued progress, while concurrently adapting to the changing dynamics of the communities being served. Leaders need to focus on leading change within their organizations and developing a culture that accepts the change process. Managing in this changing environment also will require future fire service leaders to focus on the capital management process to ensure the infrastructure is current; similarly, leaders will need to ensure the intellectual capital of personnel is managed so as to prepare them for leadership succession.

The future of the fire service will be built on system performance measurement: measuring performance, benchmarking, finding quality improvement opportunities, and measuring the agency's performance against others. Fire service leaders must be on a continual quest for best practices, which often may not lie within their own organizations. The cost benefits of where and how the local dollar is spent in fire and emergency delivery must be in sharp focus: Are we getting the best bang for our buck? Future leaders will need to study the trends in the fire service, not only within the United States but also internationally, bringing these insights into their organizations more quickly than has been done in the past.

Leaders must also begin to think how they are going to create their own opportunities for the future within their communities. Although similar types of services are being delivered throughout the world, leaders often work within significantly different communities, jurisdictions, and governance structures. Developing the framework to have fruitful dialogue on critical fire service issues locally will continue to be one of the issues fire service leaders face.

A weakness often found in leadership is sugarcoating the facts of difficult situations that chief officers face. It is a serious mistake to look no further than the immediate crisis to understand the facts of a situation, which can be hidden and manipulated for some time by using the right spin. The end result can be catastrophic. Leaders in today's world of economic challenges and continued change must not allow themselves, their departments, or this profession to fall into this trap of complacency, denial, or failing to deal with the facts at hand. The truth can be hard and the choices tough.

STOCKDALE PARADOX

In his book, *Good to Great,* Jim Collins, former faculty member at Stanford University and now head of a national management research laboratory, found that every company moving from good to great had to face and overcome significant adversity. He states, "In every case, the management team responded with a powerful psychological duality. On the one hand, they stoically accepted the brutal facts of the endgame, and a commitment to prevail as a great company despite the brutal facts." Collins came to call this duality the **Stockdale Paradox.** The name refers to Admiral Jim Stockdale, who was the highest ranking United States military officer in the "Hanoi Hilton" prisoner-of-war camp during the height of the Vietnam War. Tortured over 20 times during his eight-year imprisonment from 1965 to 1973, Stockdale lived out the war without any prisoner's rights, no set release date, and no certainty as to whether he would survive to see his family again.[44] The following is a key quote from Stockdale: "You must never confuse faith that you will prevail in the end—which you can never afford to lose—with

PEARSON
myfirekit
Visit MyFireKit
Chapter 1 for more
discussion on Any
Old Map Will Do.

Stockdale Paradox
■ Named after Admiral Jim Stockdale, a prisoner of war who stated, "You must never confuse faith that you will prevail in the end—which you can never afford to lose—with the discipline to confront the most brutal facts of your current reality, whatever they might be." Every company moving from good to great has had to face and overcome significant adversity on the road to greatness, as noted in Jim Collins's book, *Good to Great.* Management teams respond with a powerful psychological duality. On the one hand, they stoically accept the brutal facts of the endgame and, on the other hand, make a commitment to prevail as a great company despite the brutal facts.

the discipline to confront the most brutal facts of your current reality, whatever they might be."

To succeed, the individual needs to cultivate the ability—and the organization's ability—to do these two *hard* things at the same time:

1. Stay firm in your belief that you will prevail in the end.
2. Confront the brutal facts around you.

In the context of the fire profession today and of the experiences of the fire service over the last decade, these are valuable lessons. The map used by many to navigate through the past decade of challenges comprised a prop of the past, a scare tactic, a failure to be innovative, and an unwillingness to evaluate existing practices. Others used a map for the future: a map laying a foundation for economic recovery using effective dialogue, innovation, honesty, and making tough choices needed to position their organizations for the future.

The events that have unfolded the last several years have created an uncertainty at all levels of government not seen since the 1930s and 1940s. In essence, the map has been changed significantly with many roads being closed, whereas new routes have yet to open. Clearly, to lead into the future, the latest map must be used. To ensure success through its trials, the fire service must confront the difficult facts facing the profession.

Conclusion

The heritage of today's fire service is a mixture of traditions, rituals, firehouse lore, and response to numerous tragic events—all of which have helped to shape the profession into what it is today.

History not only has taught many valuable lessons but also has helped to lay the foundation for the future of the profession. With the organization of fire-control forces, including the establishment of public fire departments; changing fireground operations to create a safer environment for firefighting personnel with the deployment of resources, equipment, and training; improving the viability of a community; instituting new technology in the detection or suppression of fire; and the adoption and enforcement of new fire and building codes; the profession has seen significant improvements over the last century. This leads to the question: What will the next century bring in the advancement of the science of fire and life safety protection that will change how fire departments respond to emergencies?

In this chapter the concepts of the total cost of fire on the U.S. economy were explored; the U.S. fire experience was reviewed, including the differences by region; and the characteristics that impact the rural and metropolitan fire problems were discussed. As noted, the control of America's fire problem lies not only in effective response but also in the prevention strategies targeting the use of technology, personal accountability, research, and a focus on high-risk groups.

At the beginning of the 21st century, it is the obligation of fire service leaders to continue to lay a strong foundation for those who will follow. The fire service profession is poised like no other time in history to make a monumental impact on the safety of its citizenry. To do so will take leaders with vision, conviction, courage, and a commitment to leave the profession better than they found it!

PEARSON
myfirekit

Visit MyFireKit Chapter 1 for interviews on the Perspectives of Industry Leaders.

Review Questions

1. Describe the impact of the conflagrations of the late 1800s.
2. Research the demographic shifts within your community. What impact have they had on your department?
3. Describe the total cost of fire in the United States.
4. List, describe, and contrast the defining characteristics of the fire problem in the rural and metropolitan areas of the United States.
5. Research one of the major fires in America, and describe how that event changed the fire service with respect to operations or changes to the fire and building codes.
6. Describe the Stockdale Paradox.
7. Provide an analysis of the fire problem in the United States.
8. Provide a history of your department.

PEARSON

myfirekit

For additional review and practice tests, visit www.bradybooks.com and click on MyBradyKit to access book-specific resources for this text!

Register for MyFireKit by following directions on the MyFireKit student access card provided with this text. If there is no card, go to www.bradybooks.com and follow the MyBradyKit link to buy access from there.

References

1. Randy R. Bruegman, *Fire Administration I* (Upper Saddle River, NJ: Pearson, 2008).
2. Rodolfo Lanciani, *Ancient Rome* (New York: Houghton Mifflin, 1888), pp. 116–222.
3. Paul Robert Lyons, *Fire in America!* (Quincy, MA: National Fire Protection Association, 1976), p. xi.
4. Paul Robert Lyons, *Fire in America!* (Quincy, MA: National Fire Protection Association, 1976), p. 1.
5. Paul Hashagen, "Firefighting in Colonial America," *Firehouse Magazine*, September 1998.
6. Paul C. Ditzel, *Fire Engines, Firefighters* (New York: Crown, 1976), p. 16.
7. Ibid.
8. Ibid.
9. Ibid.
10. Paul C. Ditzel, *Fire Engines, Firefighters* (New York: Crown, 1976), pp. 18–19.
11. Ibid., p. 29.
12. Ibid.
13. Ibid.
14. Ibid.
15. Donald M. O'Brien, "A Century of Progress Through Service," *The Centennial History of the International Association of Fire Chiefs 1873–1973* (Fairfax, VA: IAFC, 1973).
16. Ibid.
17. Paul Robert Lyons, *Fire in America!* (Quincy, MA: National Fire Protection Association, 1976).
18. J. J. Apostolow, D. L. Bowers, and C. M. Sullivan III, "The Nation's Annual

Expenditure for the Prevention and Control of Fire," Project Report, Worcester Polytechnic Institute, Worcester, MA, December 21, 1978.

19. William P. Meade, *A First Pass at Computing the Cost of Fire Safety in a Modern Society,* NIST-GCR-91-592 (Gaithersburg, MD: National Institute of Standards and Technology, Building and Fire Research Laboratory, June 1991).

20. John R. Hall, Jr., *The Total Cost of Fire in the United States* (Quincy, MA: National Fire Protection Association, February 2008).

21. P. Schaenman, J. Strem, and R. Bush, "Total Cost of Fire in Canada," National Research Council of Canada Fire Research Laboratory, December 1994.

22. T. R. Miller et al., "Estimating the Costs to Society of Cigarette Fire Injuries," National Public Services Research Institute, July 1993, as published in Societal Costs of Cigarette Fires, Consumer Product Safety Commission, August 1993.

23. P. Frazier et al., *The Economic Consequences of Firefighter Injuries and Their Prevention. Final Report,* TriData Corporation, a Division of System Planning Corporation, prepared for National Institute of Standards and Technology, Gaithersburg, MD, August 2004.

24. Patricia Frazier, *Total Cost of Fire in the United States,* Fire Protection Engineering Archives, retrieved September 13, 2010, from http://www.fpemag.com/archives/article.asp?issue_id=18&i=53

25. Ibid.

26. John R. Hall, Jr., *The Total Cost of Fire in the United States,* (Quincy, MA: National Fire Protection Association, February 2008).

27. Ibid.

28. Ibid.

29. U.S. Fire Administration, *Mitigation of the Rural Fire Problem: Strategies Based on Original Research and Adaptation of Existing Best Practices,* December 2007.

30. Ibid.

31. Ibid.

32. Ibid.

33. Ibid.

34. Robert Adams, Judy Comoletti, Sharon Gamache, John Hall, and Pat Mieszala, *Urban Fire Safety Project: Report to the NFPA Board of Directors and the Metropolitan Fire Chiefs Association,* November 2007.

35. U.S. Fire Administration, *Four Years Later—A Second Needs Assessment of the U.S. Fire Service,* a cooperative study authorized by U.S. Public Law 108-767, Title XXXVI, FA-303/October 2006.

36. Robert Adams, Judy Comoletti, Sharon Gamache, John Hall, and Pat Mieszala, *Urban Fire Safety Project Report to the NFPA Board of Directors and the Metropolitan Fire Chiefs Association,* November 2007.

37. Ibid.

38. Ibid.

39. Ibid.

40. Michael J. Karter, Jr., *Fire Loss in the United States 2007* (Quincy, MA: National Fire Protection Association, August 2008).

41. Kimberly D. Rohr and John N. Hall, Jr., *U.S. Experience with Sprinklers and Other Fire Extinguishing Equipment,* Fire Analysis and Research Division (Quincy, MA: National Fire Protection Association, August 2005).

42. Ibid.

43. Ibid.

44. Jim Collins, *Good to Great* (New York: HarperCollins, 2001), pp. 12–13.

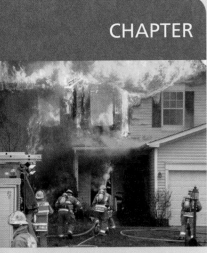

CHAPTER

2

A Community Approach to Fire Protection and Emergency Services

(Courtesy of Rick Black. Center for Public Safety Excellence)

OBJECTIVES

After completing this chapter, you should be able to:

- Describe the importance of the 1973 report *America Burning*.
- Describe the recommendations of the 2002 *America at Risk* report.
- Define the risk factors, groups at risk, and cost of fires in the United States.
- Describe the impact the adoption of a sprinkler ordinance can have on the community risk factors.
- Define the four key elements in a community approach to fire protection.
- Describe the Ten Commandments of political survival.
- Describe the importance of community engagement by the fire service.

PEARSON

For additional review and practice tests, visit www.bradybooks.com and click on MyBradyKit to access book-specific resources for this text!

Introduction

imaging the department
■ The perception people have of your department when they say its name. A department's image is composed of an infinite variety of facts, events, personal histories, marketing efforts, and goals that work together to make an impression on the public.

Every 22 seconds in America, a fire department responds to some type of fire, accounting for almost 1.5 million responses annually. As operations in the field have evolved over time due to the advancement of training, technology, and tactics, so has the fire service's approach to the stakeholder groups of elected officials, general public, special-interest groups, and employees. This chapter provides an overview of the historical documents that have laid the foundation for the discussions occurring today on deployment of resources and community risk, and it presents a nexus to what may occur in the future. Whether it is using technology, adopting stricter codes and ordinances, increasing enforcement, adding more resources to field deployment, increasing fire prevention efforts, developing positive political relationships, or **imaging the department**, it all begins with a community approach to fire protection.

America Burning

PEARSON
myfirekit

Visit MyFireKit
Chapter 2 for more
information on
History.

If fire is the purifying element, then America must rank among the most purified countries of the world. Early America, with its narrow streets and wood construction, was destined to burn and burn again, as was discussed in Chapter 1. All America seemed to be built of wood, and the danger of fire in the United States was recognized early.[1]

As President Harry S. Truman stated at The President's Conference on Fire Prevention in May 1947:

> The serious losses in life and property resulting annually from fires cause me deep concern. I am sure that such unnecessary waste can be reduced. The substantial progress made in the science of fire prevention and fire protection in this country during the past forty years convinces me that the means are available for limiting this unnecessary destruction.

If one did not know when this statement was made, one would think it could have been made today.

three Es of fire safety
■ Engineering, enforcement, and education.

In fact, President Truman's perspective is as valid today as it was then. Much of The President's Conference on Fire Prevention's report focused on the **three Es of fire safety**—*engineering, enforcement,* and *education.* The significance of the report is that it was the first time the federal government focused on and recommended a more coordinated approach to fire protection in the United States.

America Burning
■ The original report by the USFA transmitted to President Richard Nixon in May 1973 that provided an overview of the U.S. fire problem and recommendations to address the problem.

Twenty-six years later, in 1973, the National Commission on Fire Prevention and Control (NCFPC) released the report ***America Burning***. This landmark report set the course for the American fire service since that time. The report focused on the nation's fire problems and noted the following needs: (1) to provide more of an emphasis on fire prevention; (2) to promote a better trained and educated fire service; (3) to educate the American public in fire safety; (4) to understand that the designs and materials in the environment in which Americans live and work present unnecessary hazards; (5) to improve fire protection of building features; and (6) to recognize the importance of research. To encourage solutions to these problems, the *America Burning* report outlined over 90 recommendations and important tasks proposed by the

National Commission on Fire Prevention and Control. These recommendations included the following:

- Developing a comprehensive **national fire data** system to help establish priorities for research and action
- Monitoring the research in both the public and private sectors to assist in the interchange of information
- Providing grants to states, allowing local governments to develop comprehensive fire protection plans
- Establishing a national fire academy for the advancement of fire service education
- Undertaking a major effort to educate Americans in fire safety

national fire data
■ Data collected by the National Fire Data Center of the U.S. Fire Administration.

AMERICA BURNING RECOMMISSIONED

In the late summer of 1999, the director of the Federal Emergency Management Agency (FEMA) formally commissioned *America at Risk: America Burning Recommissioned.*

America Burning Recommissioned was a response to the recommendation of the Blue Ribbon Panel, which had provided its report to the director early in October 1998. The FEMA director had assembled this expert panel to give an assessment of concerns and issues with respect to the ongoing work of the U.S. Fire Administration (USFA) and to obtain recommendations on how to improve the effectiveness of this critical component of the FEMA organization. The underlying rationale of the members of the Blue Ribbon Panel, in their recommendation that *America Burning* be recommissioned, clearly was that *America Burning*, in 1973, offered important foundations and focus to the management of fire risks in America. It has resulted in the establishment of USFA's predecessor organization, and it created a blueprint for its activities. If such efforts could be adapted to current conditions, beneficial progress and results for the fire services as a whole would follow.

The FEMA director acknowledged the importance of the Blue Ribbon Panel's recommendations, and the panel's reasoning for them. He also articulated his exact purpose for recommissioning *America Burning* in his letter, and in his visits with the Commission members during their meetings in Washington, DC. In his letter to the members, the director wrote that:

. . . (the commission will fulfill an) essential role in the initiation of a much needed, and long-awaited, national effort to continue tangible reductions in our country's losses to fires. Equally importantly, it will also provide a critical framework for the evolving role of the fire services in the safety and sustainability of today's American communities. . . . Your panel will recommend an approach to an updated and renewed vision for the fire service community.[2]

In June 2002, the Federal Emergency Management Agency (FEMA) released *America at Risk, America Burning Recommissioned.* This report stated that, to a great extent, the fire problem in America remained as severe as it had been 30 years earlier. If progress is measured in terms of loss of life, then the progress in addressing the problem, which began with the first *America Burning* report in 1973, had come to a virtual standstill. The "indifference with which Americans confront the subject," which the 1973 Commission (NCFPC) found so striking, even continues today. Fire departments currently face expanded responsibilities and broader assignments than the traditional structural fire response and suppression activities of 1973.

In this 2002 report, the Recommissioned Panel reached two major conclusions. The first conclusion is that the frequency and severity of fires in America do not result from a lack of knowledge of the causes, means of prevention, or methods of suppression. America has a fire "problem" because it has failed to apply and fund known loss reduction strategies adequately. Had the past recommendations of The President's Conference on Fire Prevention (1947), *America Burning* (1973), and many subsequent reports been implemented, there would have been no need for this further work by the Recommissioned Panel. Unless the earlier recommendations and those that follow are funded and implemented, past efforts will have been an exercise in futility. The second conclusion is that the responsibilities of today's fire departments extend well beyond the traditional fire hazard. The fire service is the primary responder to almost all local hazards, protecting a community's commercial, as well as human, assets; and firehouses are the closest connection government has to disaster-threatened neighborhoods. Both firefighters, who frequently expose themselves to unnecessary risk, and the communities they serve would benefit if there were the same dedication to the avoidance of loss from fires and other hazards as there is to suppress them once they start by fire suppression and rescue personnel.

Addressing America's Fire Problem

Although each of these reports is reflective of the time in which it was written, they offer not only insightful and common themes regarding the challenges of the fire service past but also a clear vision of future opportunities. These common themes, which focus on the areas of engineering, enforcement, and education (the three Es), have been and will continue to be the building blocks of the profession's ability to address America's fire problem. While great progress has been made during the past 10 years, the opportunity exists today to create a tipping point to impact the nation's fire problem dramatically. If America's fire problem is to be addressed effectively, energies must be focused on those areas that, collectively, will create a synergy to confront the problem effectively. These areas include the following:

- Understanding that cultural orientation is unique to each community and must be addressed to be successful
- Organizing and coordinating political action toward risk reduction by an organized and coordinated effort
- Continuing to develop, improve, and implement new fire protection technology
- Creating the ability to utilize data needed by each organization to assess the impact on the outcomes the department is trying to achieve
- Continuing to refine the service delivery system so the resources deployed match the risk protected and anticipated workloads

THE CULTURAL ACCEPTANCE OF FIRE

The fire service must come to understand that the most critical contributing factors in America's high fire loss are those of public attitude, behavior, and values. A high level of safety cannot be achieved unless the views and attitudes of the American public and the American fire service are changed.

One of the significant differences that exist between America and other industrialized nations is the **cultural acceptance of fire** losses. In many countries the occurrences of fire are viewed as unacceptable from a cultural perspective. This creates a greater focus on prevention and education at the individual level, which not only is driven by local authorities but also permeates throughout the fabric of the community and often the family unit.

A strong cultural aversion to fire prevention still exists among many fire suppression personnel. Because it is much more exciting to fight fires than to prevent them, the orientation of many fire personnel has been to resist such preventive efforts. Yet, it has been witnessed in certain industrialized nations that when firefighters are fully engaged in fire prevention, dramatic results occur. Many countries have experienced double-digit drops in property loss and fire deaths when their suppression personnel are utilized for such things as conducting home safety checks on a routine basis.

Other contributing factors are the differences that exist in the application of insurance and personal economic responsibilities. In the United States, a mind-set exists that because people insure for replacement value, they need not worry much regarding fire. As has been heard on many occasions: "I didn't think it would ever happen to me." Then when it does: "Well, I'm insured so the insurance company will take care of it." In many countries if people experience a fire, the loss is theirs to bear. This surely changes the perspective of preventing a fire in the business or home.

This cultural orientation can also extend to the insurance industry. Simply stated, if premiums go down, profits go down. Insurance companies watch the trend lines of losses versus premiums, and their objective is to be profitable. This does not mean that the industry does not care about its policyholders, but that its ultimate motivation may be different from that of the fire service. Just think what would happen if this industry were to shift its cultural orientation from a purely economic view to one that promotes prevention first and if the fire service were to fully embrace prevention.

cultural acceptance
of fire
■ The tendency of
thought, public attitude, behavior, and
values, that are contributing factors to
America's high fire loss.

ORGANIZING AND COORDINATING POLITICAL ACTION TOWARD RISK REDUCTION

An organized and coordinated capability must be promoted that can identify overall problems, establish priorities, and help facilitate the fire profession's focusing on these priorities at the national level. This is the mission of the U.S. Fire Administration. Unlike many other social and economic issues that have coordinated national representation (via lobbying and public relations programs), fire safety often receives extensive public attention only after a major incident and then for only a brief period of time.

During the last decade, examples of this have ranged from the adoption of the Fire Act SAFER Grants and the Hometown Heroes Survivors Benefit Act, to several other pieces of legislation. Political action by the fire service was imperative to the passage of each. The funding for the National Fire Academy has been salvaged on many occasions due to the fire service's political engagement with its local congressional leaders advocating funding for the NFA on the issue. The recent adoption of mandatory residential sprinklers and fire codes was a result of such action. The move for adoption of these codes in each state will require the same level of commitment and political action on the part of the fire service.

The ability of the fire service to articulate clearly the needs of the service at all levels of government is an essential element in the provision of services. The fire service not only needs to speak with clarity but also needs to speak in unison. The fire service can have a much greater influence addressing these key issues when, collectively, the profession is operating from the same playbook. From national organizations to state and local associations and groups, when the fire service can focus its time and energy on specific targets using similar dialogue, it creates the momentum to influence change.

DEVELOPING AND IMPLEMENTING NEW FIRE PROTECTION TECHNOLOGY

Even though it is considered necessary to gain further knowledge in the physics and chemistry of combustion, continued research is also needed on a new generation of affordable smart detection and fire suppression systems.

The National Institute of Technology, the Worcester Polytechnic Institute, the Maryland Fire and Rescue Institute, other academic institutions, and the alarm and sprinkler industry are working on the next generation of **fire protection technology**. This is the application of the results of basic research and engineering principles to the solution of practical fire protection problems, but entailing, in its own right, research into fire phenomena and fire experience. This research will ultimately result in more cost-effective and efficient systems, which not only will be applied to new construction but also will dramatically increase the number of structures that can be cost-effectively retrofitted in the future.

USING DATA TO ASSESS OUTCOMES

The continuous and complete data collection analysis to identify fire protection problem solutions has been and still is a high priority. Unfortunately, the data collection analysis efforts being conducted within the fire community are in disarray and therefore not achieving the intended objectives.

One has to look no further than the fire department counterparts in law enforcement. Whereas the local police chief can quickly retrieve data from the Department of Justice (DOJ), Federal Bureau of Investigation (FBI), and state police, the local fire chief often collects and creates his or her own database for trend analysis. The ability of law enforcement to acquire sound data from a national perspective is a powerful tool in comparative analysis at the local level. This has provided law enforcement a means to clearly state its needs more effectively, often resulting in more resources for its programs, and has spawned numerous national research projects focused on policing issues.

Although some comparative data are available through the National Fire Incident Reporting System (NFIRS), the lack of a mandated nationalized database, which has qualified data, is the limiting factor for the U.S. fire service. NFIRS was established with the passage of the Federal Fire Prevention and Control Act of 1924 (P.L. 93-498). It authorizes the National Fire Data Center in the U.S. Fire Administration (USFA) to gather and analyze information on the magnitude of the nation's fire problem as well as the characteristics and trends occurring.

fire protection technology
■ The application of results of basic research and engineering principles to the solution of practical fire protection problems, but entailing, in its own right, research into fire phenomena and fire experience. The contribution of the practices of fire prevention is potentially much greater than that of the actual firefighting activities. Fire prevention and loss reduction measures take many forms, including fire-safe building codes, periodic inspection of premises, fire-detection and automatic fire-suppression systems in industrial and public buildings, the substitution of flame-retardant materials, and the investigation of fires of suspicious origin, serving to deter the fraudulent and illegal use of fire.

State participation in the NFIRS system is voluntary, and as a result approximately 10,000 departments do not report into the system. Compare this to the statistical information captured by law enforcement as reported through the FBI and the DOJ. As the fire service takes a closer look at the data available, the profession must be able to use these resources to provide new insights into how to address the current issues more effectively and to forecast what may occur in the future. This extends far beyond just comparative data, to having the information to conduct predictive modeling and analysis. The fire service profession needs to continue to refine this capability in the future.

CONTINUING TO REDEFINE THE SERVICE DELIVERY SYSTEM

Roles and responsibilities of the fire service will continue to evolve both to meet the changing local governmental environments and as a means to address its ever-expanding mission.

Although much has changed in the fire service over the past 50 years, the fire profession has repeatedly identified issues that will impact the citizens it serves. The question the fire service must ask itself is, Will the profession be addressing these very same issues 50 years from now?[3]

Each of the studies mentioned previously emphasizes prevention and risk reduction as critical elements of the fire service profession as it moves into the future. Research conducted by the TriData Division of System Planning Corporation for the Centers for Disease Control and Prevention (CDC) on global concepts in residential fire safety found the emphasis on prevention can have a profound impact on both the number of fires occurring and the loss of life and property. As stated in Part 3 of the report,

> National, state, and local fire agencies should consider the rich array of ideas found in this global research on best practices. There are many innovative prevention programs associated with significant decreases in residential fires and fire casualties, which are likely to have similar effect if we use them. There also are new ways to increase outreach and impact of conventional programs. All in all, there is little doubt major savings are possible in life and property loss if American fire departments use the best ideas from other Western cultures.[4]

If the fire service is successful in the incorporation of fixed fire suppression in *all* new structures and, over time, the retrofit of existing buildings, risk will be reduced. Combined with technology not yet developed in the fields of both fire and EMS, how the fire service carries out its mission will continue to develop as the use of fixed fire protection increases, population ages, and several other societal trends emerge, placing new demands upon the system.

One of the biggest future challenges the fire service may face is overcoming the traditional and cultural aspects of how resources are deployed today. Deployment strategies of the past may not work in the future. Staffing patterns, the type of apparatus and equipment purchased, shift scheduling, and training requirements may all be impacted in the future. Necessary changes will require effective planning, cooperative labor–management participation, effective community involvement, cultural reengineering in many departments, and great leadership.

As discussed in Chapter 1, one could surmise that, with the knowledge acquired over the last three hundred years, the fire problem should have been conquered by now. Yet, reviewing the annual fire statistics paints a clear picture of how far the fire service has to go to resolve this issue.

Although the fire service has seen a marked improvement in its life loss statistics from 20 years ago, it may be reaching the peak of what can be achieved with its current response and prevention practices. The fire service may also be at the apex of what it can accomplish as a reactionary force using the suppression practices, equipment, training, and public education of today, even though these are much improved from the past. As noted in his cover letter addressed to President Nixon, dated May 4, 1973, that accompanied the *America Burning* report, Richard Bland, chairman of the National Commission on Fire Prevention and Control, reports:

> The recommendations *emphasize prevention of fire* through implementation of *local* programs. This is in keeping with the very nature of the fire problem which is felt hardest at the community level. Additionally, the recommendations *emphasize built-in fire safety-measures* which can detect and extinguish fire before it grows large enough to cause a major disaster.[5]

A Community Charter for Fire Protection

In most communities today, the existence of a firehouse (whether career, combination, or volunteer) creates a feeling of safety. In fact, if a station were built with an engine in place, the lights turned on, and no responders in it, the public would still garner a feeling of safety—until someone called 911 for help, which is when this perception would turn to an expectation of performance.

The resources devoted to controlling the community's risk factors are a combination of both public and private funding sources. Public funding is the taxes that pay for the capital infrastructure and personnel to respond to the problem. Private funding is the investment made in fixed fire protection and detection, constructing buildings to code, and insurance premiums paid by property owners. Regardless of the sources of the funding, the protection of life and property is often seen as a governmental responsibility with the community members' expectation that the government will keep them safe.

The relationships among government, the insurance industry, and the building industry, and the effects of private and public expenditures for fire protection are often overlooked. Whenever the public sector invests in or the government requires higher levels of fixed fire protection, the risk to the insurance sector is reduced. An analysis conducted by the Insurance Services Office (ISO) revealed a direct correlation between the ISO grading and fire loss. Conversely, when the public sector decreases its resources or building/fire code requirements, the insurance sector and the public are at increased risk.

In every community the element of fire losses is three dimensional: life safety, responder risk, and property loss. Although the insurance industry concentrates on the property loss aspect, today's fire service organizations must focus on each dimension or else the true impact of risk cannot be quantified. The relationship between public and private funding is an important element when evaluating the effectiveness and efficiency of a fire service organization (see Figure 2.1).

Similar to all other government agencies, the fire service is not an isolated organization within a governmental structure. What other departments do within a

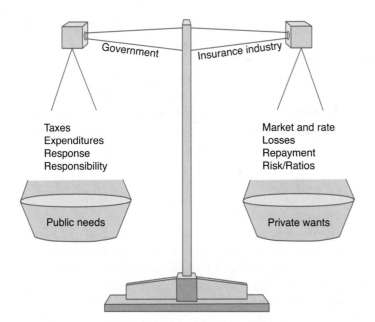

FIGURE 2.1 The Balance Between the Government and the Insurance Industry in Community Risk Management

community dramatically affects the capability of the agency to accomplish its purposes. For example, the fire service agency should have input into the policy-making process to ensure that other government operations do not adversely affect the fire service and, conversely, the fire service does not negatively affect other governmental and civic operations. An example would be the symbiotic relationship between the water department and fire department. If the water department places hydrants in areas that are difficult to access or constructs the system with undersized piping, which delivers inadequate flows, the fire protection system is impacted. If the fire department, by flowing hydrants, causes backflow problems or improperly shuts off a hydrant and damages the system, the water department is impacted. To evaluate the emergency response delivery system, the chief fire executive must assess the interrelationships among all the facets of fire protection and emergency medical services as well as the impact of actions taken on the ability of the local agency to deliver the public services.

Fire service organizations operate under an array of criteria ranging from objective requirements for performance measurements to policy decisions that establish levels of performance. This model is highly dependent upon the dialogue between the fire service and community leaders in establishing and maintaining programs and resources toward achieving a comprehensive fire/life safety policy. Often, elected and fire officials find themselves at opposite ends of the pendulum in justifying or assessing resource allocation to reduce the fire losses, in preparing for major emergencies, or in providing for an expanded emergency service mission.

How fire service leaders help balance the scale in respect to those interrelated issues that must be taken into account when developing a deployment strategy is both a challenge today and one that will continue in the future. It is easy to understand the dilemma many communities face. The resolution of the community fire problem and other general service demands almost always involves a commitment of financial resources, either public or private. Although this can be viewed

Chapter 2 A Community Approach to Fire Protection and Emergency Services   41

as either time or money, in all cases, it can be reflected in the community acceptance or rejection of specific policies and practices.

The challenge for community leadership today is finding the acceptable community balance that takes into account all the variables: level of community risk, level of public and private commitment, service level objectives, system performance, and costs. If the level of community risk is low and the commitment of public and private resources is high, the system may not be efficient in terms of cost. If the level of community risk is high and the commitment of public or private resources is inadequate, the system may be cost efficient but dangerously ineffective. Therefore, a prime objective of both fire and community leaders should be to determine the balance of resource commitment to control those risks effectively for the community in the most efficient manner.

Unfortunately, this type of dialogue does not take place in many jurisdictions. Instead, policy decisions are based on what has been done in the past, even when the community growth patterns and risk factors have changed over time. Hence, the failure to assess and deploy systematically the resources needed to meet the changing community environment often leaves agencies years behind in resource allocation and/or program development. The reverse can also be true. Communities that have undergone a reduction in population often experience a decreased demand for services. Yet, fire department resources remain the same and place a burden on the local community to pay for the historic level of service. In some cases, even when population declines, the risk actually increases due to vacant buildings and lack of appropriate maintenance found throughout the community. With no formalized risk assessment and deployment processes, organizations often cannot clearly articulate their needs during such transitions (see Figure 2.2).

FIGURE 2.2
Organizational Realities

Chapter 2 A Community Approach to Fire Protection and Emergency Services

Assessing the Benefit of the Services Provided

Elected officials must consider a variety of factors in assessing the benefit of the services provided by the fire agency. Today's elected officials and city/county managers require quantifiable data to support justifications for service level enhancement or the additions of new programs and/or projects. An element of education must also occur. Service-oriented organizations, such as public safety, have a difficult time compiling data related to the number of fires that did *not* start or the number of lives saved by prevention efforts. Although the fire service does not make widgets or does not produce a specific product that can be counted, it does provide a variety of services that can be measured. Value to the customer is found in the value of the service compared to life without those services. Among these value indicators to the customer are the following:

- Fire losses
- The nature and value of fire losses
- Overall response time, which includes alarm, handling time, turnout time, travel time, on-scene time, and intervention time
- Percentage of time actually used in providing the principal services of the department such as fire prevention, inspection, pre-emergency planning, disaster response, and training and education of personnel
- The number of complaints registered against the agency by citizens
- Comparison of data concerning the agency with that of agencies in other similar cities
- Overall productivity of department personnel in principal fire department activities

In order for elected officials to understand the operations of their fire agency better, they must consider the local factors and peculiarities affecting the agency's ability to provide service. Among these features are the following:

- Various federal and state laws and agreements governing its operations
- The geographical area served by the agency and its unique characteristics, both natural and human made
- Water distribution system serving the agency
- Staffing levels
- Degree of fire prevention efforts (which include engineering, enforcement, and education)
- Relationship between the fire agency and other governmental departments
- Type and condition of equipment used by the agency
- Various safety and personnel factors bearing on the day-to-day operation of the organization[6]

Although the fire agencies have focused their attention on fire, the reality for the majority of the fire service in the United States today has evolved to an all-hazard response service, which provides service from several different department configurations (career, combination career and volunteer, and all volunteer) (see Figures 2.3A and 2.3B).

FIGURE 2.3A HazMat Incident *(Courtesy of Rick Black. Center for Public Safety Excellence)*

FIGURE 2.3B Rescue *(Courtesy of Rick Black. Center for Public Safety Excellence)*

Many of these departments are engaged in providing the following: urban search and rescue (USAR), hazardous materials response, airport firefighting, fire prevention and education, investigations, technical rescue (confined space, trench, and high angle), WMD preparedness and response, crime scene preservation, scene security, clandestine lab operations, critical infrastructure protection, and, in some cases, management of the building code and permitting process. Because the complexity of the fire service has changed significantly over the last 30 years, so has the complexity of coherently expressing its needs.

A critical element in the assessment of any emergency service delivery system is the ability to provide adequate resources for anticipated fire combat situations, medical emergencies, and other anticipated events. Properly trained and equipped fire companies must arrive, deploy, and mitigate the event within specific time frames in order to meet successful emergency event strategies and tactical objectives. Each event (whether it is a fire, rescue operation, medical emergency, disaster response, or other situation) will require varying and unique levels of staffing and resources. For example, controlling a fire before it has reached its maximum intensity requires a rapid deployment of personnel and equipment within a given time frame: the higher the risk, the more resources needed. More resources are required for the rescue of persons trapped within a building with a high-occupant load than for a rescue in a building with a low-occupant load. More resources are required to control fires in large, heavily loaded structures than in small buildings with limited contents. Creating a level of overall service requires the emergency service system to make decisions regarding the distribution and concentration of resources in relation to the potential demands placed upon them by the level of risk in the community.

DETERMINING COMMUNITY RISK LEVEL

The objective for any local fire service/EMS provider is to have a distribution of resources that is able to reach a majority of events in the time frame stated in its service level goals. Many factors make up the community risk level, which would indicate the need for higher concentration of resources, such as the following:

- Inability of occupants to take self-preserving actions
- Construction features
- Lack of built-in fire protection
- Hazardous structures
- Lack of needed fire flow
- Nature of the occupancy or its contents, and so on

Evaluation of such factors leads to determining the number of personnel required to conduct the critical tasks necessary to contain the event in an acceptable time frame. The agency's level of service should be based on its ability to cope with the various types and sizes of emergencies that it can reasonably expect after conducting a community risk assessment. Whereas this process starts with examining the most common community risks, the potential fire problems, target hazards, critical infrastructure, and an analysis of call history data, it ends with the community and policy makers deciding what level of public services to provide and what level of responsibility the private sector and its citizens will provide. This is a new dialogue for many in the fire service and government in general. In a world in which public fire protection costs are on the rise, revenues are down, and the economy is challenged, these discussions will become much more frequent and necessary.

Increased Use of Built-In Fire Protection

One such discussion at the forefront of controlling community risk is the increasing use of residential sprinklers. The national statistics as reported to NFIRS and the research conducted by NFPA provide a clear picture of where the fire problem exists, and the incorporation of this built-in fire protection technology into the codes many years ago would have had a dramatic impact on the nation's fire problem today.

The experience with residential sprinkler technology that Scottsdale, Arizona, has had could be applied to the entire country. Because it was well established years ago that automatic sprinkler protection could have a positive impact on large risk facilities, Scottsdale posed the question regarding whether this equipment could be used in residential occupancies, proven to be the most dangerous to its citizens. Then, in July 1985, Scottsdale passed an ordinance requiring sprinklers in all new construction, even though many questions remained regarding the effectiveness and the cost of requiring built-in protection. A 15-year statistical review of this policy decision provides an overview of the impact it has had on the fire experience in Scottsdale.

The Cost Recent technology breakthroughs have made sprinklers more affordable and easier to install in homes. On the national average, sprinklers add 1.0 to 1.5 percent to the total building cost. In Scottsdale, the average cost is less than $.80 per square foot.

Based on fires from 1998 through 2001, 15-year data did not separate residential fire damage from all structures with fires. Other jurisdictions such as

SCOTTSDALE REPORT'S 15-YEAR DATA

In Scottsdale, Arizona, the sprinkler ordinance was put into service in January 1986. Ten years after the ordinance was passed, the Rural/Metro Fire Department, which provided fire service to the city at the time, published the *Scottsdale Report*. This study has since been updated to include 5 additional years of data. Currently, 41,408 homes in Scottsdale, more than 50 percent, are protected with fire sprinkler systems.

LIVES SAVED

In the 15 years, of the 598 home fires, 49 were in single-family homes with fire sprinkler systems:

- There were no deaths in sprinklered homes.
- Thirteen people died in un-sprinklered homes.
- The lives of 13 people, who would have likely died without sprinklers, were saved.

LESS FIRE DAMAGE

Less damage occurred in the homes with sprinklers:

- Average fire loss per sprinklered incident was $2,166.
- Average fire loss per un-sprinklered incident was $45,019.
- Annual fire losses in Scottsdale (2000–2001) were $3,021,225, compared to the national average of $9,144,442 for communities of the same relative size.

REDUCED WATER DAMAGE

Ninety percent of fires were contained by the operation of just one sprinkler. According to the *Scottsdale Report*, less water damage resulted in the homes with sprinklers:

- Sprinkler systems discharged an average of 341 gallons of water on the fire.
- Firefighter hoses released 2,935 gallons of water on fires in un-sprinklered structures.[7]

Prince Georges County, Maryland, and Cobb County, Georgia, report similar statistics. National statistics show the following:

- Eight out of 10 fire deaths in the United States occur in the home.
- Every 79 seconds a residence burns.
- Sprinklers and smoke alarms together reduce the risk of dying in a home fire by 82 percent relative to having neither—and thousands of lives have been saved per year.

From what is known about the benefits of residential sprinklers, one would think this would be an easy decision for all communities to make. So, why has this not happened?

Although the fire service may have become much more effective in the ability to analyze data and to review effectiveness (outcomes) and efficiencies (cost), it often struggles in its ability to engage and educate the community in a meaningful way. The debate on residential sprinklers is often clouded with misinformation and is a very political issue for the following reasons:

- Increased cost of construction
- Rejection of more government intrusion/regulation

- Power and influence of real estate and building lobbyists
- Misinformation regarding the life safety benefits
- Local political pressure by special-interest groups

To overcome these factors, the framing of the fire service message must be concise and understandable to the general public as well as a key element in the chief executive's arsenal in promoting fire and life safety.

A Community Approach to Fire Protection

It is easy for fire service administrators to get caught up with the urgent issues that must be dealt with daily. With the number of local, state, and federal regulations and mandates, budget constraints, and pressing personnel issues, fire leadership can become overwhelmed as it focuses on the most immediate needs. The administration of the fire department can become a world of plugging holes in the dike each day.

It is vital for the chief executives of the fire service to break out of this mindset, no matter how difficult doing so may seem. They must realize that the costs of being reactive instead of becoming proactive are extremely high. In reality, the fire service is in competition for funding with all other agencies in the community where money is limited, and the taxpayers in many jurisdictions are resistant to any new tax measures. Those chief executives who are just plugging holes are not spending time articulating the needs and the vision of the organization to the community. They must also realize it is the taxpayers who control the size and scope of a fire department, and it is the public officials who represent the concerns of taxpayers. If the fire service needs have not been stated in terms that the taxpayers understand and that resonate with them, the department may find it difficult to move the organization forward.

Most fire departments are part of a multiservice government agency rather than a separate and self-sufficient organization able to sustain itself independently. No matter what structure a city or county operates under, the fire department is just one vital piece in the community's delivery of service. To be successful, the chief fire executive must create an active team that works together to engage its community in the business of fire and emergency services. Stephen Covey (*The 7 Habits of Highly Effective People*) uses the metaphor of an emotional bank account to describe

> the amount of trust that has been built up in a relationship. Trust is needed for a relationship to thrive. Without trust, we may manage to accommodate and endure another person; however, it cannot be mutually satisfying in the long run.
>
> It is easy to take another person, a spouse or friend, a relative, or anyone we deal with, for granted. Yet, the level of good will that exists in the relationship determines the well-being and ease felt. It provides the foundation.[8]

This metaphor could be carried over to a community as well, although it is not so much an emotional bank account as one of equity. Developing community equity is like putting money in the bank. A fire department may not know

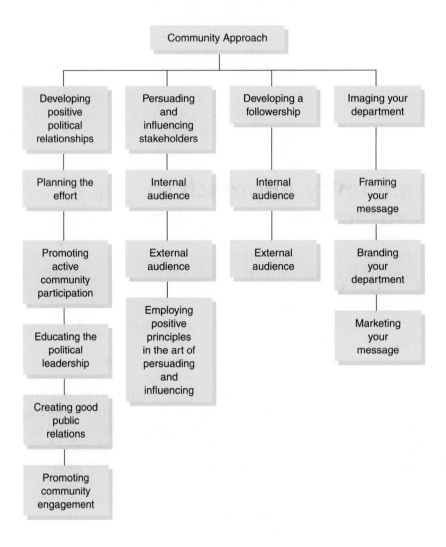

when the community will be needed for support, but if it has not invested in developing community equity, the department will be unable to draw upon it in the future.

This success is directly proportional to the chief executive and organizational team's ability to promote community involvement by engaging in the following (see Figure 2.4):

■ Developing positive political relationships
■ Persuading and influencing stakeholders
■ Developing a followership
■ Imaging your department

PLANNING

As has been noted, departments cannot operate effectively on a day-to-day basis without direction, merely being satisfied that they have made it through another day. Every department needs a strategic plan that anticipates and prepares for the future, and serves as a road map and a vision of where it should be in three to five years. Such a plan provides motivation and a good working environment for the department employees. A strategic plan has the following elements:

- Defines the community's risk factors, those of life and property.
- Creates a clear picture of the community and answers the following questions about it: What is happening? Where is it happening? When is it happening? Why is it happening?
- Establishes service level objectives based upon an analysis of community risk and historic call history.
- Identifies the resources needed to achieve the established service level objectives. This can include such things as staffing levels, fire flow requirements, apparatus requirements, facility requirements, building and fire code enhancements, and educational outreach efforts.
- Identifies alternatives for achieving the established service level objectives. These alternatives may include alterations in staffing levels and deployment options; alternative fixed fire protection models; and alternative sources of financing, funding, and revenue options.

Could a comprehensive and accurate strategic plan be written without any outside help? Would such a plan be effective if it did not fit into the overall community plan? The answer to both of these questions obviously is no. A chief executive who thinks a departmental strategic plan can be created without the help of other governmental departments and community organizations surely will not only end up frustrated but also not meet the intent of the planning process. To be effective, the department's strategic effort must be integrated into the community's general plan. It is but one piece in a jigsaw puzzle, which creates the larger community picture of fire protection. Organizational planning and decision making will be discussed in detail in Chapter 4.

PARTICIPATION

It is essential that the fire service view itself as part of its community's leadership team. As a key member of this team, the chief executive officer must see issues from a community perspective by actively contributing to all decisions that have an effect on the community, even those not specifically related to the fire or emergency medical services (EMS). The chief executive should not only attend elected officials' meetings but also participate in meetings such as those of the Rotary, Lions, and Kiwanis clubs. This is an excellent way to bridge the gap between the department, local government, and the community. When the chief executive can open lines of communication throughout the agency and the community, this creates an important link for a community approach to fire protection.

Even though those in the fire service may not always get everything they want, and are not always going to agree with public officials and key members of the community, it is essential that these fire officials remain open and reasonable. A sure road to failure is to close down avenues of communication, withdraw, and stop being an active participant in the community. To ensure the lines of communication remain open, the chief executive must be well respected by the members of the department, public officials, and the community at large. The chief executive must be well prepared when presenting proposals and be able to express views and opinions clearly and concisely. Above all, the chief executive must be honest, trustworthy, loyal, and professional. To achieve such a reputation takes time, hard work, and diligent effort; however, a chief executive who wants to be successful should be willing to devote the necessary time and effort toward achieving this goal.

POLITICAL INVOLVEMENT

The chief executive must learn quickly that being political is a large part of the job. Being political does not mean shaking a million hands, kissing babies, or endorsing a particular political candidate. Instead, it means the chief officer is involved in fire and emergency organizations and has frequent contact with public officials, writing letters or making phone calls to urge them to pass key legislation affecting the fire or emergency services. The chief executive must be proactive, not reactive. It is much easier to pass legislation and standards than to eliminate or change them. Remember, adoption of new standards or passing new regulations cannot be proposed without using some of the community equity previously mentioned.

Another way of being political is to maintain an active voice in the area of expertise. This can be accomplished by joining local, state, and national fire and emergency services organizations, such as the local fire chiefs association or emergency services association. Examples of such organizations on the state level include the state fire chiefs association, the state firefighters association, and the state emergency services association. Examples of national associations are the International Association of Fire Chiefs (IAFC) and the National Volunteer Fire Council (NVFC).

However, it is not enough to join these organizations just to have a membership card. Getting involved is essential. Local, state, and national fire and emergency services organizations promote important legislative concepts that will enhance the way the fire and emergency services operate. The chief executive must be an active participant in these organizations by serving on committees and work groups and possibly seeking to become an officer in the organizations. It is important that the fire executive lead from the front and become an agent of change in order to effect change and not be affected by change.

Being a member also offers the chief executive an excellent opportunity to network with other chiefs from various departments in the area as well as statewide and nationally. Networking is an effective avenue for discussing and influencing pending legislation that will affect the fire and emergency services, and departmental problems can be shared and solutions obtained.

PUBLIC RELATIONS

The chief executive and all members of the fire department must be concerned with public perceptions of their department. When the department works to attain a positive image in the community, it makes the department function well. If the department has a positive image, department members are proud to be associated with it; and public officials will be knocking at the department's doors to get their picture taken with the department's officers and members. The community also will respond favorably to budget requests and future improvements much more so than if the department's image is perceived negatively. As was mentioned previously, the taxpayers ultimately determine the size and scope of a

FIGURE 2.5 Public Relations *(Courtesy of Tony Escobedo)*

department. A department actively and continuously engaged in promoting a positive image will have a much better chance of gaining the public support it needs. Everyone loves a winner; looking good and walking the talk are part of a winning image. Those in the department, from the chief to field personnel, who interact with the media should be attired appropriately for every event and inject the department's message into any statements given. For example, at the scene of a residential structure fire, the department can take the opportunity to stress the importance of smoke alarms and residential sprinklers, a powerful message when done repeatedly over time (see Figure 2.5).

IMAGING THE ORGANIZATION

The importance of a positive department image will always make it easier for the department's internal audience to operate as a team. Team members having pride in their organization will result in increased productivity and improved morale.

A key element in making this happen is the appointment of a public information officer (PIO)—this person also may become the department's director of marketing—to create and promote the image of the department. In general, the PIO constantly promotes the department's image using various forms of media, such as public service announcements, newsletters, brochures, and news releases. The PIO is also responsible for overseeing the organization's Web site, Facebook, and Twitter accounts, ensuring they are attractive to the community and the information is current, accurate, educational, and relevant. The PIO interacts with the media at fire or emergency scenes, acting as a source of information so the media do not interfere with operations. It is the duty of the PIO to ensure the department complies with all laws governing the rights of the media and the public; the PIO also keeps a list of media contacts. An additional responsibility of the PIO is developing a departmental media policy.

An important aspect of interacting with the media is writing and submitting informative, timely news releases to media organizations. A news release is an extremely effective way of maintaining a department's image. Even with increased reliance on the Internet as a source of news and information, the local newspaper and television media outlets are still the best vehicles for disseminating newsworthy items about the department. In today's world of Facebook, Twitter, and a multitude of other informational exchange services (which the PIO can use to benefit the fire service), departments can push real-time information rapidly and through a variety of mediums.

The chief executive needs to do more than just be aware of the organization's image; he or she should continually monitor it by actively seeking the opinions of the organization's internal and external audiences. Every decision and statement the organization makes must be viewed from the perspective of how it will affect the internal and external images of the organization. Understanding this, one needs to know that every decision will not be viewed favorably, even though it may be the best choice for the organization. Chief executives should not be in the business of trying to please stakeholders if it is not in the best interest of their organizations.

Here are some important items to consider about a fire department's image:

- An image is what the public thinks of the organization.
- An image needs to be planned, strategically focused, and integrated throughout the entire organization.
- An image is a long-term process.
- An image gives an organization staying power.
- The majority of the department's image is often represented by the chief executive and reflected by every member.
- To create an image, an organization needs to have clearly established a mission, a set of values, and an overall vision, which all the members of the department need to know and can emulate, not just recite.
- The effectiveness of an image is measured by accuracy, consistency, repetition, and the number of people it reaches.
- A positive image will help strengthen the department's leadership voice in the community.
- A positive image will build equity within the community.
- If a good job is being done well, tell the community about it.
- Media crisis training is a necessity.
- Be proactive and create newsworthy events.
- Enlist help from within the community by increasing partner organizations.
- Select partners carefully as their image will reflect upon that of the organization.

No emergency services organization can operate in a vacuum, because it is part of the overall fabric of the community in which it operates, and each element of the community is dependent upon the others. The chief executive's relationships with the department's internal and external audiences are extremely important. To be successful, the department must be able both to identify all the members of these audiences and to communicate with them effectively.

To summarize this section, creating a positive organizational image is a critical element in maintaining high morale within a department. Image helps the department establish an excellent working relationship with key public and elected officials and community members, who will be more open to ideas toward enhancing department operations. Ultimately, this will help the department to sustain a superior level of customer service.[9]

Framing Your Message

The value of an effective message versus the result of an ineffective one can be significant. What is said is just as important as how it is said when it comes to influencing policy makers. Who is speaking the message can also be very important for making change happen. Shaping the message so that it can be understood and interpreted by the audience to whom it is being delivered is known as **framing your message**. The facts of the issue are there as the backbone of the message, whereas the frames or images created for policy makers, community leaders, and the general public make up the rest of the message. If done well, the message will educate and help to convince those on the other side of the issue to support the department's proposals. If done poorly, the cost of a weak message will diminish what the department hopes to otherwise accomplish.

Here follow some suggestions to help frame the message effectively:

- *Every issue needs a specific message.* Without a specific message, people will believe whatever they want based upon their own filters. Making sure the message is the one people hear will begin the process of educating them about the department's proposals or needs.
- *Translate an individual problem into a social issue.* Translating a problem into an issue helps others see why it is important and newsworthy. Make sure they understand the problem the department is addressing is not isolated to one or two individuals, but is a broad social issue, which could impact them on a personal basis.
- *Every message needs to be direct and understandable.* Stick to one unified concept with the message. Otherwise, multiple ideas in one message can become confusing and will ultimately weaken the message. Effectiveness often depends upon the clarity and consistency of the message, the consistency of use, and its perceived validity by the public. Remember, speak in terms they will understand.
- *Always work toward a solution.* Simply stating that something is bad or needs to be fixed is only half of a message. One must go further and tell people when and what the department wants to happen, as well as the anticipated outcomes. Never state a problem without a solution. If a solution is not presented, people may read about the issue in the newspaper, but never know what the department wanted them to do.
- *Different frames work for different groups.* Communities are often fragmented into a multitude of different special interests and concerns. The same standardized delivery will not educate, influence, or persuade everyone, so be prepared to speak differently to different audiences. Tailor the message to the specific audience(s), and utilize the most appropriate media outlet to reach them. If a picture is worth a thousand words, and the average media bite is just seconds in length or a catch phrase is just a few words, it is critical to develop compelling visuals and statements that illustrate the point.

framing your message
■ Shaping the message of a presentation so that it can be understood and interpreted by the audience to whom it is being delivered.

- *Concentrate on those audiences who might not agree with you.* Convincing those who are already in agreement about the issue is not the best use of resources, although it may be the easy thing to do. The most mileage from the department's efforts will come when the opinion can be changed of those in opposition to the proposal(s) in the first place.
- *Build a wide base of support.* The more people one can bring to the cause, the more people they will bring, and the greater effect one can have. Work toward reaching people from a variety of different groups, including various races, genders, income backgrounds, and faiths. Be as inclusive as possible, because everyone will bring a unique strength and opinion to your efforts. This synergy can be contagious and will help to propel the issue forward.

If chief officers are to be successful in the future with developing an emergency response system that is reflective of the community expectations, then framing their message is a critical element of this. This is a substantial change from what the fire service has historically done even within the last 10 to 15 years. Historically, the messaging of many departments has often been built around scare tactics to gain additional resources or forgo budget reductions. The need to continue to secure more of the general fund budget to obtain a new station (or obtain a new piece of equipment or hire more firefighters, etc.) was often driven by the organization creating a level of fear if the proposal were not approved.

In fact, while this text is being prepared, during one of the worst financial periods since the Depression, this tactic is alive and well. Due to the significant downturn of the economy, many fire departments across the United States have had to reduce staffing. The dialogue currently taking place in many communities really speaks to the issues discussed about framing of the issue.

On one side of the issue, certain departments take an aggressive stance, as discussed, saying that reducing one, two, or three firefighters will have a dramatic effect on service and will place the entire community and the firefighters at risk. On the other side of the issue, there is a leap of faith in selling to the public why it is OK to reduce services and that it will not have any impact on the residents from an emergency response standpoint. For example, the mayor of a city in Wisconsin stated the drop in fire deaths over the prior three years in his city should satisfy the public that the fire department cuts were the right thing to do. He indicated the budget decisions would have no impact on fire safety whatsoever.

Well, as many in the fire service know, significant fluctuations in fire deaths as well as fire loss from one year to the next are often attributed to one or two major events. So, looking at a one- to three-year trend and making a statement like that of the mayor of a city in Wisconsin indicate a fundamental lack of understanding as to how the fire service operates on a day-to-day basis. While making a good sound bite for the media, the statement indicates this elected official's lack of understanding of departmental operations. Comments such as these are driving a dialogue with some elected officials, labor unions, and fire service leaders that has less to do with the facts than it does with emotions. When a message is framed well, it can evoke the emotional response, but also be backed up with facts; when this occurs, the message becomes powerful.

As a rule, local governing officials today require data, a thorough analysis, cost benefit, interpretation, and expert opinion about the true impact of the investment the community makes into its fire protection and emergency services delivery system. When we move toward developing a fire protection community outreach model, it is essential that the overall objective of framing the message is to provide the ultimate level of protection in the community in keeping with local needs and circumstances. As the profession moves forward with framing the issue for the discussion, reaction, and action by the elected officials, it does so by answering a very basic question: What is the issue really about? It is an interesting question when one thinks about this from a governmental perspective, but the fact is chief executives deal with elected officials and policy makers who will ultimately make the decision on any issue brought forth. They will want to know what the controversy is, what is the human interest aspect of the issue, and what are the trends occurring not only within the region but also within the state and nationally.

These are important issues to think about as the chief executive to frame the message correctly. It is also very important to stay focused on the anatomy of winning the media. Chief executives live in a day and age where 30-second sound bites often drive public opinion. The same is true with the elected officials (as the mayor mentioned earlier was well aware of). If a possibly controversial item is brought forward, they will be watching the news as much as the chief executive is to see what the reaction is from the different stakeholders and the public. It is imperative that your message contain three things:

1. *What is the problem?*   Frame it and outline the issues or controversies and the impacts to the community or the jurisdiction served.
2. *What are the solutions?*   What are the value-added propositions to what is being proposed by the department, by the organization, and how will they impact the problem?
3. *Action.*   What are the tasks at hand that will need to be accomplished to move forward to implement the solutions to address the problem?

Often those in the fire service are much better at defining the problem than outlining the value-added solutions for the elected officials when an issue is brought forward. It is crucial for this to change. Let's return to the fire service cutbacks and the perspective of reframing the discussion, should the jurisdiction choose to take a fire company out of service. Do chief executives place the organization in a better position by articulating the impact on overall response time to emergencies, with the ability to assemble an effective firefighting force and frame the message in a way the public can understand? By educating the public to truly understand the impacts to their service from an intellectual standpoint instead of an emotional one, the fire service will ultimately be in a better position to maintain its resources in tough economic times. When the chief fire officer can frame the message to show the added value the department brings to the community and the impact such cuts and/or additions have to the organization's ability to service its residents, the officer places the department in the best position to continue to move the organization forward in a positive way.

Visit MyFireKit Chapter 2 for information on *A Time to Lead*.

On the local level, the chief executive often has the opportunity to frame the departmental message. Two examples from the Fresno Fire Department (California) illustrate the topic: one from an organizational perspective of service and the other with the specific issue of residential sprinklers. First, when the author was hired as the fire chief for the city of Fresno, it did not take long to realize the department had been significantly underinvested in for about 25 years. It took about two weeks to realize just how significant the issue was when the operations chief called and shared the news that we were closing down a company because one of the trucks had broken down and there were no reserves to replace it. This was the first time in my career this had occurred. Yet, it was indicative of how far the department had slipped behind in the depth of its resources. Equipment was old and run down, and stations were in need of significant investment of capital to bring them up to date and make them worth living and working in. We needed to add over 100 firefighters just to get back to the levels the department had been at 20 years prior.

It would have been very easy to stand up and talk about the risks to the community; that is, if the investment needed were not received, people would probably lose their lives, property would burn down, and all hell would break loose. The fact is this message had already been sent repeatedly. Instead, the objective was to frame the message(s) so it meant something not only to the council but also to the general public. Once the public moves toward an issue in a positive way, ultimately the votes will be received from the council whether they are in full agreement or not. This is typically the way government works.

The department concluded a simple and understandable message was best—"Four Minutes to Excellence!" The department chose to frame its message this way because, after discussion, it realized the reality for every resident served is when a resident dials 911 and hangs up the phone, the resident wants to hear the sirens pulling up to the front of the house. Time is a critical issue in an emergency, whether someone in the family has suffered a significant medical event or the house or business is on fire; every second seems like a minute (see Figure 2.6).

FIGURE 2.6 Four Minutes to Excellence!
(Courtesy of Fresno Fire Department)

The message was simple. If the community reinvested in the department, it could develop a plan to meet the target of placing a unit on the emergency scene in four minutes 90 percent of the time. It is known in the fire service that, with the arrival of a fire unit within four minutes, the ability to intercede prior to flashover in a structure fire and to make a positive impact on a patient having a significant medical event is much greater. Travel time is also recognized in NFPA 1710, Standard for the Organization and Deployment of Fire Suppression Operations, Emergency Medical Operations, and Special Operations to the Public by Career Fire Departments, and is found in the Commission on Fire Accreditation International accreditation processes and the ISO fire protection rating schedule. All were integrated as part of the department's planning document. The department found this theme worked, with not only our residents but also the media. In fact, when a department PIO reported from a structure fire, it did not take long for the media to begin to ask the question, How long did it take to get here and was the four minutes to excellence goal met?

The important point is the Fresno Fire Department framed a very critical and complex issue in a way the general public could understand, which ultimately led to broad community support. Of course, the message was backed up with data, charts, analysis, and all the elements important to reflect the value added to the community. Leaders' due diligence in their organizations and local government is very critical if they are to be successful. This has proven to be a successful program, which has helped the department gain a significant amount of reinvestment over the course of a five-year period for the department.

The second example involves a very specific topic that many chief officers are dealing with today in our own jurisdictions—the use of sprinklers in residential occupancies. Now, if you have been through this discussion at the local level, you know it can get pretty heated very quickly with your elected officials. As stated earlier, historically, the local building industry, in most cases, has opposed the use of sprinklers in many communities. Developers are often a major contributor to local politics and as such are very politically influential at the local level.

As many chief officers have faced, this sometimes creates a very steep hill to climb when moving forward with the incorporation of a residential sprinkler ordinance. This was the experience in Fresno as well. The department heard all the arguments on this issue: with the new building codes, buildings are built safely and do not burn; when one sprinkler goes off, they all go off and will soak the entire house; they cause additional mold concerns; the insurance rates will actually increase, not decrease; it is an economic issue; if consumers wanted sprinklers, they would ask for them and the builders would love to provide them; and there is no consumer desire for sprinklers.

To overcome the negative propaganda and images that were not factual, two messages were framed. First, a message was created to make the department's point: "sprinklers are like having a firefighter in your home." The department actually produced a public service announcement (PSA) around this concept. Because the context of the issue had become so serious, there was a need to provide some humor in the message. This PSA has gained national attention on YouTube as well as being utilized by over one hundred departments nationally. The context of the PSA is a firefighter who actually lives in a home on a 24-hour basis and provides a snapshot of what this experience would it be like for the family. The punch line is "Sprinklers are like having a firefighter in your home, maybe even better!"

The second message framed on this issue is sprinklers are a pay-me-now or pay-me-later proposition. The message was designed to drive two critical points forward: the first point was how fire protection decisions made today will ultimately result in what fire protection costs and our overall system effectiveness in the future. The second point was about controlling the community risk proactively. From a professional standpoint, the service has known that if everything had been sprinklered 30 years ago, the fire problem would be different today, as might be the deployment strategies, the type of equipment purchased, and even the fire profession. Ultimately, if everything is sprinklered, fire departments may be able to reduce their costs in some areas and expend or target those resources toward more specific problem areas as jurisdictions continue to grow.

The overall objective of any fire protection and rescue service program is to provide the optimum level of protection in the community while keeping with the local needs and circumstances of the authority having jurisdiction. Research has demonstrated a variety of factors that impact the fire department's ability and capacity to meet the objectives set forth by local, state, and national policy makers. Although a jurisdiction may pursue many different options to improve the efficiency and effectiveness of its fire protection and response systems, local circumstances profoundly affect which factors are most important and also dictate what options its chief officers are willing to explore for its overall fire protection and response systems. This is where it can become very difficult to provide for a level of service that matches the risk of the community. For the elected officials and the community at large, selecting among the various deployment options presented can be extremely complex. As such, the more education and information the department can provide to them will increase the likelihood of good decisions.

To be successful requires a combination of the specialized expertise that chief officers bring to the discussions, and the ability to clearly articulate the options available to the community. It also requires chief officers to have a thorough appreciation and in-depth understanding of the jurisdiction's budget process and future budget and growth projections. In addition, understanding the social and political circumstances of not only our communities but also our organizations is paramount in moving forward with how the department frames the message and delivers it to the various audiences throughout the community.

Persuading and Influencing

Persuading and influencing are a large part not only of the chief executive's job but also of everyone's job in the department. The chief can be seriously hamstrung by the behavior of fire department members. Credibility can be lost when negative behaviors are repeatedly played out in front of the elected officials and the media. The same is true for the department when the chief does the same thing. Such actions only make the ability to persuade and influence more difficult. In today's environment, it seems most organizations are constantly trying to effect change by introducing new ideas, concepts, policies, and procedures, as well as making an effort to secure an operating budget that will ensure the department functions in a safe, efficient, and effective manner. During this process, the chief executive and members of the department will be dealing with both internal and external audiences.

An audience may be defined as one or more people with whom a person is trying to communicate, and this communication may be written or oral. As such, it is important to know the makeup of these audiences.

DEVELOPING "FOLLOWERSHIP"

The nature of leadership may best be understood by turning the page and studying followership. Why do people follow leaders? Those in leadership positions can't just say, "follow me," and expect it to occur automatically. Leaders have to give others a good reason to follow. Leading and following entail a balance of give and take, influence, and motivation. Developing followership helps to achieve strategic objectives through a variety of professional and social means, and is a critical element in persuading and influencing.

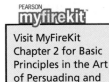

PEARSON
myfirekit
Visit MyFireKit Chapter 2 for Basic Principles in the Art of Persuading and Influencing.

THE INTERNAL AUDIENCE

The internal audience consists of one's fellow officers, firefighters, and emergency medical technicians (EMTs), as well as the office staff and maintenance personnel. It is essential to develop a team attitude in the department and realize it takes the efforts of everyone in an organization to make it successful. No one, including the chief executive, is more important than any one else in the department.

Remember, success often breeds more success. Most have experienced how good it feels to be successful at something, and how this feeling motivates us to strive harder. With this in mind, it makes sense as leaders allow the people they supervise to be successful, they will want to achieve more. A skillful leader will intentionally find ways for each and every person that he or she supervises to be successful and makes sure the follower, not the leader, receives the credit. This may not be easy at times, but creative and caring leaders strive to achieve this goal. It is the key to good leadership and management, and results in the department's becoming a healthy organization.

An essential element in developing a productive team environment is to work with the members of the organization to establish and maintain the mission, core values, and vision of the organization, as well as its functional goals and objectives. If the chief executive establishes a team environment, there is a greater chance the department will accomplish its mission in an efficient, effective, and professional manner. Maintenance of a proper team environment also will enhance the department's relationship with its external audience. When everyone feels part of the team, it is easy to go out and spread a positive message. When everyone does not, look out!

THE EXTERNAL AUDIENCE

The department's external audience consists of public officials and the community at large, as well as the career, volunteer, and combination fire and EMS departments in the region. The public officials' audience is elected and appointed members of the local, state, and national government. These people will create the legislation, which can directly affect the department. In most cases, they determine the operating budget of the department and have the final say in approving most major departmental projects.

Since the terrorist attacks of September 11, 2001, it has become obvious that included in the external audience are the other agencies the department will be working with in such disasters, such as public works, public health agencies, local and state police, the Federal Bureau of Investigation (FBI), the Department of Transportation (DOT), the Environmental Protection Agency (EPA), and the Federal Emergency Management Agency (FEMA). Although we knew this prior to 9/11, the level of motivation to engage with these groups was relatively low until then.

The community audience consists of the residents and business owners in the jurisdiction the department serves. It also includes the people who work and vacation in the community and those who pass through on its roads and highways. These people can be your greatest asset if they have a favorable image of your department, or they can be your worst nightmare if the chief officer and/or the department is held in low esteem in the community.

A large part of creating an image involves the department's ability to persuade and influence these audiences. This happens not only at planned events such as presentations to the local Rotary or homeowners association but also every day as crews are out in the public, at a school, at a neighborhood block party, or at the store, each venue providing an audience. One never knows who is sitting in the audience or whom crews may meet on an inspection or at the store! The image and the message left will resonate long after members of the department leave and will be repeated to family and friends many times. This resonance, both positive and negative, can reverberate to many throughout the community. Studies show people will repeat negative experiences more frequently than positive experiences to others.

The Community

An essential part of any department's external audience is the community. The fire department must never forget that the members of the community are the taxpayers as well as the people who use the services the department provides. With the community in support of the department, it will be much easier for the department to accomplish its goals and objectives. A successful department reaches out to its community by partnering with organizations such as the Chamber of Commerce, American Legion, Lions club, Kiwanis, Rotary club, Parent/Teacher Association, and local school district administration to enlist them as allies in accomplishing goals in providing fire prevention and life safety programs. These organizational relationships assist the department in special projects such as the passing of a sprinkler ordinance, the passing of fees, or in public safety tax initiatives.

It is also very important to invite community members to participate in the fire department's strategic planning process. This not only creates buy-in but also allows the department to draw from the pool of talent that exists in every community. Because emergency services personnel often have the attitude that they must solve their own problems because they are the people whom citizens call for help, they often are not accustomed to asking for help. The chief executive and the organization must realize and act on that certain people in the community are more than willing to offer their expertise, which will result in a more productive, professional, and successful organization, one that will reflect the community desires for service and help to create a positive community image.

Community Engagement In the fire service, most departments often find themselves in competition with the police department, as they routinely compete for public safety–related resources. Many times the police department seems to win out; of course, it has some distinct advantages over the fire department. First, most people are more frightened of crime than they are of fire. Second, as mentioned earlier in the chapter, the police department has a wealth of statistical information from the FBI and the Department of Justice that it can utilize at the local level in comparison analysis of how well its department is doing as opposed to national crime trends. With a little effort, one can access community crime maps and criminal justice statistics to determine the crimes committed and trends for a specific neighborhood, city, county, or state. However, trying to accomplish the same task as it relates to fire and EMS proves to be very difficult. Unfortunately, as of today, the fire service does not have the ability through the U.S. Fire Administration to obtain this type of in-depth comparative analysis or view fire statistics for specific areas.

Although these two factors are important, another factor often overlooked by the fire service is community engagement in the issues of policing. Community-based policing strategies have evolved over the past 25 years and really became the focus of police efforts in the 1990s.

In regard to community engagement, the fire service is in an excellent position to create a fire-based program focused on community outreach, personnel accountability, and prevention. Understanding problem solving, in respect to identifying the concerns the community has regarding fire protection, provides a great opportunity to educate the public and elected officials.

The geographic locations of fire stations throughout communities are specific to response zones, so they overlap from a travel time perspective, allowing firefighters to reach emergencies more effectively in the community. Their locations also provide an excellent means of community outreach, and many departments need to do a more effective job of utilizing this great resource than just holding open houses, pancake breakfasts, and station tours. It is critical to use the neighborhood firehouse as a means to perpetuate the focus on prevention, on increasing personal accountability, and on fire service educational efforts. As a profession, one has to ask the question, Why is the fire service not doing this?

Every department has an opportunity to create its own strategy and philosophy based on the knowledge that community interaction and support can help control the fire problem and lessen the severity of other emergencies. Being actively involved in identifying potential fire violations or opportunities to educate will ultimately reduce the loss of fire and property in our communities. One of the natural pieces to this puzzle may be the use of the Fire Corps program as the mechanism to assist with this outreach through community fire stations/districts (see Figure 2.7).

Fire Corps is a subset of Citizen Corps, which is an initiative under the Federal Emergency Management Agency (FEMA) to help coordinate volunteer activities that will make communities stronger, safer, and better prepared to

FIGURE 2.7 Fire Corps Logo *(Courtesy of National Volunteer Fire Council)*

respond to an emergency situation. Fire Corps also provides opportunities to support first responders in a wide range of measures to make the families and the homes in their communities safer from the threats of crime, terrorism, and other kinds of disasters.

Whether a department is career, combination, or volunteer, these local Fire Corps volunteers can make a difference in helping with community outreach and ultimately impact the number of people we reach in our community. By serving in a variety of nonemergency roles, these volunteers, in conjunction with the local firehouse, may provide services that volunteer and career firefighting personnel may not be able to do.

PEARSON
myfirekit

Visit MyFireKit Chapter 2 for information on Community Outreach.

The ultimate goal of any strategy utilized to develop outreach, whether it is the hiring of staff, the use of Fire Corps, or some other type of community volunteer group, is to bring together public and private organizations to elevate the level of accountability in fire and life safety.

Through a new level of community engagement, the fire service can promote the concept of personal accountability in the fire and life safety model of the future. This can be directly related to the type of incidents that the service is routinely responding to in its communities. This provides the profession with a great opportunity to educate its community, from a fire prevention standpoint, on the issues it responds to and experiences on a day-to-day basis.

Creating a Community Relationship Community relationships are established in an effort to develop positive impacts for the organization. To have this type of impact on communities, from a fire and life safety standpoint, one must approach those the fire service serves through many different avenues. One example is a professional network, which is often built upon one-on-one relationships fostered through community investment. Many other opportunities exist for the fire service to be involved in activities that provide the forum to make a difference in its communities and open the door to educating the public about what the fire department does and what the public can do to protect themselves. Community service programs such as blood pressure checks, first aid and CPR classes, residential fire safety inspections, youth firesetter counseling, and car seat programs are just a few.

Creating a Positive Relationship with Public Officials

The chief executive and the fire leadership team are constantly dealing with public officials, who are one of the key external stakeholders. There are two ways of working with public officials: the hard way and the easy way. The hard way is for the chief executive to be bullish, threatening, and stubborn. A much easier way is to open the lines of communication and understand how the officials think and the way they approach problems. Effective leaders are willing to compromise and be flexible. If patience and understanding are demonstrated, this will likely pay off in the future and is another way to elicit political equity.

A good way to begin understanding the motivating factors behind how public officials think and act is to read as many publications as possible from such organizations as the National League of Cities, U.S. Conference of Mayors, and

International City/County Management Association (ICMA). Visiting their Web sites is also an excellent way to learn about their topical issues and what is being written about them.

To open the lines of communication, the chief executive also should make an effort to get to know the public officials more personally. For example, invite them to lunch at the station or to an evening barbeque with their families and include a station tour and ride on a fire engine. Resist the temptation to ask for equipment or elicit a political conversation; just spend time together, and have fun—those discussions will come later with better results. Invite the officials to stop by occasionally for a cup of coffee, and do not *ask* for anything. Steer most of the conversation away from fire department issues. During these casual meetings, find out what you can do for them and the nature of the issues with which they are grappling. In other words, establish a comfortable relationship when you are not under pressure to ask for their support on an issue you have brought forward.

The only way public officials will fully appreciate the fire service is for the chief executive and other key members of the team to educate them about it. This starts with the relationship the chief and senior leadership team have been working to establish with the public officials. Keep them in the information loop and educate them as to how the fire service works. Remember that public officials are responsible for determining the acceptable level of service for the community. Therefore, it is the chief executive's duty to keep public officials informed regarding how the decisions the officials make may affect the acceptable level of services and to recommend ways of reducing the community's risk level.

It is essential for chief executives to work within the system and not play politics with the city/county manager's office or attempt to maneuver around them to get what they want. Creditability means everything. Keep public officials informed, but do so in an ethical and professional manner.

GETTING PUBLIC OFFICIALS TO SAY YES

When the chief executives work diligently on a proposal or budget, they sometimes forget about the psychology behind getting the public officials to say yes. All the hard work it takes to do the research and put the proposal together may be a waste of time if the request is denied or changed dramatically. Very often, the success or failure of a proposal depends on how it is presented.

When making a proposal, chief executives must be well prepared and know what they are talking about. The presentation must show how the proposal fits into the community's overall plan and how it will benefit the community.

Remember, the chief executive represents only one department in the community. Any request a chief executive makes must be justified so the public officials have no problem defending to other departments and to the community why additional funding for a project or new service is a wise expenditure of funds. As chief executives know all too well, there is only so much taxpayer money available, and public officials have a mandate to get the most for that money. This is why the choice of words in a proposal is so important in convincing public officials that a proposal warrants funding. Using terms such as *value added, mutually beneficial, cooperatively managed, interoperable, cost effective, maximizing service, with available funds, progressive,* and *quality* will go a long way in getting a favorable reaction to a proposal.

Another way of persuading public officials to say yes is to get community support behind a proposal. If the community buys into a proposal in advance, it makes it easier for the public official to agree. This is the "cashing in" process of your political and community equity.

THE POLITICAL DYNAMIC

Understanding the political arena chief executives work within is vital to the success of their departments. Unfortunately, many chief officers, although trained academically in political science, learn their most valuable lessons from on-the-job training. If these situations are mishandled, it can create issues for both the individual involved and the department.

The credibility chief executives establish at the local level and the integrity they display determine their ability to have an effective relationship with their elected officials. However, this alone is not enough. When one presents an item for consideration to an elected body, in many cases it will be judged not only on its merits but also politically. Understanding the political nature of one's position and how to interact within it is vital. Many proposals submitted to the elected body may be the right thing to do based upon good science and yet are not acted upon. This is not due to any subjective evaluation but rather to the politics surrounding the issue.

Most fire service leaders will have come through a typical career model of being evaluated and promoted based largely upon their training for and response to emergencies. This model is fairly linear, fairly black and white (something is either right or wrong), and fairly objective; and leaders tend to like it this way. It is easy for us to make a plan, set an objective, develop a course, take action, reevaluate, and move on to the next issue. The political arena, however, operates very differently. Elected officials often lean more toward the subjective or softer side of the issues. This is often a key element in their process of decision making, and it is extremely important for fire service leaders to understand it and learn to accept it. When a battle is lost because of a political versus an issue-based decision, do not take it personally; as chief, one must be able to let it go and move on to the next issue. This is part of the process and, in most cases, is not meant as a personal attack on the chief or the department. When the chief understands this and is ethical in the dealings, the chief will live to fight another battle on another day. There is much to be gained by taking one's losses with dignity and respect.

Today, fire chief survival often depends on how to get through those politically loaded encounters. It is important for the chief to continue to remind him or herself of what the outcomes and the objectives are, and what the political process is trying to achieve. Remain focused on the issue and not on the people involved in the debate. If not, the fire chief will begin to focus on the personalities and the political discussions taking place, and lose perspective as a leader in the organization and what accomplishments are to be achieved.

Many proposals get voted down, not because of a lack of political support, but because the person presenting the issues mistook the tough questioning as a personal attack. For example, instead of listening, a fire chief mounts a full assault on the council member, calling into question the ethical character of the council member. In such a scenario, one can become very vulnerable and lose one's own credibility—and, if the behavior is severe enough, one's job.

Always remain focused on the issue at hand and remember why it was presented in the first place. Remember to listen to all the options stated. Though you are the fire service professional, there are times when politically motivated suggestions are offered, and some of them might be good ideas. The chief executive has to be willing to consider them. Getting through politically loaded encounters requires fire service leaders to be focused yet at the same time flexible, politically sensitive, and open to alternatives.

How can chief officers help to create a positive political climate? The more they can deal with people face to face, the better off they will be. As has been mentioned, relationships are established best with one-on-one encounters. Increase your own tolerance for ambiguity and, when dealing with elected officials or members of a political body, remember that they may not be used to handling the black-and-white issues of the fire service. Ambiguity is sometimes the saving grace for politicians, and they often choose it over the more definitive proposal. Become more receptive to politicians' ideas as they relate to your proposals.

Take an open, receptive stance when bringing issues forward. Take a broad view, one that not only incorporates the needs of the organization but also takes into account the city or county's goals and objectives. Try to emphasize areas of mutual cooperation and benefit. Realize that buying a new ladder truck might not be the most important thing on the agenda of the city or county at a particular point in time because these officials may have more pressing needs or issues to address. Share the department's goals with your elected body; hopefully, with understanding, the public officials will begin to support what the organization is trying to achieve, on both a short- and a long-term basis. If they do not, it is a good indication that they may not be supportive of the goals and objectives of the organization, that they do not understand, or in some cases that your credibility is so low that they may not even be listening. That is an indication that you may need to look for a new job.

Remember, understanding does not always mean agreement. Understanding the political nature of the community and how it operates is very important not only to the fire chief but also to the labor president, the volunteer association, the department leadership, and the organization as a whole. This does not mean the chief executive has to agree to the final decision on an issue. Yet, as the chief executive, it is critical to realize why the decision was made and see that it is implemented. Sometimes a decision will be reached for totally political reasons that probably go against the grain of most fire service leaders, but this is the reality that most in leadership positions have faced. However, whether one agrees with the decision or not, it is one's job to support it. This builds political equity and allows survival to fight another day.

The late president of France, Charles de Gaulle, once said, "*Politics is too serious to be left to the politicians.*" This is often the case in local government. Politics in itself is neither good nor bad; it largely depends on how it is used, the people involved, and their own motivations. The political climate that exists in a given community will create an environment of trust, distrust, or skepticism. Strong loyalties can form and strange bedfellows can be made as a result of the political dynamic. All, to some degree, will influence how the political process works within and/or upon the organization. This is a reality the senior fire service leaders should not ignore if they wish to be effective, nor should they allow the art of politics to place them in a position that compromises their ethics.

The Ten Commandments
of Political Survival

George Protopapas published his version of the Ten Commandments of political engineering, which are paraphrased here in respect to the role of today's fire service leadership. The Ten Commandments of political survival for fire service leaders provide a sound road map that allows all leaders in the fire service to deal with the politics of local government today.

1. *Never show animosity toward any elected official.* Sometimes elected officials will do or say something to make us look bad in public, either at a meeting with elected officials or in front of a constituent. Always be professional. Coolness under fire can go a long way toward creating a good relationship with the entire elected body.

2. *Know your budget thoroughly.* Elected officials often refer back to specific areas of the budget that you may not have looked at for the past several months. Questions on your budget often come when you least expect them. When you are presenting an item that has a big budget, has no cost ramifications, and has already been approved, you may be asked questions about something totally unrelated, and you are expected to have an answer. So it is extremely important that you and your staff have a good understanding of everything in your budget and where it can be found at a moment's notice.

3. *If you have a proposal but know you do not have the votes to win, do not bring it to the governing body.* Give yourself time to lay the groundwork before you place it on the agenda; be patient and deliberate.

4. *For senior fire service leadership, stay out of political campaigns, especially in elections that will result in replacing people on your elected body.* It is a lose–lose situation. If the incumbent you choose to support loses the election, you will have an opposing vote on many of your proposals from the newly elected official. Conversely, if you do support a local candidate, do not think that person owes you anything in the future. Such thought can present ethical dilemmas later for both the chief executive and the elected official.

5. *Make a point of getting to know your elected officials better.* When you are attending conferences or other events together, get to know them personally. This can tell you a lot about what motivates them and is important to them as individuals, and not as council or board members.

6. *Encourage your elected officials and city and county managers to participate with you and your personnel in events, if this is allowed by the political structure of your governing body.* It is a great opportunity for them to learn more about the organization and its people and provides a way to share information.

7. *Conduct field trips for your elected officials.* The fire service has a lot of impressive equipment and great people. A good way to create a positive relationship with elected officials is to encourage them to spend some time with us observing and experiencing what we do. Allowing them to ride along and participate with the organization in prearranged events will help these officials see what the fire service does on a day-to-day basis. This can offer a great opportunity for discussing with them issues of importance to the organization. Remember, once is not enough. Some organizations hold a "Citizen's Academy" and invite

council members and other community leaders to participate one night per week for several weeks; they get to experience live fire, extrication, rappelling, mock EMS incidents, and HazMat, to name a few situations.

8. *Follow up every problem referred to you by an elected official, even if it is a minor one.* Do this immediately, return phone calls, and write to the official to confirm that the matter has been handled. Deal with the issue in concert with your city and county managers, if appropriate. Never put off decision makers.

9. *Work with your elected officials individually, before the open business session.* Often, being able to field questions from individual elected officials helps them to understand more fully the issue that you are bringing forward. It can also be a great opportunity for you to learn what they will be asking when the regular open business meeting begins. If you do this, you cannot do it for just one official; you must do it for all of them. Make sure not to offend any elected officials by leaving them out. Again, this has to be done in concert with, and with the approval of, your city and county managers, if appropriate.

10. *Sometimes your recommendations are not going to pass.* You must recognize when this occurs; do not continue to push the agenda item and end up antagonizing your elected officials. Let it go, adjust your strategy, and move on.[10]

In discussing ethics and the political dynamics of the organization, leadership, ethics, and politics are interwoven into the fabric called government. Chief fire officers should be dedicated to the concepts of effective and democratic forms of government in their response to elected officials and professional management. This is essential to achieving the objective of providing quality service. To do so, one must maintain a constructive, creative, and practical attitude toward the way local government works. Those in the various roles and positions held in the fire service have a responsibility as public servants. One can fulfill this responsibility only by being dedicated to the highest ideals of honor and integrity in public, personal, and professional relationships. The measure of success is the respect and confidence given to you by the elected officials, peers, employees, and the public.

Probably no single action that leaders can take will affect their credibility more than ensuring their behavior is ethical and professional. So how do they prepare to do just this?

PREPARATION FOR LEADERS

- Understand that the role of the fire chief is very political due to the importance of the position.
- Politics often has a negative connotation.
- Situations turn from social to political very quickly.
- Both the internal and external politics must be understood.
- The political landscape is very dynamic and can change quickly.
- Politics may often determine the posture and position one must take.
- Politics and political gamesmanship are a way of life for today's chief officer.
- Your credibility is based upon your ethical and professional behavior.

NAVIGATING THE POLITICAL ENVIRONMENT

The fire service has been reorganized toward becoming a more progressive, all hazards response profession. The roles and responsibilities throughout fire service organizations have changed dramatically over the past 25 years. The biggest shift realized may have been in the political arena. Whereas the fire chief was once the top firefighter, today the chief is a CEO, content expert, facilitator, and politician. Although not elected, the position of fire chief stands out and can become highly politicized very quickly. Those in leadership positions must understand as much about the political process as about incident command.

Whether in a small all-volunteer or combination organization or a large career metropolitan department, each chief officer will work in a political environment. It may involve the volunteer association, the local union, citizen groups, and/or elected officials.

One must have the expertise to blend one's skills, ethics, values, and organizational needs into the political process in such a manner to be professionally successful, create organizational improvements, and have positive relationships throughout the political arena. The following guidelines for political survival for the fire chief and others in leadership roles are essential to long-term job success and, in some cases, survival in the position:

- Understand that past events often dictate current situations and dynamics.
- Identify the players in the game.
- Learn the rules of the game; every community is different.
- Determine your own ethical bottom line.
- Develop a framework for making ethical decisions.
- Understand that the political process can be as important as the substance of the issue.
- Focus on the issue, not on the personalities.
- Do not take the rejection of your idea or concept personally.
- Perseverance is a must.
- The most important thing to remember is that fire chiefs and staff do not vote.[11]

Policy Parameters

Every jurisdiction operates under a specific set of laws that governs the jurisdiction and defines the basic tenets of the government's role in the community. These policy parameters and legal tenets define a political philosophy of the jurisdiction involved. These six parameters often reflect the culture of the local community and will impact how the department develops its fire protection and resource strategies.

- *Public expectations.* Does the public expect the jurisdiction to address its needs, or is there a fairly high level of personal-self reliance?
- *Service delivery strategy.* How open is the community to alternate forms of service delivery such as outsourcing or fee-for-service and financing?
- *Level of satisfaction.* Is the community satisfied with its level of fire protection and emergency service delivery, and the efficiency and effectiveness of the fire protection system?

- *Funding policies.* What impacts do your funding policies and practices have on the services you deliver? How do you account for capital expenditures? Is the jurisdiction prepared to issue debt?
- *Competing priorities.* What priority does public fire safety have in your community in comparison to the other services being provided?
- *Receptiveness to change.* Does the public recognize the need for change, and would they accept the implications of such change?

Chief officers begin to move from the development of strategic plans, which are important and will be discussed in Chapter 4, to the implementation of those strategies identified. The critical link, which is often overlooked, is the dialogue around these six policy parameters that occurs with the community stakeholders. Creating a dialogue that places everyone on the same page increases understanding of the department policies, procedures, politics, people, and services provided. External stakeholders often ask the tough questions and provide an opportunity for the organization not only to educate them but also to begin to focus on the issues important to these stakeholders.

Conclusion

The fire service has definitely evolved over the last two decades in its approach to fire protection in life safety services in each of our communities. From the days of responding only to fire calls, to expanding the fire service mission to include EMS, hazardous materials, specialized rescue services, and providing a new level of community engagement, the fire service is positioned to continue to be considered as one of the most trusted professions by the general public. This level of engagement is likely to continue to increase far beyond the aspects of emergency response. It must incorporate the effective use of our resources and deployment to emergencies driven by a comprehensive strategic plan of how to provide a community approach to emergency services. As part of this community approach, the fire service must also include programs such as those outlined in this chapter. Every community and every department should have an arsenal of programs, other than 911 response, emphasizing community risk reduction and community equity building. The ability as a profession to impact the community in nontraditional ways can happen only when a comprehensive strategic approach to community engagement is implemented. Through the development of political relationships, persuading and influencing of key stakeholder groups, developing a followership with both the external and internal audiences, effective imaging of the organization, branding the department, and marketing the messages, every department must engage the community at various levels. Making these efforts will continue to position the fire service as the go-to service in the community and ultimately help reduce fire and life loss due to emergency incidents. This will also enhance the positive perceptions the community has regarding the fire service and help develop a healthy organizational environment while providing a broad range of services to the community.

PEARSON

Visit MyFireKit Chapter 2 for Perspectives of Industry Leaders.

Review Questions

1. Discuss the six areas noted in the *America Burning* report that the fire service needs to focus on in addressing the nation's fire problem.
2. Describe the relationship between government, the insurance industry, the building industry, and its effects on expenditures for fire protection.
3. Why is a formalized community risk assessment and deployment process important?
4. Describe the four key elements of a community approach to fire protection. Why is each important?
5. The chief executive promotes community involvement in four areas. What are they and why is each important?
6. What are the basic principles in the art of persuading and influencing others?
7. What are the signs of a healthy organization? Why is each important?
8. What are the guidelines to political survival? Why is each important?
9. What is meant by imaging the organization?

10. What is meant by framing your message? Why is this important in the fire service?
11. What are the local factors that often reflect the culture of a community and make an impact on how fire protection and resource strategies are developed?
12. Describe how your department promotes community engagement.
13. Select at least five other similar organizations, and conduct a comparative data analysis on key performance indicators to include the following:

   - Population
   - Sworn firefighting personnel
   - Minimum daily staffing
   - Non-sworn safety staff
   - Civilian staff
   - Chief officers
   - Call volume
   - Fire loss
   - Daily staffing, firefighter/1,000 population
   - Total firefighter/1,000 population

PEARSON

myfirekit™

For additional review and practice tests, visit www.bradybooks.com and click on MyBradyKit to access book-specific resources for this text!

Register for MyFireKit by following directions on the MyFireKit student access card provided with this text. If there is no card, go to www.bradybooks.com and follow the MyBradyKit link to buy access from there.

References

1. Michael P. Dineen (ed.), *Great Fires of America* (Waukesha, WI: Country Beautiful Corporation, 1973).
2. Federal Emergency Management Agency, *America at Risk: America Burning Recommissioned*, FA-223 (June 2002), p. 6.
3. Randy R. Bruegman, *Fire Administration I* (Upper Saddle River, NJ: Pearson, 2008), pp. 20–21.
4. Philip Schaenman, *Global Concepts in Residential Fire Safety: Part 3—Best Practices from Canada, Puerto Rico, Mexico, and Dominican*

Republic, TriData Division, Arlington, VA, for Centers for Disease Control and Prevention and Assistance to Firefighters Grant Program, July 2009, p. xviii.

5. *America Burning,* The Report of The National Commission on Fire Prevention and Control (Washington, DC: NCFPC, 1973), p. vi.

6. Randy R. Bruegman, *Fire Administration I* (Upper Saddle River, NJ: Pearson, 2008), pp. 425–427.

7. Home Fire Sprinkler Coalition, *Scottsdale Report 15 Year Data Now Available.* Retrieved September 11, 2010, from www.homefiresprinkler.org/FS/Scottsdale15.html

8. Stephen R. Covey, *The 7 Habits of Highly Effective People: Powerful Lessons in Personal Change* (New York: Simon & Schuster, 1989), p. 188.

9. Federal Emergency Management Agency (FEMA), United States Fire Administration (USFA), and National Fire Academy (NFA), *Advanced Fire Administration,* Course Guide, August 2002.

10. George Protopapas, "Ten Commandments of Political Engineering," *County Engineers of California (CEAC) Newsletter,* 1993.

11. Randy R. Bruegman, *Fire Administration I* (Upper Saddle River, NJ: Pearson, 2008), p. 265.

The Leadership of Change

Chief Officer Badge
(Courtesy of Randy R. Bruegman)

OBJECTIVES

After completing this chapter, you should be able to:

- Describe the symptoms of organizational trouble associated with change.
- Describe how attitude impacts the change process.
- Explain how leadership styles can impact organizational changes.
- Explain how organizational culture can impact change.
- Explain why core values are important in the organizational change process.
- Explain how to create an agenda for organizational change.
- Describe why above-the-line accountability is critical to making change work.
- Describe how to overcome organizational negativity to create momentum.

PEARSON
myfirekit™

For additional review and practice tests, visit www.bradybooks.com and click on MyBradyKit to access book-specific resources for this text!

Introduction

People and culture, the human systems of the organization, are what make or break any **change** initiative. Change presents a new order of doing things or a means to undergo a transformation or transition. Who has not heard the mantra *change or perish?* Although the fire service will not perish, the changes that will be necessitated during the next several years may force many organizations to become familiar with a level of discomfort and frustration they have not previously experienced.

If it is any comfort, this problem is not new. According to Niccolò Machiavelli, who wrote *The Prince* in the 1500s, "There is nothing more difficult to carry out, nor more doubtful of success, nor more dangerous to handle, than to initiate a new order of things. For the reformer has enemies in all those who profit by the old order and only lukewarm defenders in all those who would profit by the new order."[1]

Why do some organizations seem to thrive on change whereas others stress out to the point they become dysfunctional? Organizations often become dysfunctional due to a lack of leadership, employee resistance, and entrenched **culture**, which determines what is acceptable or unacceptable, important or unimportant, right or wrong, workable or unworkable. Each of these factors can be a barrier to change; however, if two of these three are present in a department, they present an obstacle that is almost impossible to overcome quickly.

In today's times, it is especially important for leaders to prepare their organizations for future change by laying the foundation for the department to adapt to new practices and changes successfully.

As Frank Ogden, author and self-styled futurist, stated:

America has run the world for at least the past 50 years, and when you're at the top that long, you forget what it's like in the valley. There are 5+ billion people out there now who are willing to study harder, work harder for less money and be more industrious than we are. And we're linked to them by technology. With telecommuting, you can have your bookkeeping done in Madra, India, for less than it costs here. Today technology can replace whole new industries, so you have to stay flexible. To survive today, you have to be able to walk on quicksand and dance with electrons.[2]

With reengineering, relocations, political shifts, downsizing, merging/consolidations, and threats of outsourcing, public service agencies and the men and women who make them work exist on the fine line between surviving and thriving. The rules of the game in local government have changed quite dramatically in just the past decade. Therefore, the trends of the past are not good predictors of what may happen in the future. In this white-knuckle decade of change, no organization can rest on its past successes. The increased demands for service, flexibility, and quality, coupled with government cutbacks and the push for cost containment, produce a difficult challenge for even the best run organization. It is clear that the most effective way to survive future changes is to risk being part of a team that delivers value and helps create a new **paradigm** for public service. A paradigm is a set of rules based on an explicit set of assumptions that explains how things work or ought to work. Unfortunately, fire service traditions and culture often get in the way of doing just that. As J. Paul Getty said, "In times of rapid change, experience could be your worst enemy."

Whatever their position, leaders and professionals at all levels need to be motivated to make a difference for their teams and the communities they serve. This

change
■ Presents a new order of doing things or a means to undergo a transformation or transition.

culture
■ Determines what is acceptable or unacceptable, important or unimportant, right or wrong, workable or unworkable. It encompasses all learned and shared, explicit or tacit, assumptions, beliefs, knowledge, norms, and values, as well as attitudes, behavior, dress, and language.

paradigm
■ A set of rules based on an explicit set of assumptions that explains how things work or ought to work. This intellectual perception or view, accepted by an individual or a society, is a clear example, model, or pattern of how things work in the world. This term was used first by the U.S. science fiction historian Thomas Kuhn (1922–1996) in his 1962 book *The Structure of Scientific Revolution* to refer to theoretical frameworks within which all scientific thinking and practices operate.

means leaders at all levels should be action oriented, open, and flexible to change. Fire service leaders today must embrace the change journey as a never-ending opportunity for government agencies to meet the needs of the communities served.

The challenge for leaders is to develop organizations so they work smarter, are cost efficient, and are effective in promoting the safety of the citizens and firefighting personnel. Peter Drucker's maxim on change and leadership may have said it best:

> Pick the future against the past. Focus on opportunity rather than on problems. Choose your own direction rather than climb on the bandwagon. Aim high. Aim for something that will make a difference rather than for something that is "safe" and easy to do.

Although the nature of the fire service job has never been safe, organizations that fight to retain the status quo are self-destructive. In *Enlightened Leadership: Getting to the Heart of Change*, Oakley and Krug make the point that organizations that maintain the status quo are actually moving backward in relation to the organizations around them—even those that are progressing slowly. Hence, fire departments attempting to preserve the status quo are sliding backward in relation to other fire agencies around them, which are progressing amid change. The events of recent years have left a lot of people troubled about tomorrow; with the uncertainty of what is coming next and how they might be affected, many employees worry whether they will measure up to the demands that changes may bring. Will they become part of the body count in the next round of personnel cuts? And, if they do, how will they manage to get their careers back on track?

Working under the weight of such concerns wears people down, and much of their energy is spent on emotional issues, fretting about what may happen in the future. It is no wonder that this stress can affect organizational productivity dramatically. The fire service is not immune to the challenges facing local government today due to the worldwide economic issues. These challenges are forcing many organizations to rethink their deployment strategies, to do the same with much less, and to reengineer the fire department and many of its processes in a matter of months. Because the fire service as a whole is more comfortable with a more methodical approach, it does not embrace rapid change well. Unfortunately, in today's environment, the luxury of time is often gone, with significant decisions and changes having to be made in weeks and months.

Rethinking Government

In an article on reengineering government, Peter Drucker wrote:

> There have been a few organizations . . . that, without fanfare, did turn themselves around, by *rethinking* themselves. They did not start out by downsizing. In fact, they knew that the way to get control of costs is not to start by reducing expenditures but to identify the activities that are productive, that should be strengthened, promoted, and expanded. Every agency, every policy, every program, every activity, should be confronted with these questions.

- What is your mission?
- Is it still the right mission?

- Is it still worth doing?
- If we were not already doing this, would we now go into it?

. . . Rethinking government, its programs, its agencies, and its activities would not, by itself, give us . . . new political theory. But it would give us the factual information for it. And so much is already clear. The new political theory we badly need will have to rest on an analysis of what works rather than on good intentions and promises of what should work because we would like it to.

Rethinking will not give us the answers, but it might force us to ask the right questions.[3]

Drucker's thought process here is very relevant for many fire service organizations facing uncertain budget forecasts. The questions are, How many are asking the critical questions? How many are rethinking what and how they are doing business as opposed to just making cuts in an effort to balance the budget and hope revenues improve?

DON'T PLAY A NEW GAME BY THE OLD RULES

Albert Einstein once observed that the problems we face cannot be solved with the same level of thinking that created them. As an example, the significance of the change the fire service has experienced in just the past 20 years is that many aspects of business today are much different from what they were when many began their careers. With reengineering, consolidations, political shifts, lack of resources, increased service level demands, and outsourcing of services, the men and women who work in the fire service have found themselves in a perpetual state of change.

Organizational change has become a hot topic in the public sector, in local government, and in the fire service, evident by the number of books available on the subject. Just 20 to 30 years ago, changes within the fire service were often measured in terms of buying a new piece of equipment, opening a new fire station, instituting a new program, or hiring new firefighters. At present, the concept of organizational change is often measured in the expansion of the core mission and delivery of new services, in the restructuring of operations, or in the downsizing of forces due to budget shortfalls. Furthermore, instead of going through a change process, many organizations are undergoing an organizational transformation, which is often reflective of a fundamental and radical reorientation in the way organizations operate. In most cases, local government and the fire service have not gone through the degree of change experienced in the private sector until now, and they are finding it to be a very uncomfortable process.

EXTERNAL FORCES

It is important for leadership not to undertake change just for the sake of change but to approach it more strategically to accomplish an overall objective. Often a significant external force provokes organizational change: for example, an election of new leadership, funding reductions, a citywide expansion, or a dramatic need to increase productivity and enhance the service delivery system. As organizations embark on this change process, they often find themselves moving away from being an establishment that is simply reactive to the surrounding environmental forces to becoming an entrepreneurial organization looking for opportunities to leverage its strengths. This involves moving away from operating in a more stable and planned environment to working in one that is less structured and more freewheeling.

CHANGE IN LEADERSHIP

A common force for change is a change in leadership. The election of new officials or appointment of a new chief executive often promotes organization-wide change in accordance with the new leader's unique personality and leadership skill set. Whatever the driving force, typically a significant amount of resistance to change exists because people are afraid of the unknown, often do not understand the need for change, and may be cynical about the need to change (anything taking them out of their comfort zone). Any change process often creates conflicting goals in the organization, highlighting the importance of leadership throughout this process. As the fire service develops and begins to implement organization-wide changes, conflicts arise among certain aspects of the organization's culture, which are very important to the members in the organization. This is why successful organizational change is often developed around a vision that clearly articulates the desired state, yet is built upon the organization's culture and the employees' values and beliefs. Understanding this is a key component of any successful change process.

Leadership in the fire service must realize the future will not resemble the past. Leaders must focus on creating new opportunities rather than just solving today's problems. They must continually aim for something that will make a difference rather than something safe and simple. After all, isn't that what leadership is all about? For many who find themselves in positions of authority, it is difficult to make the tough decisions, to set expectations, or to demand accountability. If you have ever worked for someone who struggles to fulfill his or her leadership role, although the demands upon you as the employee were low, the overall organizational performance was less than desired. One of the problems the fire service has faced is that its heritage and traditions can often cloud the vision of what its future needs to be. In a number of organizations and for some in leadership roles from captain to chief officer, it is often difficult to step outside the established frameworks of the organizational culture to provide the leadership needed for the health of the department. It is always easier to be a buddy than a leader.

In your own organization, have you witnessed inappropriate behavior that was overlooked, tolerated, or in some cases encouraged? In a firehouse located away from administrative oversight, the acceptable behavior level is dictated by the station captain or battalion commander. If the culture of the organization is accepting of inappropriate behavior, it will become pervasive. Professionally, leaders reach only the level of performance and behavior that they envision and are willing to pursue. When the fire service's culture is like a fraternity versus a paramilitary organization, the results are predictable. In a culture that is accepting of hazing and harassment, it is anticipated that the organization is not accepting of women, minorities, or anyone who is "not like them." If a lack of *esprit de corps* is accepted in the day-to-day activities, it will eventually be reflected on the emergency scene.

This may be one of the most significant challenges for leadership in the fire service today—stepping outside a framework that has developed over time and focusing on integrating the values, ethics, and culture that will directly impact the fire service's ability to provide professional services. In many organizations, this framework has become a driving force, part of its culture, dictating how to perceive and act on information. Organizational norms become the filtration system, determining how information is shared, processed, and reacted to. The organizational culture is an unforeseen force that, if not addressed and updated, will stifle the best ideas, the most detailed plans, and the most progressive leaders.

Dealing with organizational culture can be one of the most challenging and frustrating aspects of leadership, yet it can pay the most dividends. A mark of true leadership is the ability to break down the cultural barriers existing within an organization, to provide new levels of thinking and openness. Joel Barker, a futurist, has written several books on how paradigms (frameworks) may affect both the ability to lead change and the ability of people to accept it.[4] Both organizations and individuals have conceptual structures that determine how they process information, how they communicate about it, and how they react to it. These structures, which are developed over time and are a culmination of experience and culture, (1) define or establish the boundaries in which to operate and (2) establish norms about how to behave inside these boundaries to be successful. Today the fire service faces a combination of many challenges, new service level demands, technological choices, new workforce challenges, and an economic crisis never before experienced by a workforce. Many opportunities exist to make revolutionary shifts in the fire profession, and the fire service has never before been armed with such a talented pool of firefighters and such progressive leadership. Yet if fire officers cannot recognize and understand the changing environment and help shape their departments to form an organizational culture that expects excellence, the true potential of their departments will never be realized. Paradigms often limit this.

When Complacency Becomes the Benchmark

Considerable attention has been focused on individual teams within organizations, but what about the organizational team as a whole? In the absence of leadership, organizations will arrive at a midline level of performance, driven not by the pursuit of excellence but by the preservation of the status quo. The entire team can become complacent in its **attitudes** and actions. An attitude is a complex mental state involving beliefs, feelings, values, and dispositions to act in certain ways. Complacency then becomes the benchmark, or standard, and the whole organization, from a performance perspective, is not what it should be. Such complacency promotes organizational decline. Conversely, when best in class becomes the organizational benchmark, the department is utilizing performance measures to evaluate its efficiency and effectiveness. The organization as a whole is looking for new methods and techniques to improve services and focuses on professional development at all levels of the organization. These traits have become a core value: the culture of the organization has embraced them, and they have become part of the fabric of how the department operates on a daily basis.

attitude
■ A complex mental state involving beliefs, feelings, values, and dispositions to act in certain ways.

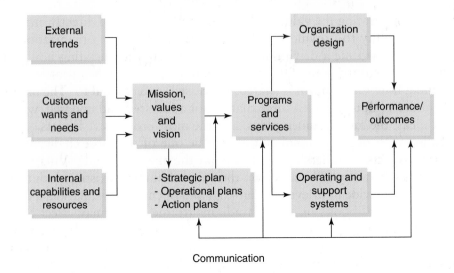

FIGURE 3.1 Influences of Organizational Change

When one looks at a fire department as a system, many forces and interrelationships have impact upon it. Each department needs to incorporate these influences, and then it can exercise a significant degree of authority because of its ability to meet its mission and provide quality services. Such influences as external trends, customer expectations, and the department's capabilities have a direct impact on the programs and services provided and are distinct factors in any organizational change process (see Figure 3.1). However, if complacency has become the norm, it leaves those in the department vulnerable and has created organizational apathy.

Organizations that are out of touch with the changing context of the profession and the community they serve, find themselves in an organizational comfort zone—a dangerous place not only for the fire service profession and its firefighters but also for individual fire companies or divisions within an organization. Experienced chiefs must avoid this especially vulnerable place, where nothing much happens because the need for bureaucracy, fellowship, or socialization is greater than the need for innovation. These organizations do not recognize the need for progress and new ideas in planning. These employees or volunteers come to work or respond to a call banking on nothing of significance happening. We all know the rest of the story! When they are called upon, their performance is less than acceptable, the outcome of the situation is not what it could have been, and the risk to those involved is increased.

Such complacency often creates an organizational culture that has minimal environmental perspectives and often cannot react when it needs to. The reason is it never sees or senses the existing threats. An analogy referred to as the boiled frog syndrome describes what organizations experience when they become complacent about what is occurring within and around them. If a frog is placed in a pan of hot water, it will immediately jump out. More often than not, it will survive the experience. If the same frog is placed in a pan of water at room temperature and then the water is heated very slowly, the frog will stay in the pan until it boils to death. The frog could have jumped out, but the change happened so slowly that the frog did not realize the danger it was in.

At times, organizations that do not respond to internal or external forces in time to avoid significant damage often find themselves in the same situation as the frog. Today's fire service has no room for such organizations that are "playing fire department." With approximately 30,000 fire agencies across the United States, the range of departmental ability can vary greatly. Unfortunately, the expectation of the community (whether a small rural community in the middle of Indiana or a major city such as Chicago) is often the same: that the department is professionally staffed, fully equipped, well trained, and ready to handle any emergency that may arise. Unfortunately, some in the fire service have lulled themselves into believing they can actually provide an all-hazard level of service to their communities, even though, in reality, they do not have the resources, training, or experience to do so. Complacency is an organizational cancer, and if left untreated, it will suck the life out of the organization and its people.

THE IMPORTANCE OF ORGANIZATIONAL CULTURE ON CHANGE

The importance of organizational culture on the ability to process change must not be overlooked. The organizations that possess the ability to facilitate change in a positive way have the following:

1. A developed, clear, and compelling purpose
2. An identified organizational mission to achieve the purpose
3. An agreed-upon set of values by which to carry out the mission
4. An adopted servant-leader attitude throughout the organization

An organization's purpose is the "why" of its existence. It is not what it does as much as what it is striving to accomplish. Its purpose is a statement of the greater good it is attempting to achieve. It answers the question: Why are we here? and helps give clarity and focus to each person in the organization. It is the yardstick by which decisions are measured.

An organization's mission is the "what" of an organization. It is a definition of and a focus on the core competencies—what the company does to achieve its stated purpose.

An organization's set of values is the "how" of an organization. It defines what a company values most in the execution of its mission. It is not an all-encompassing list of possible values as much as a statement of what the organization esteems most in its people and their conduct. It identifies behaviors and culture within an organization and helps set the guidelines of what is and is not acceptable.

At the core is the premise that the customer is the most important person to the organization. It only follows that the most important people to the customer are the frontline staff, the people customers interact with on a daily basis. This understanding leads to the philosophy that leadership's job is to support the people on the frontlines, to make their jobs as easy and effective as possible so the customer has the best experience possible. This results in an organizational chart that looks like an inverted pyramid. This servant-leader attitude focuses leaders on developing those around them, leading people to work together in a collaborative, solution-oriented environment.

How does an organization develop its purpose, mission, and values? It is critical for the organization to understand the importance of everyone within it, because the creation of purpose, mission, and values requires input from people at all levels of the organization. The purpose, mission, and values need to be relevant to everyone involved, to be consistent with one another, and to be used consistently as a yardstick for decisions and policies. There is nothing worse than a department developing values and then just paying them lip service by not putting them into effect on a daily basis. Such practice lacks integrity and actually becomes a demoralizer.[5]

SERVICE, CORE VALUES, STEWARDSHIP, AND CHANGE

Leadership is more than command and control, although in the fire service, it does mean having a command presence. Yet, leaders who have a command presence may not be effective, appreciated, or followed. The most successful leaders are those who ensure that service becomes a core value of the organization. In leading by example, these leaders strive to institutionalize the basic concept of service over self. As a consequence, the most effective organizations in the fire service and in the private sector embody this core value of service in their day-to-day operations. Think about it. Would you rather buy a product from a company whose core value is to make you happy through "best in class" customer service or one whose core value is to make a profit, no matter what? Public service organizations must have a broad perspective on what is occurring in their community. They cannot be self-absorbed, but instead must (1) understand community dynamics and politics and (2) have the flexibility and ability to change rapidly. Core values separate good organizations from those that are world class.

The fire service must ensure that its basic skills and core competencies are relevant. These core competencies focus on what has to be done from a technical standpoint; but the core value of service goes beyond the ability to respond to an emergency once the alarm is sounded in the firehouse to embody who the fire service is and the expectations those in it have for one another. The acceptance of stewardship as a core value underlying the services provided is the basis of an industry's ability to grow to its full potential. Stewardship is a responsibility for taking care of the resources, both economic and human capital, with which one has been entrusted.

Keshavan Nair, in counseling *Fortune 500* companies on leadership and decision making, claims that core values are not only policy but also something deeper and more profound. He suggests that every individual in the organization values being of service. *When service to others is valued, then trust, loyalty, and truth flourish.* These create efficiencies of trust, trust creates positive energy, and positive energy fuels the organization.

Service, leadership, and stewardship are intertwined. Successful fire service leaders, those who have the core value of stewardship, place service above self and truly believe organizational service is a core value—a self-sustaining value that has tremendous power. Keshavan Nair writes that the acceptance of a core value implies a commitment.[6] Core values within an organization and core values in service are long-term commitments. When leaders have both belief and commitment, they and their organizations can achieve great things. Core values often become the foundations of future traditions. Once leaders have articulated that service is a core value, the challenge is to determine how to influence others within the organization to act in a way that supports the department's core values. This starts with the leader leading by example to ensure he or she is doing

FIGURE 3.2
Above-the-Line—Steps
to Accountability

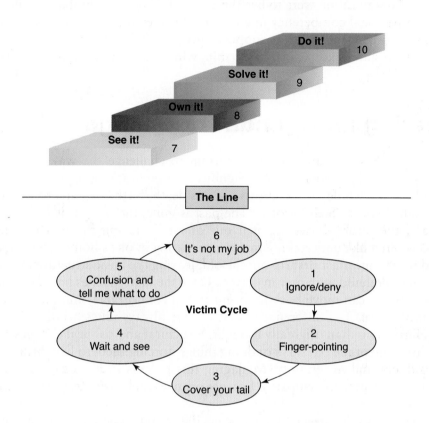

more than he or she is paid to do. The model of good leadership is the most effective way to inspire others' commitment. As Albert Schweitzer said, "Example is not the main thing in influencing others; it's the only thing."

The alignment of organizational philosophy, operations, mission, vision, and values is critical to this endeavor. This may best be summed up by the saying "We must leave it better than it was when we found it." It is vital to create in all organizations, whether public or private, a universal thought process and buy-in to the concept of **above-the-line accountability**. This term expresses the idea that "If you see it, you own it. If you own it, you can help to solve it. If you can solve it, you should just do it." Accountability only occurs above the line; and when it does, it is very powerful (see Figure 3.2).

Another aspect of organizational culture that can be very debilitating is the opposite of above-the-line accountability, referred to as the victim cycle. Those who find themselves in the victim cycle often ignore what is happening around them. They use finger-pointing to protect themselves, refuse to get involved, wait to see what happens, create confusion through starting negative rumors and other passive-aggressive behavior, and state, "It's not my job." Accountability does not result when people have placed themselves in the victim cycle. Can you think of a situation in your own department where you have seen this played out? The result would be to expend a greater degree of organizational energy to solve an entrenched problem because those involved were part of the victim cycle and did not deal with it in the first place. Just think what the organizational power,

above-the-line accountability
■ A personal choice to take the ownership necessary to solve a problem, not ignore it. The idea if you see it, you own it. If you own it, you can help to solve it. If you can solve it, you should just do it.

customer service, performance, and time saved would look like instead if the majority of an organization were to live the above-the-line accountability philosophy.

When personal competency in all areas of leadership skills is combined with an organizational culture that supports people, their development, and their success, the result is exceptional leadership, which, in turn, inspires the best effort in others.

Creating an Agenda for Change

Going through any major change is certain to challenge the way leaders view themselves and their organization. Significant organizational changes can be as stressful, on the whole, as life events such as death, birth, and divorce are for an individual. David L. Stein wrote, "The past is gone; the present is full of confusion; and the future scares the hell out of me!"[7] Living through this change process is often like undertaking major remodeling in one's home. To achieve the desired result, one must first rip out the old, leaving the basic structure intact, and then begin to rebuild with new materials. Once the final touches have been added, the homeowner can move back in, begin to feel comfortable, and once again become productive. The difference is that change always takes a little longer and costs a little more than originally thought. Structural and cultural changes require people to let go of the old ways of doing things, forcing them to live through a period of doubt and uncertainty. Managing this process in an organization takes sensitivity, patience, and empathy, because change can be very frightening and unsettling to employees.

As has been said earlier, change is inevitable. Whether someone has been on the job for 1 year or 25 years, he or she has already been witness to change. Hence, a good question is, Will we always be riding this wave of transition? The answer is yes, for without change, people and organizations would become stale and unresponsive. The challenge is to learn how to navigate through the transition as easily and creatively as possible. What helps people plot a course through this unknown territory is a map of what they can expect and information on how they can respond most effectively to challenges as they occur.

Many in the fire service today have heard the old adage regarding the profession: "two hundred years of tradition, unhampered by change" (which, by the way, was never a truism for the fire service). Yet from an organizational standpoint, one of the challenges the fire service faces is to motivate many of its firefighting personnel to look past the four walls of the fire station and understand that the world around them is changing very rapidly. As the environment that interacts with the fire profession is changing, it places significantly greater demands on the department and everyone within it.

To successfully steer through this inevitable, ever-changing environment, the leadership of the department must make a commitment to share as much information as possible with everyone within the department, and an expectation must be set that everyone will read and/or receive this information. Those in leadership roles have to understand the resistance to change; it is part of the fire service culture, and for many people it is their natural tendency. Change is easier said than done, and whether the fire service tries to alter its environment, the workplace, or the services offered, the change destabilizes organizations. In some cases, it creates fear. A sense of permanence and tradition has been the backbone of the fire service and has become a significant influence in the development of the culture of the profession. Permanence and tradition have provided stability for fire service organizations, but in today's environment they have also produced many departments that are inflexible, unwilling to change, and bureaucratic in their approach.

The need for change is evident; just look around. Almost everything in our environment is being altered; therefore, we must adapt and change with it. Remember the old saying, "If you always do what you have always done, you will always get what you have always gotten." To successfully navigate the changes that will be needed under fire leadership, chief officials must understand the importance of culture, vision, and the need to possess a global view. The ideals leaders must promote to be successful—the importance of leadership, the empowerment of others, the ability to measure performance, and understanding how change affects people and organizations—are critical in leading effective change whether one is a fire chief, a battalion commander, a captain, or a young firefighter.

Strategic Management of Change

The National Fire Academy's Executive Fire Officer Program, Strategic Management of Change, was initially developed by the Department of Defense (DOD). The fire service uses this document to provide critical thought to all aspects of the change process and to facilitate institutionalizing the change(s).

This change management model involves a systematic progression of behaviors designed to assist senior fire executives who must facilitate and adapt to rapid change in the delivery of fire and emergency services. They can use this tool to provide direction both for managing change brought upon the organization and for seeking opportunities through change. The model facilitates effective change through a systematic, four-phase process: analysis, planning, implementation, and evaluation/institutionalism.

ANALYSIS

The first phase in the change management model involves analyzing the existing situation and assessing what changes need to be made. The analysis phase is an overall needs assessment that senior fire executives perform to identify the influences creating the need for change. This identification determines the overall magnitude of change required and defines specific organizational change requirements.

Identify Organizational Conditions and Compare to Existing Mission, Standards, Values, and Norms

The first task in the analysis phase is to assess internal organizational conditions to determine whether change is needed. This is accomplished by identifying those conditions and comparing them to organizational touchstones such as quality of services, ethical standards, morale and attrition, existing culture and history of change, and any other internal indicators that may suggest the need for organizational change.

Identify Potential Destabilizing Forces

After identifying and comparing internal organizational conditions, the next task is to identify and forecast potential destabilizing forces that may influence or dictate organizational change. This focuses on the forces originating outside the organization and may address technological developments, economic pressures, social issues, and political and legal factors.

Assess the Impact of Current Organizational Conditions and Potential Destabilizing Forces

Once internal organizational conditions requiring change and external destabilizing forces are identified, they must be assessed to determine their impact on what is necessary to institute organizational change. This attempt to broadly define the need for change helps to determine the current, short-term, and long-term requirements to bring about the desired change.

Determine Organizational Change Requirements

By more specifically quantifying change requirements, senior fire executives can assess whether the change reaches all aspects of the organization or addresses only a few discrete areas. Are short-term results needed or would a long-term view be more productive? What if the organization were to simply stay its course and make no change? Answering these types of questions involves a formal determination of the types and amounts of organizational change required.

Determine the Perspective of Change

The first formal determination involves a focus on the perspective of change needed. After determining the overall type or form of change to be made, fire executives have a choice of three primary perspectives of change.

The *developmental change perspective* is appropriate when the organization needs to improve upon current skills, methods, or conditions that do not meet current expectations and standards. Over the past two decades, the impact of technology on the tools and devices used daily by firefighting personnel has changed dramatically. The adoption of new standards and legislation has also compelled changes in daily operations, forcing a developmental change.

The *transitional perspective* shifts the emphasis from developing existing methods and processes to actually replacing them. For example, introducing a new technology into a fire service organization may do away with some existing processes and systems while requiring completely new ones. Facilitating this replacement of the old with the new may necessitate several transitional steps to successfully ease the conversion—temporary workforce arrangements, pilot testing, phase-in operations, and so on. The defining characteristic of the transitional perspective is replacement through gradual evolution.

The final perspective of change, the *transformational perspective*, involves entirely replacing existing organizational characteristics. The transformational perspective differs from the transitional in the extent of replacement involved. This perspective requires developing new beliefs and systems and gaining organization-wide commitment for them. The transformational perspective envisions a radical reconceptualization of the organization's mission, culture, critical success factors, form of leadership, and/or other defining characteristics of the organization.

Determine the Magnitude of Change

The second formal determination involves deciding on the organizational change requirements, which focus on the magnitude of change required. There are three variations or aspects of the magnitude of change:

- Pace of the change
- Scope of the change
- Depth of the change

Determine the Objects of Change

The task of determining organizational change requirements focuses on the objects of the change, or simply those elements of the organization that will be changed. Typically, there are three primary objects of change:

1. Individual task behaviors
2. Strategic direction
3. Organizational culture

Change, regardless of which of the three objects it refers to, is the adoption of new behaviors. Whether the employees are learning new skills (application behaviors), basing decisions on new strategic directions (personality behaviors based on knowledge and skills guided by an understanding of the organization's strategy), or realigning their values with the organization's (culture), all reference behaviors. Change is not complete until this happens.[8]

PLANNING

The second phase in the change management model is planning. In this phase the information the fire executives gather during analysis is used to formulate a plan designed to bring about the desired change. The goal of planning is to translate the change requirements into detailed, strategically sound plans to accomplish the desired change. The planning phase generally involves developing a vision of the change; defining goals, objectives, methods, and strategies to achieve that vision; and identifying organizational characteristics and processes that can impede achievement of that vision.

Examine the Forces for and Against Change Systematically

Organizational situations are typically held in equilibrium by two sets of forces—those that stimulate change and and those that restrain or oppose change. Stimulating forces encourage change, facilitating movement in a new direction. Restraining forces hinder or prevent change, restraining movement to a new situation. In order for an organization to overcome the status quo, stimulating forces must exceed restraining forces.

Examining these forces consists of periodically investigating the stimulating forces and the restraining forces and then identifying the need to strengthen the stimulating forces and/or reduce the restraining forces.

Select Personnel to Develop a Vision of the Organizational Change

After examining these stimulating and restraining forces and identifying those that need to be strengthened or reduced, the next task is to select personnel to develop a vision of the organizational change. Selecting these people requires adopting one of three personnel strategies for vision development.

1. *Executive officer/leader strategy.* The first personnel strategy, commonly used in many organizations, is to select an executive officer or leader to develop the vision of organizational change.
2. *Executive/senior officer team strategy.* The second personnel strategy also concentrates the vision development power in the upper levels of an organization but involves selecting a team of executive or senior officers to develop the vision.
3. *Bottom-up team strategy.* The third personnel strategy involves selecting a bottom-up team. It relies on gathering input regarding the organizational change vision from lower-level personnel, such as staff employees.

Envision the Organizational Change to Be Implemented

Once personnel have been selected to develop the change vision, they must actually envision the organizational change to be implemented. The purpose of this task is to translate the change requirement formulated during the analysis phase into a vision of the desired state or position of the organization.

Generate the Desired State/Position to Be Achieved Through the Change Implementation

Envisioning organizational change means to actually produce the organizational vision. An organizational vision is the big picture statement of what a desired future for the organization would look like both internally and externally. It describes what services the public expects from the organization and how the organization will deliver those services. It becomes a guide for the organization to follow into the future. The vision must be aligned with the mission, values, and operations of the department.

Next, the articulated vision statement must be evaluated in terms of its completeness, strategic soundness, and overall feasibility.

Develop a Road Map to Achieve the Envisioned Organizational Change

It is vital to develop a structured or sequenced framework (the map) of the most important activities that must be performed to achieve the envisioned change.

Generate Ideas That Inspire and Create Emotional Appeal to Change Recipients

How the vision is presented to the organizational members affected by the change determines how to frame the vision and helps to explain the need for change. The objective of these ideas should be to inspire organizational members to support the change wholly and to do whatever is necessary to make the change a reality. Creating this connection helps individuals begin to buy into the change process emotionally.

Set and Evaluate Target Goals and Objectives of the Envisioned Change

After the desired future state of the organization is established, with the map of how to reach that state in hand, the next task is to set and evaluate goals and objectives. These goals and objectives must match the envisioned change and be consistent with the organizational change requirements identified in the analysis phase.

Select Method(s) of Change to Be Employed

Selecting the actual method or methods to use in bringing about the desired change can be done once fire executives know the fire service's destination and have developed a map for how to get there. This involves assessing the applicability of the four different types of change methods to the envisioned change:

1. *Technical method of change.* This method of change involves altering the way services are provided or how an organizational output is produced. An example of this is the total quality management (TQM) approach for improving productivity and reducing defects and errors. A similar method that is gaining popularity in the emergency services profession is system status management (SSM). The essence of both of these types of programs is to change the way an organization conducts its business.

2. *Structural method of change.* This method of change may focus on altering the structure of specific jobs or modifying organizational roles or relationships. Examples of its application to job redesign include job rotation, job enrichment, and job relationship changes. Modifying the organizational structure basically involves changes to the structure's complexity, formalization, centralization, and coordination. The following are some examples of these changes:

- *Complexity.* Number of departments, different occupational groups, highly trained specialists, layers of managerial levels, hierarchical structure (i.e., horizontal vs. vertical differentiation)
- *Formalization.* Degree to which rules and regulations govern employees' behaviors
- *Centralization.* Degree to which employees participate in making decisions
- *Coordination.* Process of integrating differentiated resources and activities in unity of effort

3. *Managerial method of change.* Examples of this method of change include modifying the evaluation process, assessing the merit systems, and enhancing cooperation between management and labor.

4. *People method of change.* This final method of change involves the organizational members themselves. Its essence is to bring about the desired change through actively engaging the people who work in the organization.

Education and training, which are aimed at upgrading people's knowledge, skills, attitudes, and beliefs, comprise one form of the people method of change.

Organizational development, another form of the people method of change, emphasizes planned interventions into various aspects of organizational life. These interventions may address individual practices, group practices, or system-wide processes, and include such activities as personal coaching and counseling, team-building activities, and process consultation.

Select Techniques to Promote the Change

Whereas the selection of change methods concentrates on *what* to do to bring about the desired change, the selection of techniques to promote change focuses on *how* to put the methods of change into place. In other words, assessing and selecting techniques to promote change deal with how to get organizational members to carry out the selected change methods. There are four types of techniques from which to choose:

1. *Facilitative techniques.* These methods of promoting change use managerial authority to facilitate the change but seek significant interaction with the group to be affected by the change. Facilitative techniques are most effective when members of the group have some sense of what they want to do but do not have all the means to do it.

2. *Informational techniques.* Change managers attempting to promote change through informational techniques demonstrate the rationale for the change by educating employees and providing them with factual information. It is assumed that, given adequate factual information regarding a change, personnel will act rationally, recognize the problem, and come to a mutually agreeable solution.

3. *Attitudinal techniques.* Attitudinal techniques rely on the use of persuasive messages and communications to ultimately produce changes in behavior. The idea is that changing personnel attitudes will cause a change in personnel behaviors. It is important to note that, whereas informational techniques rely on strictly factual information to appeal to employees' rationality, attitudinal techniques rely on persuasive information to appeal to employees' emotions or beliefs.

4. *Political techniques.* These methods depend on the use of scarce resources by giving them to or withholding them from a particular group, competing for them, or bargaining for them. These techniques are particularly variable in their nature and can run the gamut from unilateral coercion to complex maneuvering.

At one end of the political technique spectrum is the power-type technique, which is likely to depend on giving or withholding resources. This method is more effective given a change that is small in magnitude, when a change needs to be completed expeditiously, or when a change can be divided into small components.

The other end of the political technique spectrum includes artistic negotiating and aligning of groups and forces within an organization. These methods involve complex maneuvering that is likely to hinge on competing or bargaining for resources. These types of political techniques are more effective when the proposed change is extensive and cannot be easily broken into subparts.[9]

IMPLEMENTATION

The third phase in the change management model is implementation. During the implementation phase, the procedures and strategies detailed during the planning phase are executed, and behaviors most likely to ensure a successful implementation are performed. Unanticipated difficulties are most likely to occur at this stage of the change management model and process. To avoid or mitigate such difficulties, a variety of mechanisms should be put into place to facilitate implementation.

Create an Environment of Shared Vision and Common Direction

Create an environment that unites and promotes overall support of envisioned change. Such an atmosphere should engender a common direction that those at all levels of the organization perceive and support. In creating this environment and common direction, one must develop the following:

- *An appropriate communication strategy.* These strategies can vary from simple announcements by supervisors to all-hands meetings designed to celebrate the new organizational vision.
- *Political sponsorship.* This means gaining the backing of leaders within the organization, including those most receptive to the proposed change, informal leaders within the organization, and personnel who view the proposed change positively and who can instill that view in other organization members.

Minimize Initial Resistance to Change Through Effective Communications

Effective communications regarding a proposed change can go a long way toward minimizing resistance to that change. Many people seek to avoid change because they perceive it as depriving them of control over a situation. Communicating the rational implications of a proposed change, along with how the change will be instituted, can often restore this sense of control.

A number of steps can be taken to communicate these types of information to organizational members, all of which relate back to the decisions made during the analysis and planning phases. These steps may include the following:

- Describe where the organization is now, where it needs to go, and how it will get there.
- Explain the business rationale for the change.
- Explain who will implement the change and who will be affected by it.
- Describe the negative aspects and personal ramifications of the change.
- Quantify the change by explaining its success criteria, how it will be evaluated, and its related rewards.
- Describe the timing and pacing of implementation.
- Communicate key organizational activities that will not be changing.
- Convey the organizational commitment to the change.
- Explain how people will be kept informed throughout the change process.
- Utilize a diversity of communication styles and presentation formats to communicate the messages regarding the proposed changes.

Create a Sense of Urgency and Pace for the Change

It is important to focus on the change implementors, typically middle managers, who will be responsible for putting the change plan into operation. In the fire service these are the battalion chiefs, captains, and lieutenants in the field who may often perceive that the change needed is not as imminent as upper management believes or that more moderate changes will work just as well as the proposed changes. It is imperative to ensure that the personnel in these positions recognize the needed urgency and pace and to emphasize leaving behind the old way of doing things. The leadership of the department should reexamine the various strategies to support change in promoting this sense of urgency and pace.

Develop and Implement Change-Enabling Mechanisms

Lay the groundwork for the proposed change by creating new mechanisms that serve as precursors to the organizational change. These change-enabling mechanisms should relate back to the selection of change methods and strategies to promote the change.

The two types of change-enabling mechanisms are practical mechanisms and symbolic mechanisms. Practical change mechanisms involve actual workplace programs, processes, or systems. Symbolic change mechanisms may range from themes and logos promoting the change to rearrangement of office spaces. Practical mechanisms are often closely tied to the methods of change, whereas symbolic mechanisms often relate to strategies that help to promote change.

Implement Planned Change Methods and Strategies

The culmination of the initial implementation is to put into place the planned change methods and strategies and involves the following:

1. Selecting how implementation will occur—what actions are necessary to implement
2. Ensuring change strategists fully support the change implementation
3. Ensuring change implementers are aware of the microdynamics of the change effort responsibilities[10]

EVALUATION/INSTITUTIONALISM

The fourth and final phase in the change management model is evaluation/institutionalism. Once a change management plan is implemented, it must be continuously, systematically monitored to evaluate how and whether the change process is working as anticipated. An additional measure of success is that of institutionalism, which results in witnessing a fundamental change in organizational behavior, process, and the rules, systems, and procedures that were in place.

Evaluate Initial Change Implementation

Evaluate the initial change implementation using the following to measure its success:

1. *Evaluate the implementation against the initial change goals.* Assessment of whether a change management approach is working is determined by evaluating the effects of the implementation against the goals and objectives set out in the change plan. In order to evaluate the implementation in relation to the change goals effectively, those goals must be explicit, precise, and quantifiable. If the plan is not working as anticipated, the approach must be adjusted.

2. *Evaluate the implementation against the described future state.* It is at this point that the completeness of the envisioned and articulated desired organizational state becomes particularly important.

3. *Evaluate how well established, or institutionalized, the change becomes.* A change is well established if it persists without additional or excessive controls

to sustain the new behaviors, relationships, or activities it was intended to bring about.

4. *Evaluate how rapidly the change was accomplished.* The speed with which a change is accomplished must be measured against the decisions made regarding the urgency and pace of the change.

5. *Evaluate costs to individuals and the organization of conducting the change.* Costs associated with a change may be either economic or noneconomic. Organizational costs are likely to be measured in economic terms. On the other hand, costs to individuals may be economic or may simply relate to whether the implemented change is more of a benefit or a burden to those organizational members affected by it.

6. *Identify the number of unanticipated actions and occurrences the change generates.* Implementation is the phase in which the unexpected should be expected. Evaluation is the phase in which it is determined how well the unanticipated was anticipated during the planning phase.

7. *Assess initial resistance to change.* The key to assessing the initial resistance to a change is to identify attempts to maintain the status quo. These attempts may result from a lack of understanding or acceptance of the change on the part of members of the organization. Initial resistance may also arise from particular conditions within the members of the target group or within the organization as a whole that prevent organizational members from acting to implement the change.

Exercise Flexibility in the Change Management Approach

Leaders and managers must be willing to modify as necessary the initially formulated change management approach (plan) based on the results of the implementation evaluation. Modification of the change management approach is, like evaluation, an ongoing process involving several activities that relate back to earlier phases and steps in the change management model. Flexibility in the change process is a must for leaders and managers who often have to alter or modify strategies, goals, objectives, and methods to be successful.

Continue to Monitor and Institutionalize Change Implementation

It is imperative to continue monitoring and demonstrating a commitment to the change implementation. This can be accomplished by the following:

- Monitoring and reinforcing the new culture
- Promoting risk taking related to strategic change management
- Promoting incorporation of new behaviors into day-to-day operations of the organization
- Removing the means required to perform in the old way and providing only the means to act in the new way, if possible

Although institutionalism is the goal in any change management project, its achievement does not signal the end of the process. Some level of continuous monitoring is required even after institutionalism occurs.[11]

The strategic management of change model provides an excellent framework for leaders and managers to facilitate a change process. Yet, it is often the personal resistance to change and the negativity some individuals in the department demonstrate that may present the greatest challenges in the organizational change process.

THE RESPONSE TO CHANGE

During times of great change, leaders have to ask their employees to be flexible; if they do not, leaders cannot position the department to be competitive in the future. Yet for 99 percent of the fire service, there is a monopoly about who is going to provide the service. *So why the need to be competitive?* Every year at budget time, the fire service is in competition for money and resources to do its job efficiently, effectively, and safely. Therefore, the ability to maneuver and take advantage of opportunities when they arise is critical to the profession's long-term health. This has never been more evident than in the past decade, which included times of both economic vitality and crisis. The forces of change experienced in the fire service will come from both inside and outside the organization. Leaders will definitely be challenged by more demands from constituents, the constant push for cost containment, the need for quality, shifts in the political arena, and an expanded mission to include homeland security issues and traditional culture. As leaders move through such changes, they must understand their organizations, their culture, and their effect on the change process. Whether change is initiated from inside or outside an organization, culture plays a large part in determining how it will be processed, accepted, and ultimately dealt with. Cultures and the strategies for change are often in conflict. When conflict occurs internally and the organizational culture does not embrace the changes that have been initiated, the change efforts often involve a struggle and in many cases will not be successful.

In contrast, when the change is externally driven, both the organization and the fire chief can find themselves in extremely vulnerable positions. In many cases, this form of change is often the most dangerous. Whether it is political, economic, or service driven, both the organization and the people within it can find themselves in a no-win situation. Often, change caused by consolidation, by a need to save money, or by expanding service beyond the traditional mission is met by organizational resistance. In such cases the change effort often still occurs, and the organization and its culture are permanently altered.

During the most recent economic crisis, the country witnessed how an industry's culture can influence the direction it takes. The United States automobile industry, which in the 1980s controlled over 90 percent of the U.S. market share, had fallen to a point where the survivability of the major manufacturers (GM, Ford, and Chrysler) was in doubt. For years consumer surveys had been sending clear signals to corporate leadership indicating the future was in smaller, more fuel-efficient vehicles with an emphasis on quality. The American auto industry failed to recognize this shift, which has resulted in questioning the survivability of this American institution. Why did it happen? The culture of the profession, in both labor and management, was often impervious to what was occurring. Throughout the 1980s and 1990s, the leadership continued to produce the same product lines, with fewer and fewer sales. During this same period, labor contracts continued to add costs, and some would argue that government regulation and U.S. import and export policies also factored into the mix. This situation is an example of how both internal and external environmental forces, although identified, were ignored to the point the industry was placed at risk. Every industry, including the fire service profession, should take note of this. The fire service is beginning to experience the same phenomena related to the traditional methods of service delivery and increased personnel cost, specifically in the areas of pension and health care benefits. So, the fire service must recognize and correct its course, or in the future it may find its own profession where the Big Three automakers found themselves in 2008–2009.

The future fire service must promote a culture that is much more responsive, adaptive, and accepting of change, which will keep the fire service competitive. This *desired future* will be one that not only maintains the heritage and traditions of the fire service but also provides for the ability to adapt rapidly to the changing environment; offers innovation and creativity, overcoming the mind-set "because we have always done it this way"; and becomes a proactive profession that focuses time and resources on preventing rather than reacting.

WHAT CAUSES RESISTANCE TO CHANGE?

Over the years, a lot of time and effort has been spent studying the causes of failed organizational change initiatives. Research has found that employee resistance and a lack of proper training are key hindrances to organizational change. People know that realignment of the hierarchy within an organization or division means that, no matter what type of change occurs, someone or some group will ultimately lose or gain "power" as a result. Until the costs/benefits are made clear, staff will anxiously await the effect the change will have on them personally.

Here are some of the common employee concerns about change:

- *Fear of job loss.* A common fear during a change initiative is whether positions will become redundant or jobs will become unnecessary.
- *Fear of increased responsibility.* Some staff may question whether they will have more responsibilities and/or accountabilities as a result of a change.
- *Frustration with the process.* If personnel have not been consulted before, during, or after an organizational change, they will likely become disheartened, particularly if the change directly impacts their jobs. If they were consulted and their positions or suggestions were not incorporated, they will also become frustrated, unless their managers can satisfactorily show their input was not ignored but, rather, discussed and shelved for logical reasons[12] (see Figure 3.3).

The five patterns of behavior and response to change are innovators, explorers, late adopters, resisters, and refusers. It is helpful as a leader to understand these reactions as a change process is implemented.

Innovators

People who have a firm commitment to the organization and are motivated to stay on top professionally are usually on a continual quest to be at the cutting edge in everything they do. These *innovators* are often the most creative people in the organization, the first to embrace new ideas. Their commitment to achieve a competitive edge through knowledge is very high. These people, referred to as "lead ducks," are often shot at because they will be the first out of the water on an issue, to the point where they can make their peers uncomfortable with their

FIGURE 3.3
Individual Response
to Change

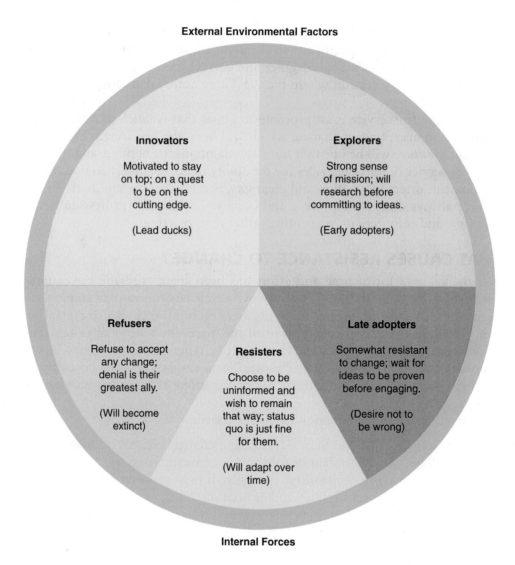

External Environmental Factors

Innovators

Motivated to stay
on top; on a quest
to be on the
cutting edge.

(Lead ducks)

Explorers

Strong sense
of mission; will
research before
committing to ideas.

(Early adopters)

Refusers

Refuse to accept
any change;
denial is their
greatest ally.

(Will become
extinct)

Resisters

Choose to be
uninformed and
wish to remain
that way; status
quo is just fine
for them.

(Will adapt over
time)

Late adopters

Somewhat resistant
to change; wait for
ideas to be proven
before engaging.

(Desire not to
be wrong)

Internal Forces

exuberance. They are usually the earliest to arrive to the appointed destination; then they are off to the next adventure.

Explorers

The next category is those referred to as the *explorers,* or *early adopters.* These people in organizations have a high degree of curiosity and a strong sense of mission. They are on the leading edge of change and do not mind taking a few risks along the way. They tend not to be lead ducks but are definitely near the front of the formation. They will research the project or issue before they commit to it, but once they do, they will be a strong part of the team.

Late Adopters

The next category consists of the *late adopters,* who are somewhat resistant to change, slightly timid, and desire never to be wrong. They wait for ideas to be well proven before they jump on the bandwagon. They like to be in the safe zone, with the objective of reducing their vulnerability and risk. They can be very strong performers but join only after some degree of validation of a concept or new idea has occurred.

Resisters

The next category is the *resisters,* who are uninformed and want to stay that way. They work hard to maintain the status quo based on their apathy, negativity, or simple ignorance. Change can overcome the resisters and, in most cases, they will adapt over time. However, it often takes an enormous amount of organizational energy to do so.

Refusers

This last category, in which too many people and, in some cases, organizations are found, is the *refusers.* They have chosen not to progress. They refuse to accept any change within the organization or in the world around them. This is why they are often referred to as about to become extinct. Choosing to follow the road of the past leads them to a community called obsolescence. They live in a time warp, and the only way to deal with their denial of the change occurring is to wait for or motivate them to retire.[13]

The Negativity Factor

Who has not experienced bouts of negativity about the changes occurring at work? Who has not become negative when a colleague or friend loses his or her job? Who has not been put off by the individual who complains about everything the department does?

Everyone can find a reason to become negative about his or her work or organization from time to time. However, **the negativity factor** is the personal chemistry of some people whose expression is pessimistic, gloomy, hostile, and detrimental. When negativity becomes a routine posture for the entire company, the employee, or others around the employee, it can take hold like a virus and begin to eat away at everyone's performance.

the negativity factor
■ The personal chemistry of some people whose expression is pessimistic, gloomy, hostile, and detrimental.

U.S. companies lose billons of dollars a year to the effects of negativity, according to the Bureau of Labor Statistics. The fire service has a serious problem when negativity affects productivity, revenue, expenses, employee and customer relationships, medical claim rates, absences, or turnover. Negativity can stem from organizational negativity and individual negativity, and each can impact the attitude of many in the organization. As such, this virus must continually be attacked, or it may create long-term damage to the organization.

In the case of organizational negativity, the department may affect the attitudes of employees in negative ways through its procedures, policies, structure, culture, and management style. Major causes of organizational negativity are mismanaged change, inappropriate norms, problems related to levels of trust, and employees who are unhappy in their personal lives. Individual negativity is the virus that spreads from one individual to another. The negativity virus can quickly spread in a matter of days or weeks and, once transmitted, is not easily cured.

NEGATIVITY AND THE VICTIM CYCLE

A key element that drives the below-the-line victim cycle mentality (Figure 3.2) is negativity in the workplace. Some of this can be related to the changes an organization is going through. Negativity can quickly spread through a department with devastating impact. It took me a few years' experience as a fire chief to realize some people are not happy with their lives and want everyone to join in their misery. Their cloud of negativity can easily envelope those around them and make them negative as well. This can be seen in the firehouse when one negative person infects an entire crew; the crew can infect another crew and, if not stopped, the battalion and ultimately the department. I have seen negativity move through a department based solely upon a rumor or misinformation. It is amazing how many will jump onto the negativity bandwagon based upon misinformation, and along the way they damage relationships and harm reputations. It just shows how much easier it is to turn on others and be negative rather than to build others up and be positive. Allowing this virus to proceed to a point that the entire organization is "infected" is also a manifestation of poor and uncourageous leaders at *all* levels of the organization. Leaders who allow negative influences to run rampant without correcting those influences are not leading.

A 1998 article by Gary Topchik outlines an interesting perspective on managing complex change in respect to controlling negativity. He states that it is not change itself but rather how the company handles change that determines whether it is viewed negatively or positively. In one scenario, the workforce responded quite negatively to a new computer system that replaced the one in use for the previous seven years. On the surface, it seemed employees were displeased with the system itself. In reality, it was the way in which the change had been handled that led to their negative attitudes. I have found myself in this situation on a couple of occasions due to lack of adequate planning and poorly executing the implementation of the change.

Here are some lessons to be learned from the experiences:

1. *Communicate.* If you do not, those in the firehouse will fill the void with information they believe to be correct or, in some cases, make up scenarios to test the system.
2. *Don't spin the facts.* Change is often a result of an environmental force outside of the chief's control. Many changes may be political and sensitive. If you spin, you will not win.
3. *Understand you will not appease everyone.* If you try to satisfy everyone's needs or concerns, the change will not happen. Change will upset some people no matter what the change is.

MANAGING THE ELEMENTS OF CHANGE

A method of managing organizational change that can dramatically reduce negativity includes the following five elements. If any one element is missing, it impacts the change initiative negatively (see Table 3.1).

1. *Vision.* The better people understand the reason for change, the more they will remain positive about it.
2. *Incentives.* If individuals believe a change will benefit them or the organization, negativity is likely to subside.
3. *Skills.* People become anxious—and anxious people become negative—when they feel they lack the skills needed to perform or are placed outside of their comfort zone.

| V | I | S | A | R | RESULT/IMPACT |
VISION	INCENTIVES	SKILLS	ACTION PLAN	RESOURCES	(IF MISSING)
X	X	X	X	X	Positive change
	X	X	X	X	Likely confusion
X	X		X	X	Higher anxiety
X		X	X	X	Gradual change
X	X	X	X		Heightened level of frustration
X	X	X		X	False starts on projects

TABLE 3.1 Managing the Elements of Change

4. *Action plan.*   Individuals like to know the specific steps involved in a change. The more they know about how it will be implemented, the more positive they may remain.
5. *Resources.*   Employees need time, tools, money, and other resources to implement change in a positive manner.

The greater the number of elements that are effectively addressed, the more successful the change initiative will be.

Both organizational negativity and individual negativity occur when the people with whom you work spread the negativity virus to others through their words, actions, and behaviors. The "vaccination" is courageous leaders and leadership.

Such negativity usually results from a lack (or loss) of one of the three Cs: confidence, control, or community. More than 80 percent of negative people do not know they have a negative impact on others, according to interviews with about 500 managers conducted at the University of California at Los Angeles. No one ever gives them feedback about this aspect of their performance. In fact, 75 percent of managers interviewed say they avoid dealing with the negative attitudes of others.

Four strategies have proved to be effective in contending with most negativists. One or more of these may be needed to overcome the negativity of a particular virus carrier, and these strategies are not listed in order of use. The personalities and the issues involved determine the best strategy to use.

1. *Avoid becoming defensive.*   Do not take negativists' words, actions, or behaviors personally as they may not be directed toward you. It may be more of an attack on themselves. If you become defensive, they will become more negative.
2. *Focus on solutions.*   Negativists get into the habit of enumerating all the reasons why something will not work. So by asking them why something should not be done, you become trapped in their spider's web. Instead, ask them for suggestions about what they would do differently. If negativists cannot provide suggestions, end the conversation.

3. *Use the "AAA" list.*   Tell negativists about their *abilities, accomplishments,* and *positive attitudes* when they shine through. Their thoughts about themselves may become more positive over time; and these positive thoughts may influence their actions, words, and behaviors in the future.
4. *Confront the negativist.*   You cannot—must not—ignore the negative behaviors. You will have no choice but to confront unacceptable behaviors. Eventually, negativity will affect the performance of the individual or others, and it may have an impact on the entire organization.

In these cases, treat negativity as you would any other performance or behavior problem. Focus on changing the person's behaviors or actions, not his or her personality; listen to the negativist to identify the reasons for this behavior (this interaction may be enough to motivate some negativists to change); decide what alternative behaviors or actions are needed; and monitor the change. Provide positive feedback when the negative behavior changes, and determine consequences when it does not.[14]

The Impact of Leadership on Change

Having been hired by five different organizations either to lead a change process or to take the organization to the next level, my experience has taught me that leading a cultural or organizational change involves four key areas, which must be addressed to be successful.

1. *Information.*   What is the change?
2. *Inspiration.*   Why is it needed?
3. *Implementation.*   How will it be done, both individually and organizationally?
4. *Institutionalism.*   How will the organization know whether it has succeeded?

Over the past three decades, the fire service has experienced numerous significant changes as the fire profession has strived to meet new objectives, an expanded mission, more extensive educational requirements, and enhanced safety for the public and our firefighters. This new environment has forced leaders to rethink many of the established leadership and managerial processes. At times, leaders have been reluctant to make needed changes in their organizations, often as a direct result of their failure to understand how to perceive, understand, and interpret their environment. Many discussions regarding leadership and group dynamics can be summed up by using the concept of above-the-line accountability within our organizations, as discussed earlier. However, unless organizations realize its importance, they will never achieve it. Although a significant amount of time is spent training personnel in the technical areas of the job, leaders often fail to focus their efforts on the specific areas that can make an organization reach a level of individual accountability, which in turn can build an exceptional organization. However, when leaders do promote above-the-line accountability, organizational excellence will not be a sometime thing, but an everyday thing.

Leaders must never underestimate the power of their own actions. Even the smallest gesture can change another person's life, for better or for worse. The prime goal of leaders must be to create conditions that produce commitment and creative action by the people within their organizations. Their efforts will be most effective when they can create ownership using above-the-line accountability. Mark Twain may have stated it best: "Always do what is right. It will gratify most of the people and astonish the rest."

In local government today there is also a focus on creating organizations that can quickly adapt to the changing environment. The nature of bureaucracy is the foundation of its processes, procedures, regulations, laws, and governing structure that produces stability. This stability can also create difficulty for many organizations to react quickly to address threats or to take advantage of an opportunity. An inertia in government exists that can be difficult to overcome. The future success or, in some cases, economic viability of local government may rest on its ability to overcome this inertia. Creating organizations that adapt quickly, are flexible, and are entrepreneurial will help maintain the stability required of the governmental structure.

Whether an agenda for change is motivated by internal or external forces, leaders and managers must address certain basic requirements in order to succeed:

1. *Top management* must be involved. It needs to set the example and be active in the change process so others in the organization recognize its commitment.

2. *Measurement systems* must be used to track the progress of the change both at the upper level of the organization and in day-to-day operations.

3. Leaders need to *set the bar high* and push their organization, their division, or their company.

4. Leaders must understand the need to provide education on how and why the change has to occur and the route they plan to take.

5. If an implemented change has been successful in the organization, *spread the story* within the organization and to the entire fire service.

6. Do not ignore the *people side* of the "change movement." The practice of the change management is a combination of the methods used by people (usually management teams) within organizations to ensure organizational transition is completed efficiently and effectively. It is extremely important that leadership and management teams consider not only the technical process but also the human element of any organizational change. This makes sense, because the field of change management is described as the study of "approaches" or "processes" an organization follows when moving from its current state to a desired state. Many academic studies discuss how changes to structures, processes, policies, and technologies will improve efficiencies. The buzzwords used to describe this type of organizational change include *organizational reorganization, corporate restructuring, process reengineering,* and *resource reallocation.* However, for any of these change processes to work, the impact they have on people cannot be overlooked or discounted or the change will likely fail.

7. Hone interpersonal and communication skills. In the past managers were told to focus their efforts on managing resistance to their change initiatives. The most common suggestion was for managers to hone their interpersonal and communication skills so they could help their staff overcome the pains associated with change. Today courses such as Global Knowledge's *Management and Leadership Skills for New Managers* and *People Skills for Project Managers* offer managers specific training in the types of interpersonal and management skills needed to help staff deal with change. These skills include motivational techniques, team building, coaching, feedback, setting priorities, negotiating priorities,

stress management, dealing with conflict, systematic problem solving, and effective delegation.

8. Stop thinking of change management as a stand-alone initiative and start accepting it as an everyday reality. Managers need to accept the fact that organizational change is inevitable and its pace is quickening. Basic economics dictates an organization must constantly adapt or risk failure. The introduction of factors such as new technologies, economic constraints, new legislation, new management, new ideas, and new geopolitical shocks/crises may force organizations to change how they conduct their operations.[15] A survey done by the American Management Association in 2006 suggested:

> 1,400 executives and managers found that 82 percent of them reported the pace of change experienced by their organizations has increased compared with five years ago. Further, 7 out of 10 noted their organizations experienced disruptive change during the last year.[16]

9. Anticipate what and where the resistance will be and plan for it accordingly. Given that change is an ever-present reality in today's workplace, it is safe to say that behavior resistant to it is inevitable in most organizations. Managers need to identify this behavior and help staff manage it by utilizing the proper interpersonal and communication skills, although management training alone will not guarantee the success of organizational change initiatives. Students of organizational change today are looking at "managing resistance" as a reactive answer to change management. To manage a successful organizational change, leaders instead need to be more proactive in their approach by asking, Who is responsible for making transitions successful? The answer is clear—their leadership and management team. Therefore, both executives and managers need to minimize resistance by accepting responsibility to ensure employees are ready and willing to embrace change.

10. Become a change promoter. Global Knowledge's Change Management Implementation Survey found employees were still confused about the new requirements and unprepared to handle them. This end-user confusion and training avoidance demonstrate that managing change for employees is still not fully understood.

Managing in today's organizations means one must constantly identify when and where change is needed and become an advocate for these changes to occur. To be effective, when a change opportunity exists, it has to be communicated up, down, and out. This requires a set of skills that may be new to many, namely, skills related to planning and managing departmental communications. If managers see a way for change to improve a business function, they need to be able to brief their superiors about the effects on employees and, in many cases, do a cost–benefit analysis. Good managers will already have discussed the change idea with their staff and heard their professional input and personal concerns. If their change initiative is accepted, they will need to work with other affected managers to develop a project plan, including a plan for regular communications to staff to inform them of the changes, to explain the rationale, and to ensure buy-in (i.e., ease resistance).[17]

Confronting this resistance to change can be a challenging yet exhilarating leadership experience when one overcomes the resistance. Although no checklist will guarantee success, several strategies can prove very effective for everyone involved.

- Tell people the truth and give as much information as you can. Keep giving information as soon or as often as possible.
- Give them time to digest the news. Do not expect buy-in initially.
- Give them time to vent—there may be anger, but this is normal.
- Listen and empathize! This means fully listen to employees and their concerns! Do not interrupt or try to define; instead, take notes and summarize what they have said. Empathizing does not mean you agree but that you understand they are upset. Say so!
- After you have sensed employees are moving past anger, move to the next step. This always takes longer than you think, so be patient.
- Start getting them involved by asking implementation questions. Determine what needs to be done, when, and why by defining the objectives, constraints, and expectations. The Golden Rule in times of change: let employees determine how the change gets implemented! When possible, give them control to design and implement the change. This helps to eliminate the fear employees may have about losing control of what they know.
- Build on their ideas.
- Work with them to determine reasonable implementation plans, which should include achievable tasks and timelines.

Overcoming the Inertia

Inertia is the tendency of an object in motion to remain in motion or of an object at rest to remain at rest, unless acted upon by a force. This concept was quantified in Newton's First Law of Motion. Organizations also demonstrate inertia, and those in the public sector tend to require a bit more force to get them to move.

In their 1992 book, *Reinventing Government: How the Entrepreneurial Spirit Is Transforming the Public Sector,* David Osborne and Ted Gaebler provided a perspective on the need for government to become more entrepreneurial in its approach. According to Osborne and Gaebler, governments do not work well because they are tall, sluggish, overcentralized, and preoccupied with rules and regulations.

> We designed public agencies to protect the public against politicians and bureaucrats gaining too much power or misusing public money. In making it difficult to steal the public's money, we made it virtually impossible to manage the public's money. . . . In attempting to control virtually everything, we became so obsessed with dictating how things should be done—regulating the process, controlling the inputs—that we ignored the outcomes, the results.

Osborne and Gaebler recommend *entrepreneurial government:* a government that can and must compete with for-profit businesses, nonprofit agencies, and other units of government.

PEARSON
myfirekit
Visit MyFireKit Chapter 3 for more information on The Impact of Leadership on Change.

OSBORNE AND GAEBLER'S 10 PRINCIPLES OF REINVENTION

According to Osborne and Gaebler, the entrepreneurial government should follow these 10 principles of reinvention:

1. *Catalytic.*   Steering rather than rowing
2. *Community-owned.*   Empowering rather than serving

3. *Competitive.*   Injecting competition into service delivery
4. *Mission-driven.*   Transforming rule-driven organizations
5. *Results-oriented.*   Funding outcomes, not inputs
6. *Customer-driven.*   Meeting the needs of the customer, not the bureaucracy
7. *Enterprising.*   Earning rather than spending
8. *Anticipatory.*   Prevention rather than cure
9. *Decentralized.*   From hierarchy toward participation and teamwork
10. *Market-oriented.*   Leveraging change through the market[18]

David Osborne and Ted Gaebler's model for entrepreneurial government led to the initiation of the National Performance Review (NPR) guided by Vice President Al Gore in 1994.

Fast-forward 15 years from the release of *Reinventing Government*: What has changed? In an excerpt from a speech to the United Nations in January 2007, David Osborne stated:

> Reinventing public institutions is Herculean work. To succeed, you must find strategies that set off chain reactions in your organization or system, dominoes that will set all others falling. In a phrase, you must be strategic. By strategy, I do not mean detailed plans. There is no recipe you can follow to reinvent government, no step-by-step progression to which you must slavishly adhere. The process is not linear, and it is certainly not orderly. Things rarely go as planned; reinventors must constantly adjust their approaches in response to the resistance and opportunities they encounter. Rather, by strategy, I mean the use of key leverage points to make fundamental changes that ripple throughout the bureaucracy, changing everything else. Reinvention is large-scale combat. It requires intense, prolonged struggle in the political arena, in the institutions of government, and in the community and society. Given the enormity of the task and the resistance that must be overcome, the reinventor's challenge is to leverage small resources into big changes. Being strategic means using the levers available to you to change the underlying dynamics in a system, in a way that changes everyone's behavior.

CREATING MOMENTUM

A critical factor in successful change is the creation of positive organizational momentum around the change. This is a significant element in overcoming the inertia in which many organizations find themselves. Momentum creates energy, energy creates forces, and it is these forces that can move the organization forward. Remember, an organization in motion will stay in motion and an organization at rest will stay at rest.

In his book *The Tipping Point: How Little Things Can Make a Big Difference*, Malcolm Gladwell explores the concept of how ideas, products, messages, behaviors, industry shifts, and national perspectives spread similarly to a virus. Gladwell points out three particular characteristics that, when combined with one another, can dramatically impact the world in which we live.

These three characteristics—(1) contagiousness, (2) the fact that little causes can have big effects, and (3) change happens not gradually but at one dramatic moment—are the same principles that define how the common cold moves through a grade-school classroom or the flu attacks every winter. Of the three, the third trait—the idea that epidemics can rise or fall in one dramatic moment—is the most important, because it makes sense of the first two and permits the greatest

PEARSON
myfirekit
Visit MyFireKit
Chapter 3 for more
information on
Reinvention.

insight into why modern change happens the way it does. The name given to the one dramatic moment in an epidemic when everything can change all at once is the tipping point.[19]

So, as the fire service shapes its future, it has to contemplate what its tipping points may be in the near future: changes that establish the framework by which the fire profession will change and the direction the profession may travel as a result of these changes. In fact, no matter the type of business or organization one leads, one of the key elements that will determine the level of success is momentum, defined as "the force of movement." It can help decide how successful one's department will be and how quickly that will happen. At the same time, momentum can often prove illusive, something that seems to be here today and yet is gone tomorrow.

What Does Momentum Look Like?

John Maxwell calls momentum "the big mo." In his book, *The 21 Irrefutable Laws of Leadership,* he says, "Momentum is really a leader's best friend. Sometimes it is the only difference between winning and losing."

People know they have momentum when they run over obstacles in their path as though they were nothing. Momentum is when things happen easily, when one success follows another and forward growth comes quickly. Momentum allows leaders to move past mistakes quickly, and it makes any kind of change possible. People throughout the organization are motivated to achieve more and at a higher level.

Achieving momentum starts with creating forward progress. Getting started is the most difficult part, similar to the law of inertia: an object in motion tends to stay in motion, and an object at rest tends to stay at rest.

Momentum starts with the leaders and then moves outward and impacts the entire team. Building momentum requires fast action. The most important part of the race is often leaving the starting block. Building momentum requires following one success with another. Success breeds tremendous positive energy and confidence. Take advantage of it and take the next step.

Momentum is everything![20]

The establishment of a team creates an infectious drive, which feels more successful. To do so requires leaders to carry out the following:

- Align employees' goals with organizational goals.
- Properly frame the issues.
- Consider the audience and present accordingly.
- Seek employee input.
- Delegate tasks throughout the organization.

All of these are included in the organization strategic plan, which drives the organization forward.

PEARSON
myfirekit

Visit MyFireKit
Chapter 3 for more
information on
Momentum.

Conclusion

Today's fire service leaders must embrace a much larger community interest than they have recognized historically. Not only must they learn how to deal with zoning and land use issues, private sector business interests, and neighborhood safety concerns; but they have entered an era that includes regulatory mandates, which necessitate coordination among law, health, planning, and public works interests. Today's fire service leaders need to employ the same planning tools and business logic as are in the private sector's contracts and agreements. They must develop the who, what, why, where, and how of management with an understanding of the political realities and the multidisciplinary functions now embedded in almost every aspect of what fire service leaders carry out on a day-to-day basis. To do so, they need to coordinate with other governmental organizations to develop successful and creative innovations that lead to safe and healthy communities. They must have a clear understanding of how to manage today's emergency services and how to plan for what needs to occur in the future.

The role of fire service leaders is to provide the policy makers (governing board) with the correct information so they can make quality policy decisions and develop strategic methods for spending the taxpayers' money. Effective and well-directed communication is critical for a successful overall coordination. The ability to manage any organization depends largely on the quantity of the resources available to accomplish the desired result and the quality of their leadership and management. In addition, the legal and regulatory requirements in which fire organizations operate, the parameters of the organizational foundation, and the budgetary challenges affect strategic planning and the vision about how to manage their own future. Understanding the dynamics of how to manage a fire service organization and clarity of vision are critical if fire service leaders are to be successful in designing the future of their organizations. At all levels of the organization, fire service leaders must thoroughly understand budget development; how funds are allocated; and how general fund, enterprise fund, capital fund, and asset replacement funds are utilized. They need to have a clear understanding of the liabilities and regulatory mandates that not only have an ongoing cost but also require them to change their operational procedures and structures to meet those mandates. The connections among strategies, mission, goals, objectives, action plans, values, and vision take into account the elements critical for successful change. The relationships among budgeting organizational design, labor–management relations, discipline, and how people are treated are all pieces of the puzzle of how to lead and manage organizations into the future.

Allan Kaye, formerly of Apple Computers, was once quoted as saying, "The best way to predict the future is to invent it."[21] Fire service leaders can do so only through effective strategies built upon a clear understanding of how organizations work and effective change strategies, which will position their organizations to be successful in the future. Whatever strategy is utilized, a critical component is the inclusiveness of the workforce, elected officials, and the public.

Leading change in the fire service is not an easy task. The traditions of this profession run deep in every organization and can create barriers to overcome and engrained mind-sets on how things should be done. The work schedule of 24 to 48 hours on duty creates a bond and a connection like no other profession, except the military, with the fire department becoming a second family for most in its employ. This extended work period creates the opportunity for personnel to analyze everything the leader does, to contemplate all the dimensions of a decision, and, in many cases, to come to a conclusion based on minimal facts and possibly misinformation. This often plays out in the firehouse as a rumor that will start and spread across the organization like wildfire, a much more common occurrence in the fire service than in most other professions.

The bureaucracies in the public sector were designed to create stability, process, procedure, and systems to provide service delivery. The bureaucratic template was not structured for rapid change. Yet, many in government and the fire service lead effective change by providing inspiration and a clarity of vision of where the organization needs to move in the future. This is accomplished through effective information sharing via multiple communication media to help raise the level of understanding for the employees and to reduce the fear and speculation that accompany any change of significance. Through the design of an implementation process, which is inclusive, fear is reduced, better ideas are created, and ownership in the change grows. Leading change requires leaders to focus on attitude, organizational culture, addressing negativity, attacking below-the-line accountability, and pushing the envelope of the old frameworks (paradigms), which many departments still operate within today. To lead a successful change strategy, one must inspire, inform, be inclusive, and have an effective implementation plan.

Simply stated:

PEARSON

Visit MyFireKit Chapter 3 for Perspectives of Industry Leaders.

WORKING TOGETHER WORKS

Working together can never be a policy.
It can only be an idea.
It can never be a code of rules.
It can only be a way of looking at the world.
We can say, 'This is mine,' and be good,
or we can add, 'This is ours,' and become better.
We can think, 'I do my share,' and be satisfied,
or we can ask, 'Can I do more?' and become prosperous.
We can work alongside each other and function,
or we can work with each other and grow.
As we have.
Our country's history
makes it clear that combining all efforts into one
has been the only way to achieve
that progress and that strength we take such pride in . . .
pride not only in what we've achieved
but pride in knowing that we've achieved it together,
with our own work and our own visions.
That's really the key.
Because when all is said and done,
working together doesn't only bring out the best in all of us,
it brings out the best in each of us.

Unknown Author

Review Questions

1. What type of change has occurred in the last five years within the jurisdiction you serve and in your department?
2. Why are core values important for an organization?
3. Describe above-the-line accountability. How can you promote this with those you work with?
4. How does your organization react to change? Why do you think this is the case?
5. How does organizational culture impact the change process?
6. What three things would you change in your department and how would you implement/process each?
7. What is meant by changing your leadership style to the organizational need?
8. Why is organizational momentum important when leading change?
9. What causes resistance to change?

PEARSON

myfirekit™

For additional review and practice tests, visit www.bradybooks.com and click on MyBradyKit to access book-specific resources for this text!

Register for MyFireKit by following directions on the MyFireKit student access card provided with this text. If there is no card, go to www.bradybooks.com and follow the MyBradyKit link to buy access from there.

References

1. Niccolò Machiavelli, *The Prince*, ed. Quentin Skinner and Russell Price (New York: Cambridge University Press, 1988), pp. 20–21.
2. Frank Ogden, *Americans and America*. Retrieved October 14, 2010, from http://www.cybernation.com/quotationcenter/quoteshow.php?type=author&id=6643
3. Peter F. Drucker, "*Really* Reinventing Government," *The Atlantic Monthly*, February 1995, pp. 49–61.
4. Joel Barker, *Discovering the Future: The Business of Paradigms* (Lake Elmo, MN: ILI Press, 1985).
5. Michael Beck, *Is Your Leadership Effective?* May 6, 2005. Retrieved October 14, 2010, from www.buzzle.com/editorials/5-6-2005-69563.asp
6. Keshavan Nair, *Beyond Winning: The Handbook for the Leadership Revolution* (Hamilton, Ontario, Canada: Paradox Press, 1990).
7. Bright Quotes. *Future*. Retrieved October 14, 2010, from www.brightquotes.com/fut_fr.html
8. P. E. Connor and L. K. Lake, *Managing Organizational Change*, 2nd ed. (Westport, CT: Praeger Publishers, 1994); R. W. Stump, "Change Requires More than Just Having a Vision," *HR Focus*, 71, no. 1 (1994), 34; J. Wright, "Vision and Positive Image," *The New Bureaucrat*, 23 (Winter 1994), 55–56.
9. Ibid.
10. Ibid.
11. Ibid.
12. American Management Association, *Agility and Resilience in the Face of Continuous*

Change—A Global Study of Current Trends and Future Possibilities, 2006–2016 (New York: American Management Association, 2006).

13. Randy R. Bruegman, *Fire Administration I* (Upper Saddle River, NJ: Pearson, 2008).
14. Gary S. Topchik, "Attacking the Negativity Virus," *Management Review*, 87, no. 8 (1998), 61–64.
15. Ibid.
16. Steve Lemmex, *Managing Change in Organizations: It's Management's Responsibility*, Global Knowledge_{TM}, Expert Reference Series of White Papers, 2007. Retrieved September 16, 2010, from http://www.cba.co.nz/download/ManageChange.pdf
17. American Management Association, *Agility and Resilience in the Face of Continuous Change—A Global Study of Current Trends and Future Possibilities, 2006–2016* (New York: American Management Association, 2006).
18. David Osborne and Ted Gaebler, *Reinventing Government: How the Entrepreneurial Spirit Is Transforming the Public Sector* (Reading, MA: Addison-Wesley, 1992).
19. Malcolm Gladwell, *The Tipping Point: How Little Things Can Make a Big Difference* (Boston: Back Bay Books, 2002).
20. Danny Gamache, *Creating Momentum: A Leader's Best Friend*. Retrieved September 9, 2010, from www.powerhomebiz.com/vol43/momentum.htm
21. The Quotations Page, *Quotation Details*. Retrieved October 14, 2010, from www.quotationspage.com/quote/1423.html

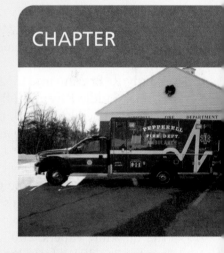

Planning for Organizational Achievement

OBJECTIVES

After completing this chapter, you should be able to:

- Describe the various types of planning models used at the local level.
- Describe the differences between comprehensive, master, strategic, operational, and action planning.
- Describe the defining requirements for the development of an effective strategic plan.
- Explain how a SWOT analysis is used in the planning process.
- Describe the importance of using a structural decision process in effective planning.
- Explain the various analytical tools that may help facilitate group decision making.

PEARSON

myfirekit™

For additional review and practice tests, visit www.bradybooks.com and click on MyBradyKit to access book-specific resources for this text!

Introduction

Organizational **planning** was formally introduced in the fire service as a recommendation of the 1973 report to Congress, *America Burning: The Report of the National Commission on Fire Prevention and Control* (NCFPC). Planning is the process of establishing objectives and suitable courses of action before taking action. The term *master planning* also arose from this report. Its Recommendation 10 states, "The Commission recommends every local fire jurisdiction prepare a **master plan** designed to meet the community's present and future needs in fire protection, to serve as a basis for program budgeting, and to identify and implement the optimum cost benefit solutions in fire protection."

Master planning makes a statement of community policy, and to be successful, it must involve the community in the decision-making process to determine the desired level of fire protection. The NCFPC's recommendation provided for an analytical approach to fire defense planning. In fact, with the creation of the National Fire Academy, one of the first courses taught was on how to develop a long-range master plan for community fire protection.

Planning and Public Policy

Planning in organizations and in formulating public policy involves both the organizational process of creating and maintaining a plan, and the psychological process of thinking about the activities required to achieve a desired goal in the future. These thought processes are essential to the creation and refinement of a plan and its integration with other community plans, which combine forecasting of future development with the anticipated scenarios to which the department may be called upon to respond.

PLANNING CYCLES

Effective organizations are proactive in their efforts to plan and implement current and future operations and direction through the use of proper operational and strategic planning. A planning cycle can range from annually, such as for the budget, to 20 years, a typical length for community growth management plans.

Planning should be viewed as a continuum. At one end are plans to complete the next thing on the to-do list; at the other are plans for the distant future. As such, planning has many levels. People plan for a variety of reasons and then use those plans to make and implement their decisions. Few people do anything without some form of plan. It may be easy to think of responding to fire calls as reactive actions, but significant preplanning is involved before the alarm even goes off in the firehouse. Fire personnel may not know the time of day, day of the week, or location of the next fire, but they plan their strategies and tactics well in advance of the call. Because the best fire departments plan in this way, they create a higher level of effectiveness.

planning
■ The process of establishing objectives and suitable courses of action before taking action. Planning is the first step toward an effective implementation of a decision to achieve the desired outcome. Managerial function that determines in advance what organizations, subunits, or individuals should do and how they will do it.

master plan
■ A variety of different planning processes may be called a master plan. However, it is most often linked to land use community growth plans for a specific development project or particular element of a community's comprehensive plan. For example, a master plan created for a specific area to be developed may include research and surveys on population and job projections, infrastructure, traffic, services required, and estimated timeline for completion.

PEARSON
myfirekit™

Visit MyFireKit Chapter 4 for information on Why Strategic Planning.

Types of Planning

Although the concept of master planning was not globally accepted in the fire service, the idea of planning has been embraced and has continued to evolve and refine itself over the past two decades. In the public sector, there often exist planning terminologies with different applications for various governmental operations, which can be applied locally in several ways. Although these common terminologies and applications are not universal in their use, it is beneficial to understand how they are used at the local level to describe various planning methodologies. The terms used frequently include *comprehensive, master, strategic, operational,* and *action planning.*

COMPREHENSIVE PLAN

comprehensive plan
■ An official public document, created through a public participation process. This document is intended to be a guide in making decisions concerning future land use, extensions of community services and facilities, parks and open space, designation of environmentally sensitive areas, and desirable urban design elements of a city.

For many communities a **comprehensive plan** is a 20-year growth document, adopted by the local governing body, that outlines what type of growth will occur in a community.

In government circles, the term *comprehensive planning* is often heard in regard to development. The primary job responsibility shared by planning commissions and city and county governments across the nation involves the design and development of a comprehensive plan. Whether the plan is labeled comprehensive, master, or general, in certain instances, it may describe the same thing: putting down on paper the aspirations a community holds for itself, and capturing in words and pictures what a community hopes to become in the future. Preparing a comprehensive plan involves a number of technical, political, legal, and managerial considerations; and even though how this is accomplished will vary from one community to the next, three common phases exist: planning the process, plan preparation, and plan implementation. Bear in mind, however, that effective comprehensive planning is actually more like a continuous loop, because feedback from monitoring implementation of the plan's recommendations ideally should be used to initiate needed changes to the plan itself. Comprehensive planning often provides the foundation and guidelines for community growth. During the planning process, the community and its governing body define the scope of the project, develop an achievable project plan, assign resources, and provide the technical support to complete the plan.

The comprehensive plan is an official public document, created through a public participation process. It is intended to be a guide in making decisions concerning future land use, extensions of community services and facilities, parks and open space, designation of environmentally sensitive areas, and desirable urban design elements of a city. Also, comprehensive plan documents are used as the basis for developing a city's development regulations to ensure consistency between policies and regulations and that development meets the intent of the plan.

As a broad statement of community goals and policies, a comprehensive plan directs the orderly and coordinated physical development of a city into the future. The plan serves as a guideline for designating land uses and infrastructure development as well as developing community services. It anticipates that the necessary changes will occur in the future and provides specific guidance for future legislative and administrative actions to accomplish them. It reflects citizen involvement and input, technical analysis, the input of local decision makers, and the adoption by the governing body. The maps, goals, and policies of the plan provide the basis for the adoption of regulations, programs, and services that implement the plan.

Certain states mandate that every municipality adopt a plan addressing specific subject matters, one of which is land use. Certain states, such as Oregon and Washington, go even further, requiring that the tax and zoning maps for a city, as well as other development regulations, be consistent with the land use map and policies contained in a separate document entitled "comprehensive plan." In such states, zoning mandates and development decisions cannot be approved unless such consistency is achieved. The majority of states that have adopted mandatory planning are those with vast undeveloped land or an abundance of natural resources, either of which may be subject to development and related impacts associated with growth.

MASTER PLAN

The term *master planning* can be used to describe a variety of different planning processes. However, it is most often linked to land use community growth plans for a specific development project or a particular element of a community's comprehensive plan. For example, a master plan created for a specific area to be developed may include research and surveys on population and job projections, infrastructure, traffic, services required, and estimated timeline for completion.

Under classic zoning theory, a community's zoning ordinance must be predicated on a comprehensive plan, which takes into account existing land use and anticipates future growth and development. The master plan contains those elements, which relate to a specific land use, employment, population density, public facilities, infrastructure (such as water and sewer service and streets), and housing. However, master plans are usually developed through broad-based community involvement. The purpose of master planning is to integrate various public policies in order to create and implement a vision for the future growth and well-being of the community in respect to a specific area, project, or infrastructure need.

Most important, master planning provides a road map for future growth of a specific site, complex, or area of the community. For example, many projects will trim budgets to a bare minimum and sacrifice land usage, which may impact future land development or make it more expensive. The goal of a master plan is to provide reasonable current development while keeping in mind future developments, which need to align with the community's comprehensive plan.

A master plan presents a conceptual layout and looks at historically placed buildings, the next phase of growth, and the future growth. It outlines a logical phased growth plan and indicates the maximum potential usage of a site. Master plans often include a topographic survey of the property analyzed with respect to streets, easements, buffers, zoning, setbacks, floodplains, and natural features. The land remaining is the actual usable acreage.[1] Depending upon the community, the terms *master planning* and *comprehensive planning* may be used interchangeably.

THE COMMUNITY STRATEGIC PLAN

A community strategic plan is typically a five-year plan that brings together various community elements and departmental strategic plans into one document.

This helps to focus on the community's comprehensive growth plan and other priorities (key objectives) as set forth by the governing body.

Most public strategic planning processes are born out of a strong desire to influence the future direction of the community proactively and positively. The essence of the public strategic planning process involves finding what the community can realistically become, and then planning and implementing actions to make it happen. This is a critical element because it takes the various departments, working separately yet in concert with one another, to accomplish the community's objectives.

The public strategic planning process is a systematic way to manage, change, and help create the best possible future for a jurisdiction. It is a creative process for identifying and accomplishing the most important actions in view of strengths, weaknesses, opportunities, and threats (SWOTs, discussed later in the chapter).

strategic plan
■ A document by which the members of an organization envision its future and develop the necessary procedures and operations to achieve that future. Strategic plans are designed to define and achieve the long-term objectives of the organization.

Strategic planning, though not always labeled "strategic," can be distinguished from other kinds of planning by its specific methodology:

- It is a focused process that concentrates on selected visions or issues.
- It explicitly considers major events and changes occurring outside the organization and jurisdiction.
- It considers resource availability, not only financial but also organizational and political.
- It assesses the strengths and weaknesses of the community and local government.
- It is action oriented, with strong emphasis on practical results.

Whereas traditional kinds of planning consider the "destination," expressed in goals and projects, strategic planning also takes into account the "itinerary" and the "road map" to assure all variables required to arrive successfully at the destination are met. Strategic planning puts the "how" into planning. It is a process of articulating and achieving a dream, tempered by the reality that exists in every community.

Strategic planning does not replace other annual functional planning efforts or effective budgeting. Instead, it can help to integrate activities, assuring that a common purpose and a sense of direction guide them. It helps tie all the agency's planning efforts together.

At the least, strategic planning can give the department and the community a clearer picture of their own unique identity and potential. At its best, strategic planning is planning for results, and implementing the plan will have positive results now and for decades to come.

Although strategic plans can be developed in varied ways, most variations include several basic steps that can be found in other planning efforts:

1. *Scanning the environment* to identify key factors and trends important for the future and then assessing how these forces will influence the community and impact the organization
2. *Conducting a resource audit* to take inventory of the managerial, operational, fiscal, and political resources of the local government and community, as well as the strengths, weaknesses, opportunities, and threats within the operating environment and the community at large
3. *Developing a vision* of the desired future and "beginning with the end in mind" to establish the direction for strategy development

4. *Selecting issues and developing strategies* to achieve the desired vision tempered by external and internal analyses of the forces, opportunities, and constraints affecting realization of the vision
5. *Developing an implementation plan* to consider timetables, resource allocation, and responsibilities for carrying out strategic actions
6. *Monitoring results* to ensure strategies are carried out and adjusting them as necessary to achieve success

THE DEPARTMENTAL STRATEGIC PLAN

A departmental strategic plan is typically a five-year plan focusing on a department's goals and objectives and linking them to the community's comprehensive and/or strategic plan and the annual budget process.

For most departments, strategic planning is the process of defining the requirements for delivering high-payoff results. Strategic planning also helps to identify how to shift from the current reality to future goals that add value to society at a macro level. It is not a rigid process, but rather a self-correcting set of defining requirements and relationships for stating *what is* today in terms of results and what will it take to move to *what it should be* from a macro perspective, not just from the daily operational viewpoint.

Departmental strategic planning involves formally asking these questions:

1. What anticipated shifts will influence the future of the department?
2. What will be the direction of and response to these shifts?
3. To meet the anticipated shifts, what elements must be addressed and why?
4. How will the desired *results* be defined in terms of measurable performance?
5. What are the best ways and means to get there?
6. How will progress be measured?
7. How will success be measured?
8. How will the plan be revised?

Strategic planning is the formal process for documenting the results identified by strategic thinking. At the minimum, strategic planning develops, creates, and records the following results:

1. The ideal vision to help create the department of the future
2. The organizational mission or purpose
3. The strategic objectives for achieving the desired results
4. The operational and action plans needed to achieve the desired results
5. The needs assessment based on organizational priorities and anticipated demands
6. The solutions (methods and means) for delivering internal and external results

Strategic planning for the department formally documents the results and contributions of strategic thinking, namely, the results an organization, customers, suppliers, coworkers, and society want to achieve in a specified period of time. Strategic planning, properly defined and accomplished, involves a process for creating and describing a better future in measurable terms and the selection of the best means to achieve the results desired.

Formal strategic planning is an explicit written process to determine the department's long-range objectives, the generation of alternative strategies for achieving these objectives, the evaluation of these strategies, and a systematic procedure for monitoring results. Each step of the planning process should be accompanied by a process for gaining commitment from both internal and external stakeholder groups.

To be effective, strategic planning should be a continuous and systematic process. This is accomplished when the members of an organization engage outside stakeholders to assist in making decisions about its future, to develop the necessary procedures and operations to achieve that future, and to determine how to define success in the strategic planning process.

In many cases departments that develop a strategic plan do so informally and seek no outside input. Unfortunately, this can produce a myopic focus on what the plan addresses. The value of opening the process to outside stakeholders is that they will bring knowledge and perspectives not found within the organization. Community-based strategic planning can be extremely effective when its focus is the following:

- Oriented toward the future and focused on the anticipated future. It sees how the environment around the organization could be different 5 to 10 years from now. It is aimed at creating the organization's future based on what this future is apt to look like.
- Based on thorough analysis of foreseen or predicted trends and scenarios of the possible alternative futures, as well as the analysis of internal and external data.
- Flexible and oriented toward the big picture. It aligns an organization with its environment, establishing a *context* for accomplishing goals and providing a *framework* and *direction* to achieve the organization's desired future.
- Creates a framework for achieving competitive advantage by thoroughly analyzing the organization, its internal and external environment, and its potential.
- Enables the organization to respond to emerging trends, events, challenges, and opportunities within the framework of its mission, vision, and core values.
- A qualitative, idea-driven process, which can integrate "soft" data that are not always supported quantitatively; provides for involvement of the organization in the ongoing dialogue; and aims to provide a clear organizational vision and focus.
- Allows the organization to focus, because it is a process of dynamic and continuous self-assessment.
- An ongoing, continuous learning process, creating an organizational dialogue that extends beyond attaining a set of predetermined goals. It aims not only to change the way an organization thinks and operates but also to create a learning organization.
- Influences all areas of operations by becoming a part of the organization's philosophy and culture, and impacting the perspective of the community and elected officials.

operational plan
■ Those identified key departmental initiatives that support the departmental strategic plan. Examples include the building of a fire station, equipment purchases, and the adoption of a fire protection ordinance.

OPERATIONAL PLAN

Operational plans are those identified key departmental initiatives that support the departmental strategic plan. Examples include the building of a fire station, equipment purchases, and the adoption of a fire protection ordinance.

Unlike comprehensive, master, and strategic plans, operational plans are designed to achieve their goals within one to two years. Most departments develop an operational plan in order to predict needed expenditures and to develop a project implementation schedule. A community that has limited growth has a much easier time developing operational plans than does a community that is growing at a faster rate. Rapid-growth communities are highly affected by the economy of the local area, and growth in such communities can have dramatic peaks and valleys even over a period of two years.

For many fire departments, operational plans have been the focus, and sometimes even the extent, of their planning process. If a plan is measurable, has a definitive time frame of less than two years, and includes results alerting the department when the outcome has been achieved, it is an operational plan. An operational plan is still an operational plan, even if it is called a strategic plan.[2] Although many fire departments may have conducted some of the components of a strategic plan, they have never actually completed one.

The difference between an operational plan and a strategic plan lies with the anticipated outcome. On the one hand, because an operational plan has measurable outcomes, it is a combination of goals, objectives, and action plans. Operational plans focus on those strategic goals and objectives and then develop the actions necessary for their completion. The different types of operational plans are determined by the length of time they cover or the length of time it takes to reach the desired outcome or goal.[3] An example of an operational plan is the department's annual budget. Once the budget is adopted, numerous action plans will be developed and implemented during the 12- to 24-month operational cycle. On the other hand, a strategic plan gives direction and focuses on those global issues faced by the department. It reflects a broader perspective on where the organization is and where it needs to be in the future.

ACTION PLAN

Supporting the operational plan is the **action plan**. Action plans outline the specific steps and actions to be taken to accomplish a particular operational plan.

Action plans generally are considered to have a targeted completion date of less than 12 months. Action plans are more specific than longer-range plans because of the greater degree of uncertainty that exists the further one goes out into the future. These plans should be very precise in defining the intended outcomes and goals, and the definite actions required to complete the task in an identified time frame.

An example of the community planning process through all levels is provided in Figure 4.1. This shows an overview of how the various planning processes used in a community can be integrated to form a dynamic planning process.

As planning has been analyzed, described, redescribed, and reformulated over the years, many planning processes have been developed. A good number of them often begin with analyzing the facts of a current project or service. If you do not know where you are going, any road will take you somewhere. However, if you have a destination in mind, you must have set goals to get there. Establishing the goals of the plan is an important step in realizing what must be done to achieve them, where to place the primary emphasis of resources, and what must be accomplished throughout the process.

action plan
■ Action plans outline the specific steps and actions to be taken to accomplish a specific operational plan. A basic project management tool defining clearly the actions to be taken, the responsibilities, the goals, the date to start, and the date for completion. Action plan timelines are typically less than 12 months.

PEARSON myfirekit

Visit MyFireKit Chapter 4 for information on Tracking Success.

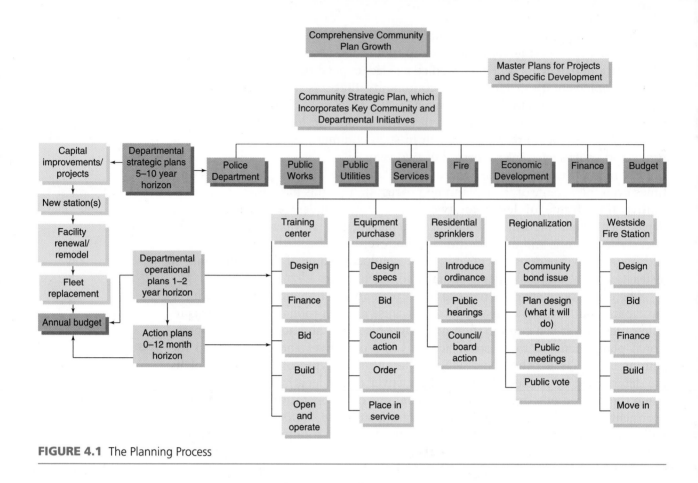

FIGURE 4.1 The Planning Process

Seldom is there only one way to accomplish a goal; this often leads to searching for and examining alternative courses of action. However, when responding to emergency incidents, incident commanders (ICs) normally use a rapid decision process instead of examining alternative actions in depth as described in traditional planning processes.[4] This mind-set often encroaches on the strategic process, as those in the fire service are trained to react quickly and go with the initial assessments. Although this proves effective on the fireground, it is not a good model in the strategic planning process.

In any planning process, the need to establish specific objectives, benchmarks, or milestones by which to measure the progress of the project or service is critical to the plan's success. When there is no mechanism of measurement or evaluation, how can progress be measured? Measurement also serves as a feedback mechanism, which will help to validate the goals and objectives established and to expose areas for improvement.

Goals and Objectives

Whether it is a strategic, operational, or action planning process, it must include defining the goals of the plan. Goal statements typically start with the phrase, "The goal of this plan is to. . . ." The planning team must take it from there and make the necessary decisions. The type of plan utilized and the scope of work determine how broadly or how narrowly a plan will be designed.

From a strategic perspective, departments typically focus on six to eight organizational goals for the time period of a plan. These goals are the organizational compass, providing the direction the department will be taking during the time the plan is intended to cover. Each goal is supported by reaching one or more objectives, which can become an operational plan and/or an action plan, depending on the complexity of the organizational goal and length of the time allowed to complete it. A goal is a dream with a deadline. A goal is observable, measurable, supported by objectives, and it is to be achieved in a given time frame.

An objective is a specific, achievable, measurable event or accomplishment. Objectives provide milestones for the implementation of the plan by fulfilling goals. Objectives should be described in quantifiable terms, such as quality, quantity, time, and measurement of cost. Accomplishment of the objectives associated with a particular goal should result in the achievement of the goal. Objectives should have the following characteristics: be realistic and attainable, have target dates for their completion, and be compatible with other objectives, which are components of the same goal. Objectives are the results to be achieved by the steps described in the department's action plans.

As such, action plans consist of specific action steps that work toward accomplishing a specific objective of the department. They are the implementation techniques designed to create the desired outcome. Action plans may be designed to accomplish a goal from start to finish and include all of the goals and objectives.

Development of effective action steps will focus on the following:

1. Identifying the results needed to accomplish the operational plan
2. Determining the most critical results needed to achieve the operational plan
3. Providing an agreement on the approach to achieving the operational plan
4. Documenting the results of each task
5. Inviting commentary from the department's various stakeholders who will implement the program
6. Providing final documentation of the action plan, including how to modify the plan as necessary
7. Implementing the action plan and evaluating the results[5]

The action plans developed may also support various components of several operational plans.

The Impact of the Environment on Planning

Every fire department operates within two distinctly different environments: the internal and the external environments. The internal environment is the portion of the fire department that it has some control over. Most of the internal environment exists within the confines of the fire department itself. Every organization has its own set of distinctive competencies and its culture, which create strengths and weaknesses within the organization. An understanding of where an organization is strong and where it needs to improve is a very critical component of any planning process. Without an accurate assessment of its strengths and weaknesses, setting the strategic direction of the organization becomes very difficult.

The external environment consists of the community and its unique makeup, elected officials, city/county managers, and a host of other stakeholders and special-interest groups.

THE INTERNAL ENVIRONMENT

Organizational culture plays a vital role in any planning effort. As discussed in Chapter 3, the phenomenon often called the culture of an organization refers to its patterns of development as reflected in its system of knowledge, ideology, values, laws, and day-to-day rituals. The culture of an organization helps define its characteristics and is shaped by the values of the employees, both past and present. Every fire department has its own culture, the characteristics of which are evident in the patterns of interactions observed between individuals, the language used, the images and themes explored in nondirected conversations, and the various rituals of daily routine. Whether or not cultural characteristics are recognized or articulated, they exert an important influence on the department's overall ability to deal with the daily and weekly challenges presented and often exert great weight on the strategic decisions made, which will have long-term consequences.

The most positive impacts that effective strategic planning can have on an organization are that it facilitates the workforce's higher level understanding of the challenges and opportunities the department will encounter, and it helps to integrate the values of the organization into the planning process. Values may be shaped by the culture but they are held by each individual. Many believe that in order for a person to be successful within an organization, he or she must hold a set of values that are shared by the other members of the organization as well as the organization as a whole. One of the first and most basic steps of strategic planning is the identification and development of the value system, which drives everything within the department, including its daily operations and individual actions.

Strategic planning is not necessarily effective simply because the organization wants it to be. First, a department and its culture must be ready to plan strategically. In the "planning to plan" step, the organization ensures all mechanisms needed for strategic planning and effective communication are in place. Part of this is accounting for the culture that exists in the organization at that time and facilitating a basic understanding of where it will need to evolve in the future.

Every organization operates according to a definable philosophy often reflected in its mission, core values, and vision. When an organization is understood by and has buy-in of its members, its philosophy of operations is apparent as a combination of many factors and reflects its approach to delivering its service. Taking the time to define and understand a department's philosophy of operations will pay big dividends in the long run and go a long way toward developing an understanding of the strategic issues a department faces. How can you tell? You can see it every day in the way the department members deliver services, how they treat the customers, and how they treat others within the organization.

THE EXTERNAL ENVIRONMENT

The external environment also impacts every organization as certain issues and influences exist outside the department's control. The external environment presents potential opportunities and a variety of challenges to a department. The department must understand the pertinent issues and assess them thoroughly before being able to take advantage of the opportunities and to meet the challenges, which may impact the organization in the future. However, if an organization does not identify them, it probably will not be able to address them adequately.

To be effective in today's rapidly changing environment, successful organizations often create a matrix to consider the relationships among the various components of their internal and external environments. Each component must be analyzed in relation to every aspect of the organization. The internal environment of an organization, such as a fire department, is controlled by its strengths and weaknesses, culture, work ethic, experience, and the talent of its leadership. Its external environment is often controlled and motivated by people who may have very little knowledge of the work done and little understanding of the service impact on them.

Because every fire department exists to serve the public, the external environment should be assessed prior to the internal environment. The community's comprehensive and strategic plans and its wants and desires, not the fire department's culture, should drive the direction and operations of the fire department.[6] Although a department can influence this direction, ultimately the community through its elected officials will make the decisions.

As the department develops a strategic document, it will begin to have a clear understanding of the strategic issues it faces or will face in the future. This happens as those in the fire service answer questions about how the department will handle specific issues in the future. Those answers are then translated into departmental strategies that define its direction and its vision of the ideal future. However, strategies and visions remain simply dreams unless they are translated into actions. Whatever type of planning process the organization undertakes, the focus is to provide it with strategic direction to be successful. Planning provides the strategic framework to help shape the future.

The Decision Process

Any planning process is only as good as the information used in it. Those in the fire service often use their intuition process based upon field experience to direct their analysis. Although this selective process can have merit, their personal and cognitive biases influence their direction, which may or may not be based upon the relative facts of the issue being analyzed. Another interesting aspect of the fire profession is how culture can impact the decision process. The norms of fire service organizations can either open doors or present barriers to effective decision making. Often the extent of a department's culture can have as much influence on the decision process as the performance analytics used to reach the best organizational decisions. As such, it is important to use multiple processes to provide analysis and make decisions on the direction of the organization.

DECISION-MAKING ANALYTICAL TOOLS

Leaders make decisions to solve problems and achieve desired outcomes for the department. A problem may or may not be something catastrophic; many times the word *problem* has a rather different and more specific meaning than that. A problem may include a complex situation that requires action, and a decision must

decision-making analytical tools
■ A model or methodology that provides knowledge-based systems that will support decision making; examples include core analytics, application-centered data warehousing, analysis and exploration tools, alerting and notification systems, scorecards, dashboards, and various integrated reporting tools and queries on system performance.

be made about what action to take. Or, a problem may consist of something that has gone wrong or is the cause of an undesirable situation or action taken by a person within an organization, and a decision involves making choices to resolve the situation. A problem may involve making a difficult choice or may be the anticipation of future trouble and deciding how to mitigate or prevent its effects. It may simply be the problem of deciding between two options that have both positive and negative effects on the desired outcome. If there is clearly one best option, decisions are easy; however, most of the time the situation is not so clear cut.

As organizational planning is conducted, leaders focus not only on addressing the problem areas in their organizations but also on laying the foundation to meet future demands for service or changes in the jurisdiction served. Within these discussions, challenges, problems, opportunities, and barriers emerge; each requires analysis to provide options either to address an issue or to take advantage of one. The beauty of having this type of strategic dialogue is it opens organizational possibilities not seen before. Another positive result is it prepares leaders to have substantive dialogue with their community and elected officials. All this starts with good analysis.

SWOT ANALYSIS

A scan of the internal and external environments is an important part of the strategic planning process. Environmental factors internal to the firm usually can be classified as strengths (S) or weaknesses (W), and those external to the firm can be classified as opportunities (O) or threats (T). Such an analysis of the strategic environment is referred to as a **SWOT analysis**.

SWOT analysis
■ Analysis of the strategic environmental factors internal to the firm usually can be classified as strengths (S) or weaknesses (W), and those external to the firm can be classified as opportunities (O) or threats (T).

As work on development of an effective environmental assessment continues, some managers may discourage an organization from considering anything coming from its public, considering it as a threat. However, the organization exists to serve the community and must be aware of its needs and desires. Therefore, some departments use the term *challenges* to replace *threats* in the strategic assessment of the external environment of a community. Certain communities may choose the challenges, opportunities, weaknesses, and strengths (COWS) analysis as a more acceptable presentation.

The SWOT analysis provides information helpful in matching the department's resources and capabilities to the competitive environment in which it operates. As such, it is instrumental in strategy formulation and selection.

SWOT (strengths, weaknesses, opportunities, and threats) analysis is a method of assessing a department, its resources, and its environment to better understand the department. The essence of the SWOT analysis is to discover what the department does well; how it could improve; whether it is making the most of the opportunities around it; and whether any changes exist in its environment—such as technological developments, mergers of services, or unreliability of suppliers—that may require corresponding changes in its daily operations.

The SWOT process focuses on the internal strengths and weaknesses of leaders, their staff, their products, and their business. At the same time, it recognizes the external opportunities and threats, which may have an impact on the business, such as market and consumer trends, changes in technology, legislation, and financial issues.

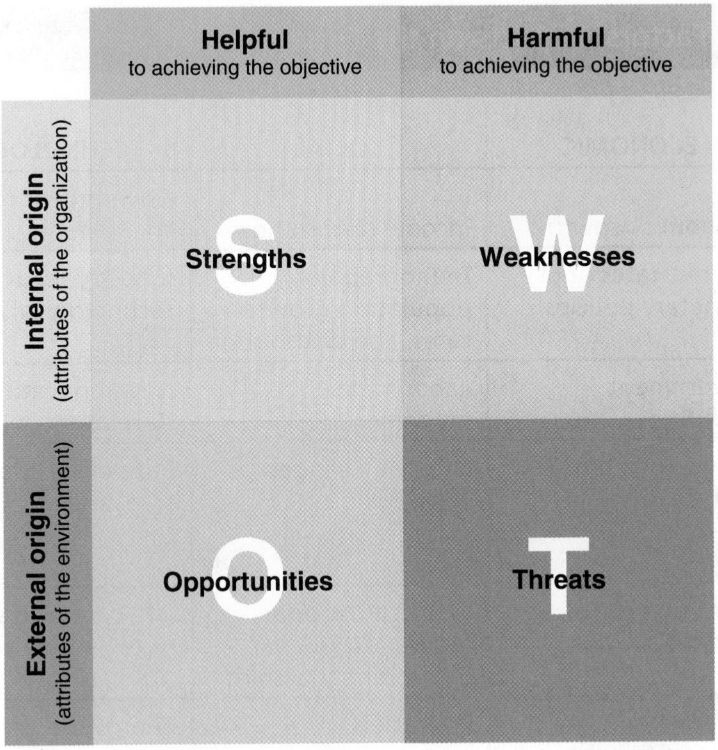

FIGURE 4.2 Sample SWOT Analysis

The traditional approach to completing SWOT analysis is to produce a blank grid of four boxes—one each for strengths, weaknesses, opportunities, and threats—and then list relevant factors beneath the appropriate heading. Do not worry if some factors appear in more than one box, and remember that a factor that appears to be a threat could also represent a potential opportunity.

Completing a SWOT analysis enables leaders to pinpoint core activities and identify what they do well and why. It also points them toward their greatest opportunities and highlights areas where changes need to be made to make the most of the resources, including the talent, in their organization (see Figure 4.2).

PEARSON
myfirekit
Visit MyFireKit Chapter 4 for information on Making It Happen SWOT.

PEST ANALYSIS

The **PEST analysis**, which stands for political, economic, social, and technological analysis, describes a framework to assess the macro-environmental factors for use in the environmental scanning component of strategic management. As a part of the external analysis when conducting a strategic analysis or doing market research, PEST analysis gives a certain overview of the different macro-environmental factors, which a department must then take into consideration. It is a useful strategic tool for understanding growth or decline, organization position, potential, and direction for operations.

The growing importance of environmental or ecological factors in the first decade of the 21st century has given rise to green businesses and encouraged widespread use of an updated version of the PEST framework. As such, analysis systematically considers sociocultural, technological, economic, ecological, and regulatory factors, which are also relevant for the fire service.

PEST analysis
■ Political, economic, social, and technological analysis, which describes a framework to assess the macro-environmental factors for use in the environmental scanning component of strategic management.

TABLE 4.1 | External Factors' Impacts on the Organization

POLITICAL (INCLUDING LEGAL)	ECONOMIC	SOCIAL	TECHNOLOGICAL
Environmental regulations and protection	Economic growth	Income distribution	Government research spending
Tax policies	Interest rates and monetary policies	Demographics, population growth rates, age distribution	Industry focus on technological effort
Federal, state, local regulations and restrictions	Government spending	Labor/social mobility	New innovations and development
Contracting law, consumer protection law, governmental transparency	Unemployment policy	Lifestyle changes	Rate of technology transfer
Employment laws	Taxation/local revenue	Work/career and leisure attitudes, entrepreneurial spirit	Life cycle and speed of technological obsolescence
Government organization/attitude	Federal and state revenue take backs, unfunded mandates	Education	Energy use and costs
Administrative/growth regulations	Inflation rates	Community expectations of government	(Changes in) information, technology
Political stability	Community growth patterns	Heath consciousness and welfare, feelings of safety	(Changes in) Internet
Safety regulations	Consumer confidence	Living conditions	(Changes in) mobile, technology

The PEST analysis has proved to be a flexible and easy-to-understand tool in the context of strategic planning. As with all tools, the real value of the PEST analysis depends on how its information is used.

The PEST analysis is a useful starting point for the examination of an organization's external environment and the forces at work there. Different opinions exist in literature about the inclusion of legal and ecological factors, because their importance differs from industry to industry. In case such factors are of high relevance to an industry, they should be analyzed separately. In industries that are less influenced by legal and ecological factors, they could be allocated to the other categories; for example, legislation as a political factor, as ecological awareness, or as a sociocultural factor. In any case, it is important to include only external factors that will influence the organization.[7] Table 4.1 shows some examples of political, economic, social, and technological factors.

Completing a PEST analysis is relatively simple and can vary from department to department depending upon the aforementioned factors to be evaluated.

PEARSON
myfirekit™

Visit MyFireKit Chapter 4 for information on PEST Analysis.

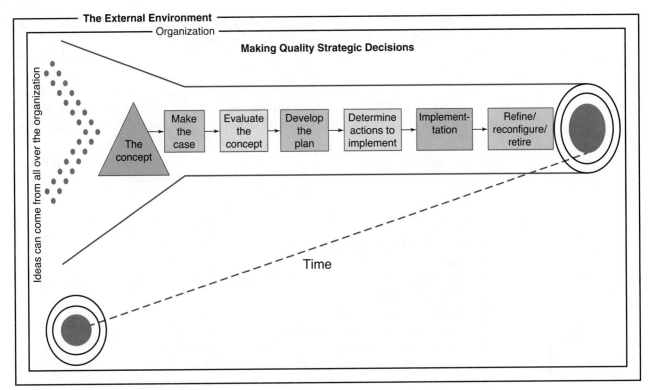

FIGURE 4.3 The Structural Decision Process   *(Adapted from a model of the Strategic Decision Group)*

STRATEGIC DECISION-MAKING PROCESS

At times strategic decisions are not clear-cut; there may be uncertainties or ambiguities about using a particular strategy. As data are analyzed and collected, one often finds that the relevant competing options or ideas have merit. Hence, choosing a direction can be difficult. For example, building a station in the wrong location or beginning a service that gains immediate acceptance but does not have the funds to continue may have long-lasting impacts. Such decisions can produce an irretrievable commitment of resources, creating indirect impacts throughout the organization. Like throwing a rock in the pond and watching the ripple effect across the water, a poor strategic decision can ripple through the organization and cause unintended consequences. The importance of planning based upon sound use of the **decision-making process options** and analysis helps to minimize the unintended consequences.

Several processes exist that provide a structured approach to making critical decisions. One of them is called the **structural decision process** (see Figure 4.3).

Implementation of a structured decision process means all new ideas will be processed through specific decision points. The purpose is to make quality strategic decisions by selecting the most viable ideas and concepts, which have the highest capability to build value to the organization and those being served by the department. New ideas will be assigned a "champion," or owner, who will take responsibility for moving the ideas through the strategic decision process.

The executive team of the department, or another group that is formed for this purpose, may act as the "gatekeeper." Its objective is to make "pass or do not

decision-making process options
■ A position, opinion, or judgment reached after consideration. Choosing between alternative courses of action using cognitive processes—memory, thinking, evaluation, and so on. The process of mapping the likely consequences of decisions, working out the importance of individual factors, and choosing the best course of action to take.

structural decision process
■ A systematic approach to analysis and decision making.

pass" decisions based on the owner's input, balanced against the strategic plan of the organization, and other organizational considerations. Moving through the process is determined by the quality input to the group the owner defines as the gatekeeper and its output in the form of pass or do not pass decisions. Multiple concepts can be "in the process" at any given time. The process is not rigid and the team can modify it, based on lessons learned, as the process is used. As the intensity and complexity of organizational issues have grown in recent years, so has the need to refine our decision processes. The structural decision process provides six decision points from which to evaluate a project, from inception to reconfiguration of the product, service, or process that is being considered for implementation or the retirement of one already in existence.

Decision Point 1—Make the Case

- Identify the opportunity.
  - Assign ownership.
  - Develop the customer and associated economic model.
  - Define the objectives.
  - Define the value-added proposition.
  - Formulate the customer value proposition.
  - Define the investment required to make it happen.
  - Determine whether it fits the department's strategic plan.
  - Report the lessons learned from this step to be shared with others.

Decision Point 2—Evaluate the Concept

- Define what the success criteria are.
- Initiate a customer feedback loop to define the customer value proposition.
- Initiate a process for employee feedback, which may be affected by or will implement the concept.
- Understand the challenges and the risks.
- Review decision alternatives, and quantify the risks of and rewards for each alternative.
- Report the lessons learned from this step to be shared with others.
- Determine whether the resources (staff time and money) are available to move on to the next step.

Decision Point 3—Develop the Plan

- Develop a plan that includes the following:
  - Create a staffing and funding plan.
  - Establish the core team.
  - Define the strategy to implement and define what are the organization's value propositions: what's in it for the organization.
- Report the lessons learned from this step to be shared with others.
- Determine whether the resources (staff time and money) are available to move on to the next step.

Decision Point 4—Determine Actions to Implement

- Define the actions required to implement the program.
  - Who, what, when, where, how.

- Field-test, if possible, before full implementation.
  - Determine refinement requirements.
  - Use the lessons learned to refine the program/process and incorporate before implementation.
- Report the lessons learned from this step to be shared with others.
- Determine whether the resources (staff time and money) are available to move on to the next step.

Decision Point 5—Implementation

- Implement the actions needed to launch the program/service.
- Measure the implementation against predetermined performance success factors.
- Run through the customer feedback loop and modify as needed.
- Report the lessons learned from this step to be shared with others.
- Determine whether the resources (staff time and money) are available to move on to the next step.

If the success factors are not met, the concept/program will move to decision point 6.

Decision Point 6—Refine/Reconfigure/Retire

- Modify the product based on customer feedback and/or lessons learned.
- Define the trigger points to refine, reconfigure, or retire the program.[8]

WHAT IS THE PROBLEM?

Solving problems is seldom easy. Whenever a department identifies a problem, it tends to focus on actions that will reduce or correct the negative effects of the situation. Yet sometimes those actions do not correct the problem. It may be discovered later that the true cause of the problem was different than first believed after an initial decision had already been made about the corrective action.

When a problem arises, the perception that there is a need for a quick, effective decision to solve it can be compelling. However, many times quick decisions result in unintended consequences that may include confusion, wasted efforts, and failure. Several pitfalls are common, which include the following:

- *Jumping to conclusions about the cause.* Based on limited information, people assume they know the cause. However, when additional information is applied to the decision theory, it does not support the choice made.
- *Failure to define the problem.* Some people find it difficult to sort out the important facts from large amounts of information related to the problem. Information may be so complex or varied that at times decision makers can be diverted from the original objective that began the analysis in the first place.
- *Action overkill.* This is the "ready, fire, aim" pitfall. When strong action is believed necessary, decision makers act before they know whether or not the action will correct the problem.

To avoid these and other drawbacks, decision makers need an effective way of defining the problem, so they can identify and collect relevant facts and analyze the available information prior to making the decision.

A **problem analysis** is important in assessing the following:

- Performance is below or has not reached an expected standard; that is, the actual situation deviates from the expected norm.
- The cause of this deviation is unknown.
- The cause needs to be determined so that corrective action can be taken.

Problem analysis is intended to gather information to determine the cause and effect of a problem before committing to an action that will solve it. Problem analysis uses three processes:

1. *Describing the problem thoroughly by gathering specific information about it.* Start with a clear understanding of the problem before searching for information needed to solve it. A deviation statement provides a point of reference for the problem solver. Irrelevant information must be discarded. Getting agreement on a statement of the observed deviation helps keep everyone on track. Then, thoroughly describe the problem by asking what, where, when, and similar questions to determine the problem's extent and to provide a complete picture.

2. *Developing possible causes of the real problem using experiences and/or analysis.* Once the problem is specified, possible causes are developed. Experience and common sense may suggest some possibilities. As these possibilities are noted, test them against the information in the problem description if these causes fail to explain the deviation, or if plausible causes have not been developed. First, identify the distinctions. Identify what *is* and *is not* the problem to create clear boundaries around it. Second, identify changes to determine why there is a deviation from a normal standard. If the situation has been reaching normal expectations and then suddenly fails, what has changed? By isolating relevant changes among the many one can expect to find, one should be able to determine the cause. Changes in, around, or about these distinctions may have caused the problem. Once changes that may be relevant to the problem have been identified, one can move from gathering facts to constructing hypotheses. By building on past experiences, knowledge, and common sense, it is possible to identify how each change could have caused the problem. Next, the task is to determine which cause seems most probable and whether it is, in fact, the actual cause.

3. *Confirming the most probable cause.* Test the possible causes to determine which one best explains what is observed and by taking the steps necessary to verify critical information. At times one may find that multiple causes contribute to the same problem.

By examining each hypothesis to see whether it explains the observed deviation, the probable cause can be tested. This can be done by asking how each cause is or is not a dimension of the identified problem statement. That is, test how well the hypothesis fits the known facts against each line of the problem statement. To substantiate the true cause, confirm it by verifying any questionable information through observation, simulation, and experimentation.[9]

Making Good Decisions

Sometimes problem solvers need to assess a recommendation presented for approval, and they approach decisions with their minds already made up. Or people may have a favorite alternative they tend to choose. Analysis in these instances tends to be biased. *Decision analysis* is a management technique in which various decision processes are applied to assist in determining the most advantageous alternative under the current circumstances, and it can help shift the focus from alternatives to objectives. It is beneficial to define the decision carefully before jumping on solutions or to conclusions. People must ensure their decisions are based on a thorough analysis of the relevant information. By using the decision analysis approach, they can expect to make more carefully reasoned decisions based on information and analysis.

At least three difficulties are encountered in such situations:

1. *Focus on a favorite alternative.* The presentation or discussion concentrates on the strengths of one alternative and provides no means for comparison with other possible choices. With group decisions, failure to explore other alternatives can lead to frustration and conflict.
2. *Ignore the consequences of a particular choice.* At times, a particular alternative seems so attractive that little or no thought is given to its risks.
3. *Rely on inadequate information.* Either people make the decision with a limited amount of information, or they allow assumptions to take the place of facts.

These difficulties occur especially when pressure exists to make a decision quickly. Problems may also crop up when the responsibility for making an important decision lies with a group of people who have differing perspectives on the risks and benefits of the possible solutions, or when the information needed to make solid decisions is unclear or dispersed throughout the group unequally.

The decision analysis approach is intended to overcome these common difficulties by first examining the purpose of the decision, and what it is expected to achieve, before selecting a decision alternative. This method provides a systematic framework as each step follows the previous one logically.

Next, after the purpose of the decision is understood, it is time to consider a sequence of four fundamentals.

1. *Determine the objectives from the results the decision is expected to achieve.* Describe the outcomes that are desired as a result of the decision to be made. In many decisions, not all of the objectives are of equal importance.

Objectives can be categorized into musts and wants and should be prioritized as they relate to the success of the decision. The must objectives will cause the decision to fail if they are not achieved, so they need to be stated in measurable terms. Decision makers need to recognize when an alternative fulfills the qualifications of a must objective and when it does not. For example, the departmental strategic plan calls for a tiered transport system to be put into place, and the elected body has approved the plan (see Figure 4.4).

FIGURE 4.4
Transport Unit

The following goals and objectives would be relevant for implementation.

Goal. To improve the fire department's capability to provide emergency ambulance transports through a tiered transport system.

Objective. To upgrade current practice of utilizing qualified dispatchers to triage patients for ALS and BLS transport by January 1, 2012.

Objective. To hire the needed number of qualified paramedics and EMTs required to implement the system by February 1, 2012.

Objective. To provide statistical updates to the elected body on the first six months of operations by September 1, 2012.

Some decisions will not have must objectives. Those objectives that are not essential to the success of the decision but remain desirable are called want objectives. They are not usually equally important in a decision and need additional classification. One method to classify them effectively is to provide weight to the want objectives on a 1 to 10 scale, indicating their relative importance. The most important want is designated as a 10 and becomes the benchmark for other objectives. Objectives assessed to be of equal importance are given the same number. An objective assigned a weight of 5, for example, has half the influence of an objective assigned a weight of 10.

When weighing objectives seems too difficult, it may simply be necessary to gather more information and assess how each objective relates to the specific decision. All identified objectives should pertain to the decision statement, short- and long-term issues should be considered, and the objectives should be stated clearly so that musts are measurable and wants are well defined.

The set of objectives for assessing the alternative decisions should now be clear. Alternatives selected need to satisfy the must objectives in order to fulfill the purpose of the decision. The musts serve as a screen for eliminating alternatives that fail to meet mandatory criteria. The wants provide a selection profile for the remaining alternatives and can help to distinguish among the alternatives in terms of their ability to meet the purpose of the decision.

2. *Consider the alternatives that seem to satisfy the objectives, and identify the alternative that best fits the goals of the situation.* Rather than considering only the obvious alternatives, conduct a wider search for other possibilities. Then gather information about each alternative and assess it against the objectives. The decision statement clarifies the purpose of the choice and defines the acceptable range of alternatives. If problem solvers feel too constrained by the decision as stated, they may need to test their perception of its fundamental purpose. Raising the level of a decision enables consideration of a much broader range of alternatives. The objectives themselves often provide a source of alternatives, and each objective can be used to generate possible choices.

Generating alternatives is a creative step in the decision-making process. Any process that enables, releases, or fosters creativity helps ensure that the range of alternatives includes the best ones. Once the possible alternatives have been identified, the process of assessing them against the objectives begins. Problem solvers gather the best available factual information about each alternative, their best projections, and the opinion of experts. They then use the "musts" and "wants" to screen out the alternatives that fail to meet minimum requirements.

If all alternatives are eliminated when screened through the must objectives, problem solvers must develop additional alternatives or review both the decision statement and the must objectives to see whether they can make a realistic choice.

3. *Assess the adverse consequences (risks) of an alternative before making a final decision.* Decision analysis is not complete without a careful assessment of the risks of choosing each alternative. Once the risks are identified, the impact of each is assessed: What problems might be encountered? What effects might the risks have on the decision's outcome? Is it likely that the negative risks will actually cause problems? If problems are encountered, how serious will they be? Overall assessment of the risks will depend on the degree of the total threat rather than the number of adverse consequences. Committing to an alternative that indicates serious adverse consequences invites disaster. Once the probability and seriousness of the consequences of each alternative are known, it is possible to make a decision with the full knowledge of the risks involved.

4. *Assess the benefits of an alternative before making a final decision.* Decision analysis benefits are anticipated once a decision is made. Often an alternative is chosen solely based upon the minimization of risk, and then once implemented, no quantifiable benefit is realized. The overall benefits to be derived must be based instead upon the organizational goals, the objectives to be obtained, and they should include a cost/benefit assessment. What is the value proposition to the organization—improved service, reduced cost, more effective service delivery, or a more effective internal process? Whatever the case may be, assessing the benefits is as important as assessing the risks.

POTENTIAL PROBLEM ANALYSIS

There is no way to be certain that decisions will be as effective as anticipated once they are implemented, especially in the fire service, where decisions must be put into operation in a rapidly changing environment and during emergencies. Many times the department experiences difficulty and it must adjust decisions to succeed. Experience indicates that, when leaders are strategically planning for the department, it is better to take the time needed to analyze and improve the plan before implementing it.

Potential problem analysis is a conscious approach that anticipates future difficulties and takes actions to prevent them. It also is designed to consider contingencies for successfully implementing a decision.

Potential problem analysis is best used to do the following:

- Complete a plan for implementing a decision.
- Monitor a plan in progress or a decision while it is being implemented.
- React to internal or external changes that affect the organization.
- Improve current operations.

The amount of time required for this approach will depend upon the situation. In certain circumstances, asking a few key questions is all that is needed. When the situation is complex or unfamiliar, consider more formally all steps of potential problem analysis, and rely on experienced personnel within the organization for help. Also, create a complete implementation plan for the decision. As its execution proceeds, monitor it closely and modify it as needed.

Using the potential problem analysis approach systematically reduces the level of risk in any decision. This approach considers cause-and-effect relationships to identify what might go wrong when implementing a decision, plan, or action. In its most complete form, potential problem analysis includes the following components:

- A statement of action or plan
- Anticipation of likely causes
- Preventive and contingent actions[10]

RATIONAL PROCESS

Rational process incorporates the logical, commonsense ideas typically used to resolve concerns. The strength of rational process lies in providing a consistent, systematic framework for gathering, organizing, and evaluating information. The process builds on the underlying principle of cause and effect to answer three key questions:

- Why did this happen (problem analysis)?
- What course of action should be taken (decision analysis)?
- What lies ahead and what can be done about it now (potential problem analysis)?

Solving problems and making decisions that involve processes, machines, and non-personnel often are much easier than those involving personnel. Using rational process helps overcome the common pitfalls of personnel decisions. It is critical to use a careful, systematic approach that emphasizes sound judgment. The common problems that occur when decisions involve people rather than machines or systems follow:

- Communications between people are more difficult because of biases and hidden objectives, leading to misinterpretation and misunderstanding.
- People have feelings and emotions that affect their ability to work with others.

- It is much more difficult to experiment when people are directly involved. For example, it is challenging to reduce job responsibilities once they are increased. Once someone is promoted, it is difficult to return him or her to a lower position.
- People tend to jump to conclusions when dealing with personnel problems. Blame and questionable assumptions cause added problems.
- People have a tendency to take on responsibilities that exceed their skills and training. Leaders can easily find themselves acting as a psychologist in situations involving people.
- Discussions of problems tend to depict the problem as broader than it really is to avoid making criticism personal. People talk in vague terms, such as about "the morale problem" or "communications difficulties," without really defining the specifics. Although this provides a sense of security, it hinders their ability to resolve the real concerns.
- People often take an opinion and/or rumor as fact, depending upon information that is heard rather than observed.

Rational process provides a framework for resolving personnel issues effectively and helps leaders avoid falling into these common traps. Scientists have long used a six-step process called the scientific method to make decisions, analyze a problem, and reach the proper conclusion. Its steps include the following:

1. Recognizing the need
2. Defining the problem
3. Collecting data
4. Analyzing the data (inductive reasoning)
5. Developing a hypothesis
6. Testing the hypothesis (deductive reasoning)

People have adapted this method many times and in many ways for specific applications. One of those applications is the situational appraisal, which applies the scientific method used for effective decision making. Decisions of all kinds are made on a daily basis, and time is often a critical factor in the way they are made. The less time there is to make a decision, the more people rely on recognition-primed decisions. If people have a relatively long time before having to make a decision, they rely on some form of the scientific method. Decisions can also be sorted by the relative degree of risk involved. Decision makers face some types of decisions routinely and others infrequently. A matrix based on the risk and the frequency of decisions can be used to display decisions.

For simplicity, people recognize that any decision is either high or low risk. Those decisions that have few or no consequences are low-risk decisions. On the other hand, those decisions that will have catastrophic results or consequences if the wrong choices are made are high-risk decisions. High-risk decisions made frequently have a high degree of consistent outcomes due to experience. High-risk decisions that are not often encountered (low frequency) become the most critical decisions in the fire service.

At the same time, most decisions that can be delayed for a period of time until sufficient time exists to consider the options are called non-time-constrained decisions. In the fire service, ICs must make decisions without sufficient time to

FIGURE 4.5 Risk of Decision *(Source: Graham & Associates, 2001)*

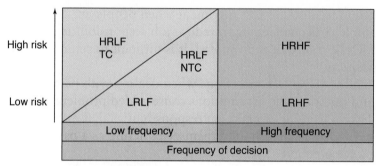

explore all of the options and without delay. These are called time-critical decisions (see Figure 4.5).

Low-risk, high-frequency decisions are the easiest to make. As long as high-risk decisions are made often, the outcomes usually are successful because they are based on experience. Time is seldom a significant factor. Low-risk decisions made infrequently present more challenges; successful outcomes are often the result of careful analysis. The critical decisions of an organization are those with a high degree of risk but low frequency. Time is a significant factor in these decisions. If the decision is not time constrained, situation appraisal may be the best choice; however, when a high-risk/low-frequency decision is severely time constrained, such as a decision faced during emergency incidents, the success of whichever process is selected depends primarily on the experience of the decision maker.[11] (If you get the chance to hear Gordon Graham's presentation on this topic, do so.)

PEARSON
myfirekit

Visit MyFireKit Chapter 4 for information on the Six Thinking Hats.

The Problem-Solving Toolbox

PEARSON
myfirekit

Visit MyFireKit Chapter 4 for a more detailed discussion on Problem-Solving Tool Kit.

A key objective in the decision-making process is to effectively apply the right analytical tools at the right time in order to arrive at the best choice. The tools featured in Figure 4.6 can be used to help facilitate a group to arrive at an appropriate decision and to build consensus. These tools are commonly part of the quality improvement process that many corporate businesses and public sector agencies utilize.

Conclusion

Looking at the landscape of the fire service today, those departments that are viewed as progressive have an effective, up-to-date strategic plan that provides a road map for the organization into the future. Several factors will continue to impact local fire and EMS providers in the future; and without an effective planning process, organizations will struggle to meet the increasing demands for service and deal with the complexities of the changing fire profession.

■ *The changing role of government.* Historically, government is supposed to have been structured to solve problems at the most basic level—the human

Cause-and-effect diagram	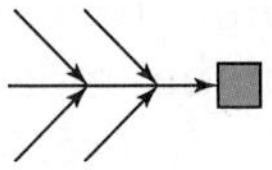	To identify, in an orderly way, all the possible causes that may be contributing to a specific problem
Pareto chart	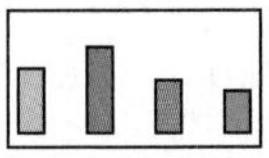	A simple graphical technique for ranking/ordering causes from the most to the least important
Force-field analysis	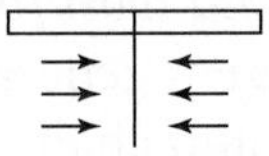	To identify the forces affecting a situation or a problem
Affinity analysis	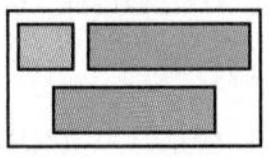	To organize and group ideas based on natural relationships
Multivoting	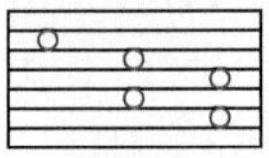	To reduce to a manageable number the improvement opportunities/problems a team has identified
Histogram	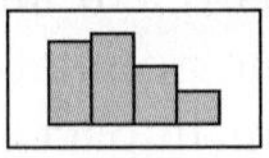	To help display or discover the distribution of measured or counted data
Stratification diagram	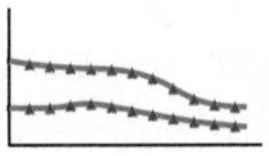	To find and isolate improvement opportunities by breaking down data into meaningful categories
Scatter diagram	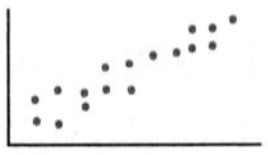	To study or to test the relationship between two variables
Modified nominal group technique	MNGT	To make a common and participative decision

FIGURE 4.6 Problem Tool Matrix

need. However, in the past two decades, one of the prevalent governmental shifts has been devolution—the process of moving power or, in many cases, responsibility from the federal to state and local levels. This devolution has been mainly financial with a number of unfunded mandates driven to the local level by the federal and state governments, still with no attached funding.

■ *The global marketplace.* Globalization has dramatically impacted the way Americans think of and prepare for security. While the world's economic interest overlays and is entwined in a global marketplace, the ease of communication and travel has resulted in new security threats for the United States and all branches of government.

As this globalization continues, its influence on the fire service will be the introduction of technology, processes, and procedures used by our fire service partners in other parts of the world. Yes, the American fire service is waking up to the fact that there are some great ideas, concepts, and technologies that will work here and will have a tremendous influence on the fire profession in the United States over the next 25 years. Our global market is beginning to share what works, and the U.S. fire service is starting to see this networking with the international fire service as having a positive influence worldwide.

■ *Technology.* Given the way technology has evolved thus far, where will it take us in the future? The speed of computing and the amount of data that can be processed will allow computing to achieve milestones that are currently unimaginable. Every facet of the fire profession will be impacted by the emerging technology of today and what will occur in the future. Technology may very well change the way we do business, and if the fire service wants to take advantage of and prepare our organizations for such a transition, it had better have a plan to do so.

■ ***The changing demographics, values, and workforce.*** Today, we have four generations in the workforce, each with different generational values. Leading and managing the workforce today is a complex and challenging task. The population is aging at an unprecedented rate, which is reshaping the very structure of society and fire service demands. This change also affords a great opportunity to tap into the desire of this aging population to be part of a "belonging network"—the fire service family. Belonging and being part of the community will become much more important for this group due to the sheer numbers that are entering retirement. Is there a place for them to help out?

The percentage of the population that is Asian and Hispanic is also rapidly increasing. By the year 2050 no single ethnic population will dominate the U.S. population. Such changes in ethnic makeup impact the societal, political, and economic climate of the nation.

■ ***The changing fire service.*** The fire service today is facing a more diverse set of service demands and organizational challenges than those seen in the past.

- A mounting body of evidence confirms that in recent years, urban–wildland interface fires have increased in frequency and severity; increased fuel loads; become bigger and faster spreading; and are growing more lethal, destructive, and expensive to fight.
- Today's wildland fires can cost taxpayers over a billion dollars annually in fire loss and costs to suppress and for recovery.
- The heavy dependence on oil products will generate new fuel-efficient engines and alternatively fueled vehicles, which present new challenges to first responders.
- Cost-effectiveness will play a central role in achieving the modernization of the fire service. It will be applied not only to support services but also to frontline services such as firefighting and EMS.
- The public will be increasingly consulted on their expectations about the level and quality of the services, requiring demonstrated proof that these are being provided at the best possible price.
- Even though the fire service has continued to enhance its capability and performance during emergency incidents, tremendous improvement opportunities still exist.
- Although today's workforce is the most educated in the history of the fire service, more and more line officers in more and more departments are advancing in rank without significant operational fire backgrounds.
- Fewer fires are being fought with less field-experienced fire officers, which means the potential for disastrous results.
- Many fire departments have and, in all likelihood, will continue to have serious deficiencies in their initial fire attack capability. One fact never changes: the safest and least costly fires are those that either never start or receive strong initial attack to contain and suppress while still small.

- Changes, advances, technological issues, and a global influence will significantly impact fire prevention programs of the future. Each of the following will continue to impact the profession in the years to come:

  Risk-based response
  Weapons of mass destruction (WMD)/terrorism
  Critical infrastructure protection
  Federal mandates for certification and training
  Electronic reporting
  Digital incident documentation
  Self-inspection
  Community outreach through the Web

- It is estimated that a significant number of the profession's experienced fire officers will be eligible for retirement in 5 to 10 years. This is due to a "bubble" of rapid expansion of fire departments in the 1970s and early 1990s.

Many of the changes noted have come to pass. Others are still in play today. What does your list have on it for the changes that will occur at the local level? How do you plan to deal with each? Organizational planning is a must for the profession's future. To cope with the rate of change, the fire service has to adjust the way it plans. It has to be dynamic, flexible, and able to update the plan annually; and it must be driven both from the top down and from the bottom up. Everyone has to understand its importance and be involved.

The importance of a good plan may have been summed up best by one of America's most famous inventors, Thomas Alva Edison, who said, "Good fortune is what happens when opportunity meets with planning."

PEARSON
myfirekit

Visit MyFireKit
Chapter 4 for the
Perspectives of
Industry Leaders.

Review Questions

1. Explain the differences between strategic, operational, and action planning.
2. Explain the following decision analysis tools: SWOT, PEST, and the strategic decision-making process.
3. Conduct a SWOT or PEST analysis of your department.
4. What are the eight fundamental questions in developing a department strategic plan?
5. Explain the three processes of problem analysis.
6. How may organizational culture impact the planning process?
7. Explain the six basic steps found in the strategic decision-making process.
8. Provide an overview of the various analytical tools in the problem-solving toolbox. How could each be used to build consensus on an issue?
9. Evaluate the plans utilized in your community and department. Explain how they are linked.

PEARSON

myfirekit™

For additional review and practice tests, visit www.bradybooks.com and click on MyBradyKit to access book-specific resources for this text!
Register for MyFireKit by following directions on the MyFireKit student access card provided with this text. If there is no card, go to www.bradybooks.com and follow the MyBradyKit link to buy access from there.

References

1. *Partnering for Fire Defense and Emergency Services Planning,* FEMA Course R508, September 2007.
2. *Unit 4: Planning and Implementation,* National Fire Academy Degrees at a Distance Program, p. 69.
3. Ibid.
4. *Unit 4: Planning and Implementation,* National Fire Academy Degrees at a Distance Program, p. 71.
5. *Unit 4: Planning and Implementation,* National Fire Academy Degrees at a Distance Program, p. 72.
6. *Unit 4: Planning and Implementation,* National Fire Academy Degrees at a Distance Program, pp. 61–65.
7. 12Manage: The Executive Fast Track, *PEST Analysis*. Retrieved September 20, 2010, from www.12manage.com/methods_PEST_analysis.html
8. *Converting Strategy into Action,* Stanford Center for Professional Development and the Strategic Design Group, Palo Alto, CA, 2008.
9. *Unit 4: Planning and Implementation,* National Fire Academy Degrees at a Distance Program, pp. 24–28.
10. *Unit 4: Planning and Implementation,* National Fire Academy Degrees at a Distance Program, pp. 22–24.
11. *Unit 4: Planning and Implementation,* National Fire Academy Degrees at a Distance Program, pp. 31–33.

(Courtesy of Tony Escobedo)

An Integrated Approach to Community Risk Management

KEY TERMS

OBJECTIVES

After completing this chapter, you should be able to:

- Describe the history of resource deployment.
- Explain the term *standard of cover*.
- Define the three key considerations in developing a community risk management plan.
- Describe the relationship among the CFAI self-assessment process, the ISO Fire Suppression Rating Schedule, NFPA 1710, and NFPA 1720.
- Define the terms *probability* and *consequence,* and describe why they are important in the design of a community risk model.
- Define the four classifications used for building risk assessment.
- Describe what is meant by effective response force.

PEARSON

myfirekit™

For additional review and practice tests, visit www.bradybooks.com and click on MyBradyKit to access book-specific resources for this text!

Introduction

As we progress in the 21st century, one of the areas that will evolve over the next 50 years will be the deployment of resources. As technology continues to progress, new codes are adopted and integrated into daily life; a reduced risk factor in most communities should be the result.

In this environment, it will be imperative for fire service leaders to utilize business analytics and provide continual education as to the efficiencies and effectiveness of the changes that may occur. The deployment of firefighting resources is often one of the most emotional, political, and controversial issues many communities face. Opening, relocating, or closing fire stations or relocating fire companies can significantly impact a community and firefighting personnel.

The history and evolution of the fire service—beginning with the founding of Jamestown, the conflagrations of the late 1800s, the development of full-time fire departments, and the creation of the board of fire underwriters—provided a systematic methodology about how to distribute fire resources throughout a community.

In the early days of this country, the idea was to locate and staff fire stations based on what was considered a neighborhood, the location of available volunteers, and the available means for hauling heavy equipment. The concept of multiple fire stations emerged only as cities grew and the need developed to space them sufficiently apart to be able to protect the entire community. Previously, the fire service had been built, not upon response time, but on the availability of personnel who could pull the hose carts and staff the bucket brigade. Response was based on the use of horse teams, which would arrive at an emergency in a relatively short period of time; for example, a good team of fire horses could haul a steamer about a mile and a half in approximately five minutes.

Then, with the conflagrations of the 1800s, it became evident the American fire service needed a more precise system to evaluate the effectiveness of its fire suppression efforts. The formation of the National Board of Fire Underwriters established an evaluation system based on science, past practice, and experience. This system had an immediate, significant impact on the formation of the fire service when it was created, and it has continued to guide the application of response and deployment for over a hundred years.

The **ISO** grading schedule has evolved (covered in more detail later in this chapter) from the National Board of Fire Underwriters, and other methodologies have emerged in the fire service to assist in the development and application of a more comprehensive, integrated approach to managing risk.

The Development of National Standards of Cover

During World War II, elected officials in England were scrambling to prepare the populace for everything, including attacks on the homeland. The German invasions of countries on the European and African continents had been viewed from afar, but the **consequences** were becoming alarmingly evident. What steps could England take to protect the populace and the assets of the country? How should

ISO
- Insurance Services Office. A national organization that evaluates public fire protection and provides rating information to insurance companies. Insurers use this rating to evaluate setting basic premiums for fire insurance.

consequence
- Something that logically or naturally follows from an action or condition. The relation of a result to its cause. A logical conclusion or inference. Importance in rank or position, significance, or importance.

it position emergency services and prepare for attacks from the air, potentially from within as acts of espionage, or from the sea, which had traditionally provided a natural block to invaders? In 1936, Great Britain was very conscious that war with Nazi Germany was extremely likely and potentially imminent.

In 1936, the British government formed the Riverdale Committee to contemplate the possible effects of aerial attacks upon the United Kingdom, particularly in light of similar events used in 1930 during the Spanish Civil War and the Japanese attacks on China. In these earlier conflicts, deliberate aerial assaults on undefended towns and cities had occurred for no other reasons than to terrorize the occupants, break the morale of the nation, and overwhelm the civilian infrastructure. All this served as a backdrop as the United Kingdom was at war in 1939 and, by 1940, was under serious, sustained air assault. A group was formed to develop a strategy to combat the risk of fire as a result of such aerial attacks.

The national standards that the Riverdale Committee recommended in 1936 were further refined in postwar years. As a result of a comprehensive review in 1958, the development of a **standard of cover** was completed for Great Britain. This provided a mechanism for the British fire service over the course of the past 70 years to adopt a series of standards of cover dealing with a wide variety of conditions ranging from rural to urban settings.[1]

standard of cover
■ A written document; an assessment of community risk; analysis of current performance that states service level objectives with the distribution and concentration of a fire agency's resources to provide service.

INTEGRATED RISK MANAGEMENT

The concept of standard of cover has evolved over time, with the most recent edition of it unveiled in a May 2004 report entitled *Integrated Risk Management Planning: The National Document*.

This new approach, *integrated risk management planning* (IRMP), seeks to extend the traditional standard of cover approach one step further in protecting the safety of not only citizens but also firefighters. It attempts to create a strategic methodology that not only answers the fire call but also seeks to intervene *before* the alarm is sounded. IRMP's additional component not only addresses prevention in code enforcement and mitigation but also incorporates threats such as a chemical, biological, radiological, nuclear, or explosive (CBRNE) attack occurring on British soil. The London subway and bus bombings in 2005 demonstrated the consequence of this new methodology by the ease and relative calm that appeared to prevail in the wake of those terrorist attacks. The new IRMP requires agencies not only to plan the response for anticipated situations before they occur but also to design standardized response to defined incidents on a national basis.

IMPACT ON AMERICAN DEPLOYMENT

Think of this standardized British approach in comparison to the variations of response, training, and equipment existing in the American fire service. Local performance is guided by various state training requirements and over 30,000 units of local government, which has resulted in a kaleidoscope of response and capability. Beginning in 1987, efforts have been made and continue to be refined by the Center for Public Safety Excellence (CPSE) to adapt the work by the British fire service and the body of knowledge acquired to date for use in the United States. The research done by the CPSE, as the entity overseeing the fire service agency accreditation project, has led to the development of an American-based standard of cover process, which many jurisdictions in the United States and

NFPA 1710
■ Standard for the Organization and Deployment of Fire Suppression Operations, Emergency Medical Operations, and Special Operations to the Public by Career Fire Departments.

Canada utilize today. In addition, the influence of this work can be seen in aspects of the National Fire Protection Association's **NFPA 1710** and 1720 in respect to response, reporting, and firefighter safety.

Prior to these efforts, the history of deployment analysis from the 1920s until 1968 indicates little action in respect to examining the response methodology within the American fire service other than updating the National Board of Fire Underwriters (ISO) grading schedule. In 1968, the RAND Institute developed a research project to study the variables of response patterns and the community impact on fire station location, which included a review of the factors of both time and distance. The complexity of the RAND studies often made it difficult for local government and fire service personnel not only to understand the results but also to translate them to their local elected officials in a meaningful way. Several academic principles were expressed in these studies, such as travel time and distance models, probability models on fire company availability, and dispatching and analyzing the demand for department services. At the time this research had little widespread impact either on the operational fire service or in the context of the insurance industry application at the local level.

During this same time frame, the International City/County Management Association (ICMA) began a series of exchanges with the insurance industry regarding its concerns that the insurance industry criteria were antiquated and inconsistent with contemporary issues facing local government and the fire service. As a result of these discussions, a fire station location package was developed utilizing actual street networks and grids. The station location package, developed by Public Technology Incorporated (PTI), was reflective of the RAND studies' methodology and was first made available to local governments in 1971. Although this system required a larger amount of computing data not readily available at the time to local government, many communities subscribed to the service and conducted individual studies with PTI.

FIRE AND EMERGENCY SELF-ASSESSMENT PROCESS

In the mid-1980s, the International Association of Fire Chiefs (IAFC) approached the ICMA regarding the creation of a more comprehensive approach to the evaluation of the local fire service. In 1986, the IAFC signed a memorandum of understanding with the ICMA for the development of such a concept of fire department self-assessment. Over the course of the next decade, a working group of more than 250 individuals representing all facets of the fire service industry endeavored to create a self-assessment process. As a result, in 1997 the IAFC and the ICMA signed an agreement to form the Commission on Fire Accreditation International (**CFAI**) (see Figure 5.1), which is now known as the Center for Public Safety Excellence (CPSE).

CFAI
■ Commission on Fire Accreditation International; now known as the Center for Public Safety Excellence (CPSE).

FIGURE 5.1 CFAI Logo *(Courtesy of Center for Public Safety Excellence)*

With the CFAI's publication of the first edition of the **Fire and Emergency Service Self-Assessment Manual (FESSAM)** and the implementation of the standard of cover process, numerous fire agencies began to develop documentation for their individual departments in order to achieve fire agency accreditation. FESSAM is a document that provides a comprehensive self-assessment and evaluation model for fire department use that enables fire and emergency service organizations to examine past, current, and future service levels and performance, and compare them to industry best practices. This process leads to improved service delivery by helping fire departments determine community risk and safety needs, evaluate the performance of the department, and establish a method for achieving continuous organizational improvement. A companion document, *Standard of Cover (SOC) Manual,* provides a detailed methodology in the evaluation and planning the deployment of resources.

The development of this process has resulted in a significant amount of research in the methodology and application of resource deployment and how to define community risk.[2] The influence of the CPSE's work on standard of response and development of integrated risk management methods in the United States can also be seen in associated works such as NFPA 1710 and 1720 and the update of the fire suppression rating.

Fire and Emergency Service Self-Assessment Manual (FESSAM)
■ A document providing a comprehensive self-assessment and evaluation model for fire department use that enables fire and emergency service organizations to examine past, current, and future service levels and performance, and compare them to industry best practices. This process leads to improved service delivery by helping fire departments determine community risk and safety needs, evaluate the performance of the department, and establish a method for achieving continuous organizational improvement.

Community Risk Management

The methods that fire departments traditionally use to assess risks and prepare to respond are applicable to all hazards. Each community and/or jurisdiction should assess all of its risks and determine its ability to respond to each one. Considerations in such an assessment follow:

1. *Life safety.* What is the hazard to life? What events threaten injury or death? What is the likelihood that the hazard will cause death or injury?
2. *Responder risk.* What is the potential risk to responders? What can be done to protect responders from harm? What are reasonable risks for firefighters to take? What steps or actions can be taken that would mitigate an event should it occur and at what cost?
3. *Property loss.* What is the community potential for property damage or loss?[3]

LIFE SAFETY

The most important aspect of any risk assessment is assessing potential harm to individuals. Humans are constantly making choices, either conscious or unconscious, about risks they are exposed to as part of their everyday lives. Likewise, communities must make choices on how to respond to the potential **life safety** risk to the lives of occupants from life-threatening situations including fire and EMS in their jurisdictions. Fire risk assessments evaluate items that pose a threat to life. For example, there are important life safety concerns to address in hospitals, nursing homes, and other non-ambulatory facilities. As a result, building and fire codes for such facilities are more restrictive than those for other types of facilities. Once a life safety threat is identified, new codes are developed or existing codes are modified, accordingly.

life safety
■ Risk to the lives of occupants from life-threatening situations including fire and EMS in their jurisdictions.

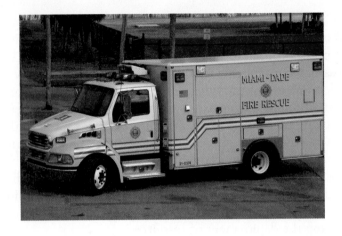

Because local jurisdictions will not always be able to identify all the life safety threats facing them, they often use historical events or past instances in the adoption of new codes. Whereas an obvious risk results in definitive action, the more subtle threats may or may not receive immediate attention. It is the fire service's responsibility to serve as an expert, raising public awareness and enabling local policy makers to make informed decisions.

Because the role of fire departments has expanded, there is now more to be done than just fight fires. Yet, no matter what job is assigned to the fire department, it always has a life safety element: there are lives to be saved. Think of the types of service a fire agency provides: EMS, special rescue, hazardous materials response, response to terrorist acts, and whatever else falls within the realm of emergency response. Regardless of the emergency, the fire service's first priority is to save lives.

Risk assessments related to life safety can be divided into two types: individual (or small group) risks and potential mass-casualty events. The majority of incidents in most communities are of the first type, such as medical emergencies, vehicle accidents, and even residential fires that present risks to individuals or small groups of people. Fire departments should be well prepared to respond to and function at all types of incidents.

Most fire departments handling EMS report 50 to 80 percent of their responses are related to medical treatment (see Figure 5.2).

It is important for those in the fire service to answer these questions: How does your department assess risks with respect to EMS? Has there been an evaluation of potential calls and a plan for the response? Because emergency medical service was not seen as a primary function of the fire department in the past, the EMS response system, in many cases, has evolved without much forethought or planning. Traditionally, the choice for many fire station locations was based on such factors as whether the land was available and affordable and/or whether the site allowed equipment to reach fires in accordance with insurance standards. However, because the same forethought does not usually enter into the equation when adding EMS response, this challenges the paradigms of any organization and indicates the importance of risk assessing and planning an integrated emergency response.

Although some medical calls are not easy to anticipate, many present an evident risk. Because senior-living centers are a growth industry, due to the aging

FIGURE 5.3
Residential Structure
Fire *(Courtesy of Rick
Black. Center for Public
Safety Excellence)*

population, they are a predictable source of increased call volume. In addition, interstate highways, rural roads, and high-incident intersections are predictable locations of accidents and subsequent injuries; yet many departments have not calculated their impact on the deployment matrix. The ability to identify the major causes of EMS incidents in a community has become part of the risk assessment for departments. One challenge fire departments must address is the ability to access and evaluate available data. Oftentimes, crash data are readily collected through police agencies, but are not shared with the fire department.

Another acknowledged community risk is single-family residence hazards related to fire, on which many fire departments focus their staffing and equipment. Individual and small group risk incidents often define the makeup of services provided by the local fire department. Although all communities should be able to handle these types of incidents with the resources they have, many cannot. Of course, communities can do certain things with respect to building and fire code enforcement, fire alarms, automatic sprinklers, and education (see Figure 5.3).

The second type of risk assessment related to life safety is potential mass-casualty events. Even though many fire departments have a base understanding of the common risks found in their community, they often discount incidents involving multiple or mass casualties, which present a much greater challenge to local departments, because of the "it will not happen here" mentality. Because many communities choose to ignore the possibility of these incidents until they happen, they often fumble through the incident. Realistically, it is not possible for all communities to prepare for these massive events or to have sufficient local resources necessary to mitigate all possible events; however, they should at least analyze the possibilities, conduct a risk assessment, and have a response plan in place.

Such catastrophic events can be human-made or natural disasters: weather-related events, accidental occurrences, or terrorist acts. Since the events of September 11, 2001, virtually all communities have an awareness of the potential for terrorist incidents. The inclusion of critical infrastructure vulnerability and key resource analysis has now become part of risk assessment. Hurricane Katrina, on August 29, 2005, and Hurricane Rita, on September 24, 2005, once again displayed the vulnerability of communities to respond effectively to large-scale natural disasters.

RESPONDER RISK

One risk to a community that is often ignored is the risk to response personnel. **Responder risk** is the risk encountered by emergency personnel when responding to and mitigating an emergency. Think of the tremendous negative impact on a community when a firefighter is severely injured or, worse, killed in the line of duty. Obviously, adding to the intangible damage done, tremendous financial and operational fallout result. In preparing for serious events, it is vital for fire departments to evaluate the risks in the community, plan how to respond to the risks, and ensure that firefighters or first responders have the necessary training and equipment to respond safely and effectively.

As has been mentioned earlier, work that has been done in the United Kingdom resulting in the expansion of the traditional standards of cover to an integrated risk management plan approach showed that, without mitigation and prevention, the safety of the firefighter and citizens would plateau. Aggressive prevention and mitigation efforts can truly impact the safety of the civilian population and firefighting personnel. Once communities identify particular risks and hazards, they must prepare their employees accordingly. It is not acceptable to ignore firefighter safety when considering **community risk management**, those policies, procedures, and practices involved in identification, analysis, assessment, control, avoidance, minimization, or elimination of unacceptable risks.

The events of September 11, 2001, raised the level of the public's awareness of the dangers facing first responders. Aside from the obvious deaths and injuries that occurred on that day and during the immediate search and rescue action, longer-term health impacts and detrimental effects on the psychological welfare of the workforce must be addressed. Such situations not only have a negative impact on the departments and the jurisdictions involved but also often have a long-term effect on service levels, recovery efforts, and the economic health of the community (see Figure 5.4).

Recognizing the unusual and unprecedented nature of these types of events, it would have been difficult (at least up until that point in time) to have anticipated the magnitude of 9/11 and have made the necessary and reasonable preparations. However, many risks can be anticipated, and consideration given to the safety of responders must become engrained in the thought process about how to manage emergency incidents.

The development of standards such as the following are designed to reduce the risk of harm to firefighters and other responders: NFPA 1500 (Standard on Fire Department Occupational Safety and Health Program), NFPA 1710 (Standard for the Organization and Deployment of Fire Suppression Operations, Emergency Medical Operations, and Special Operations to the Public by Career

responder risk
■ The risk encountered by emergency personnel when responding to and mitigating an emergency.

community risk management
■ Those policies, procedures, and practices involved in identification, analysis, assessment, control, avoidance, minimization, or elimination of unacceptable risks. A community may use risk control, risk avoidance, risk minimization, risk transfer, or any other strategies (or combination of strategies) in the proper management of future events.

PEARSON
myfirekit™
Visit MyFireKit Chapter 5 for The Physical Protection of Critical Infrastructures and Key Assets.

FIGURE 5.4 World Trade Center, 9/11 *(Courtesy of Michael Reiger/FEMA News Photo)*

Fire Departments), and **NFPA 1720** (Standard for the Organization and Deployment of Fire Suppression Operations, Emergency Medical Operations, and Special Operations to the Public by Volunteer Departments). As risks are identified, threats to responders must be addressed to minimize the potential impact on their safety. The analysis of risks faced by firefighters and first responders must become part of the community-base model in respect to the effectiveness of the actions taken, programs instituted, and other risk reduction efforts that have been implemented.

The integrated risk management planning (IRMP) model, released in 2004 by the United Kingdom, clearly states that risk to responders reaches a plateau that will not be further impacted unless mitigation and prevention are made priorities before incidents occur. The effects of failure to conduct and implement preplans and risk assessment strategies can be evidenced by the outcome of efforts during Hurricane Katrina. Its multibillion-dollar cleanup far surpassed earlier dollar estimates of preventative measures such as the strengthening of levies and other proactive steps to diminish risk. As another example, the Federal Highway Administration, utilizing data from asset management research conducted in Australia and New Zealand, has found that every $1 spent on prevention can save $4 in reconstruction of roadways.

NFPA 1720
■ Standard for the Organization and Deployment of Fire Suppression Operations, Emergency Medical Operations, and Special Operations to the Public by Volunteer Fire Departments.

PROPERTY LOSS

Much has been written about assessing the risk related to fire threats in terms of potential damage and fire department deployment. For example, the ISO's Fire Suppression Rating Schedule, NFPA Standards 1710 and 1720, and the CPSE SOC process assess fire risks and community response capabilities. Although each has a slightly different methodology, collectively they focus attention on the need for a comprehensive analysis in respect to community risk and the subsequent deployment of resources to address these risks.

However, the risks to a community go beyond the immediate result of a disastrous fire or other catastrophic event because the community lives with these effects long after the fire is extinguished or the incident is over. What happens when a building burns or another significant event occurs in a community? It is not easy to measure the resulting loss of jobs, loss of tax revenue, negative impact of burned-out or damaged structures in neighborhoods, commitment of resources following any significant loss, and psychological damage to those involved. For many small communities, the economic engine for the entire region may be built on a single employment center. Therefore, special consideration must be given to such facilities as their loss can result in devastating consequences for the community.

As community growth has encroached on wildland and forested areas, a significant risk that has emerged in many communities is wildfires crossing into suburban and urban areas known as the **urban interface**—the line, area, or zone where structures and other human development meet or intermingle with undeveloped wildland or vegetative fuel (see Figure 5.5). These fires have become modern-day conflagrations; since the 1960s, many instances have occurred of significant portions of communities burning down as a result of urban interface fires. The severity of and associated losses from these fires are staggering in property and life losses and the cost to extinguish the fire. The primary cause of these large fires is the urban encroachment on the wildland, although some believe the impact of global warming and climatic changes may be a contributing factor. As codes and prevention and suppression efforts have improved, the frequency of such burnouts has not declined due to the continued expansion of the wildland–urban interface area. Yet many communities have continued to build in urban interface areas and to experience the incidence of heavy losses. Although

urban interface
■ The line, area, or zone where structures and other human development meet or intermingle with undeveloped wildland or vegetative fuel.

FIGURE 5.5 Wildfire
(Source: www.blm.gov)

such hazards are easy to identify, they are often more difficult to prevent due to freedom of choice and lack of zoning and planned development, as well as the motivation of elected officials to create a larger tax base. Still other communities are constantly making efforts to reduce the risk of fire occurrence and improve response capabilities should a fire occur.

As a way to prevent property loss, a major component of building codes is to protect buildings from fire risks. The fire service needs to assess the following: Are buildings constructed to withstand a fire? Is there built-in fire protection? Is the building stock segregated to keep any fire relatively small and within the capabilities of the fire department? Geographically specific codes have been developed to cope with earthquakes, hurricanes, and other natural disasters. For example, California would be more prepared for earthquakes and Florida more prepared for hurricanes. However, the adoption of national (or international) codes to protect communities is a critical element of community risk management. Some people still believe the insurance companies exert the most influence on code compliance. However, the fire chief and local building officials remain the lead professionals in most localities for identifying and protecting against hazards in the community.

PEARSON
myfirekit
Visit MyFireKit Chapter 5 for more information on Causes and Circumstances of Fire.

RISK AND PLANNING

Today, the identification of community risks is not enough. Risk assessment is effective only when subsequent planning and implementation of risk reduction take place as a result. This planning needs to occur on two levels: strategic and operational (see Chapter 4 for details). First of all, strategic planning involves a conscious effort to identify risks and design the community response to address those risks. Operational plans are those planning efforts, usually one to two years in duration, that support the strategic plan initiatives. As these risks are identified, local governing bodies make policy choices concerning issues such as standard of cover and the **level of service** to be provided. This entails the resources needed to meet the stated service level objective(s). Level of service is defined only in terms of what resources are provided and not in terms of effectiveness or quality. Deployment policies should be based upon the risk present and the community's expectation of service.

level of service
■ The resources needed to meet the stated service level objective(s). Level of service is defined only in terms of what resources are provided and not in terms of effectiveness or quality.

Local elected officials ultimately determine the policies to be implemented including the decisions to prevent or mitigate, respond or react, or in some cases just plain ignore. They do this by designing laws, codes, regulations, and standards to eliminate or minimize threats. If the prevention and mitigation efforts do not work, then a response is necessitated to mitigate the problem. For a safe and adequate response, the goal should be to send sufficient resources to the scene of an emergency in a defined period of time to affect a positive outcome.

With the recent increase in terrorist acts (and threats of such acts), governments at all levels are taking action to prevent or mitigate such acts. For example, airports and mass transportation providers are increasing security and changing the ways they do business in the interest of providing a safe environment for travel. Some of their actions are mandated by the federal government and others result from national, state, or local initiatives. Today's risk planning and community risk assessment have taken on a broader meaning using more precise measurement to determine what defines success or failure.

Success is also measured when a community incorporates these planning processes as part of its long-term plan. Unfortunately, too often community risk analysis, standard of cover, and long-term departmental strategic plans are made but are not utilized. As a result, as the community grows, redevelops, or in some cases contracts, the fire service is left in a reactive mode. When this occurs, the fire service finds itself with either a lack of resources deployed in a timely fashion, or stations and equipment that may be in the wrong location to provide an effective standard of cover. This is why it is critical to incorporate the key elements of the department's strategic plan into the community's general plan.

The Community Risk Model

As has been discussed, a critical element in the assessment of any emergency service delivery system is its ability to provide adequate resources for anticipated fire combat situations, medical emergencies, and other events that are likely to occur in the community. This assessment is called a community risk model. Properly trained and equipped fire companies must arrive, deploy, and mitigate the event within specific time frames in order to meet successful emergency event strategies and tactical objectives.

distribution
■ The number of resources placed throughout a community.

concentration
■ The number of resources needed in a given area in the community based upon the risk, workload, call volume, and adopted service level objectives.

Each event—fire, rescue operation, major medical emergency, disaster response, and other situations—requires varying and unique levels of staffing and resources. For example, controlling a fire before it has reached its maximum intensity requires a rapid deployment of personnel and equipment in a given time frame. The higher the risk, the more resources needed. More resources are required for the rescue of persons trapped within a high-risk building with a high occupant load than for a low-risk building with a low occupant load. More resources are required to control fires in large, heavily loaded structures than in small buildings with limited contents. Creating a level of service requires making decisions regarding the **distribution** and **concentration** of resources in relation to the potential demand placed upon them by the community's level of risk.

PROBABILITY AND CONSEQUENCES

probability
■ A measure of how likely it is that some event will occur; the quality of being probable; a probable event or the most probable event; probability is a way of expressing knowledge or belief that an event will occur.

Jurisdictions should assess risks based upon the potential frequency (**probability** of occurrence) and the potential damage, should they even occur. For example, although a terrorist act has a low probability, the potential damage and psychological impact to the country of such an occurrence are potentially very high. This same outlook in regard to risk assessment can also be applied to natural disasters. For example, tornadoes generally do not hit the same communities every year, but if they do strike, the damage can be great. Conversely, medical emergencies happen every day. The overall potential damage from medical emergencies to the community as a whole is not nearly as significant as that from a tornado or other natural disaster (although these individual incidents greatly affect those requiring the service).

Organizations should be able to compare the probability and potential damage of events that may affect their community. Communities of all sizes need to conduct this type of analysis. Unfortunately, there is often a knee-jerk reaction to a disaster: tornado sirens are installed after a tornado devastates an area; fire codes get tougher after a major fire; Congress appropriates more funding for the war on terrorism after the World Trade Center and Pentagon attacks. Although there are many ways

TABLE 5.1	Probability Matrix
High Probability Low Consequence	High Probability High Consequence
Low Probability Low Consequence	Low Probability High Consequence

to view risk management, it simply comes down to assessing the chances of an event occurring and the damage that could result from the event (see Table 5.1).

The combinations of probability and consequence are pictured in the four quadrants of this matrix. For example, structure fires are relatively frequent in comparison to other types of incidents in a community; and the subsequent dollar loss, loss of irreplaceable items, and loss of business or jobs make the consequences of such fires high. In some communities, activation of automatic fire alarms is high probability with low consequence. Tornadoes and hurricanes may be infrequent but represent a large potential loss to life and property. Comparatively, dumpster fires may be high probability as they happen often, yet have little consequence outside the fire response.

With an understanding of the different levels of probability and consequences, fire organizations can develop deployment strategies in respect to overall community risk management. If the only threat in a community is dumpster fires, there may not be much need for a fire department. If structure fires are a daily event, the fire department should be appropriately staffed, equipped, and deployed throughout the community to handle such events. Begin thinking of the probability matrix of potential threats in your own community. What are your call volume, breakdown of calls, risk factors, critical and key assets, and current standards of cover?

COMMUNITY RESPONSE TO RISKS

If it is easy to identify risks, probabilities, and outcomes, then why are all communities not prepared? First of all, many communities and fire departments may not assess risks; instead of being proactive, some communities and/or fire departments wait for problems to occur before taking any action. Others proceed as if nothing of significance will happen in their communities (just as individuals do, because fires happen to others, not us!). Certain communities do not view risk assessment as a priority, just as certain fire departments are virtually inactive in fire prevention programs.

What about those communities that are good at assessing the risks they face? Why might they still not be prepared when something goes wrong? Unfortunately, just because their fire professionals and other risk managers can identify potential problems does not guarantee that these communities take the necessary steps to invest in the resources to address these risks. Why? This is because communities have to make choices, just as individuals do, and these choices are often based upon financial and political considerations. The community may elect not to spend funds at the level needed to address its risk properly or may choose not to adopt codes and standards due to the political risk and pressures exerted from special-interest groups.

Many communities that have not conducted any type of risk assessment simply are unaware that their current deployment of resources may not be sufficient to handle the risk at hand. Although this can be a conscious choice, in many cases it is a result of the "we have always done it this way" mentality. This may occur in a community with an insufficient tax base to address fully all the potential risks identified or one with an equally dangerous mind-set of a "one size fits all" response. The level of deployed resources, both physical and mechanical, to handle a multistory fire should be recognized as different from that for a 2,000-square-foot residential structure. The resources necessary to respond to a dumpster fire should be different from the response to a five-story, wooden-frame warehouse fire in an old industrial neighborhood.

Those responsible for setting policy and budget plans may not consider community risk management to be a critical issue. Think about how you make decisions as an individual regarding risk analysis; policy makers do the same thing by consciously choosing based upon their willingness or ability to pay and their perception of the risk factors in play. An example of this, on both the personal and community levels, is having flood insurance. You may never need it, but should it flood, then it is too late to do anything about it.

The challenge to community risk management lies not only in the work necessary to assess the probabilities of an emergency event in a community but also in the political arena. It is the policy makers who will determine the level of service to provide, and, in many instances, these people may not be aware of the community risks. It is vital for policy makers to assess what risks a community addresses and which others it ignores. Only through their active participation in the analysis of community risk and deployment strategies can fire service agencies affect policy through education of their community and elected officials.

Unfortunately, in many jurisdictions, both large and small, these discussions never occur. This level of leadership is the responsibility of the chief executive and senior leadership of the department. However, a level of complacency often exists in these organizations because they have not experienced a significant event in a long time. In some cases the complacency is within the department leadership that does not want to put forth the effort either to address the issue or to initiate the needed changes. The destructive results of this complacency have been seen in fires with larger losses and civilian and firefighter deaths and injuries.

COMMUNITY RISK ASSESSMENT—FIRE SUPPRESSION

As has been discussed, to be effective, the evaluation of fire risks must take into account the frequency and severity of fires and other significant incidents. Determining risk by analyzing the real-world factors in the area served is essential to the development of an emergency services strategic plan. As noted in the probability matrix, risk assessment is divided into four quadrants, and each imposes different requirements for commitment of resources (see Table 5.2).

The relationship between probability and consequence determines the level of risk and the needed distribution and concentration of resources. Distribution is

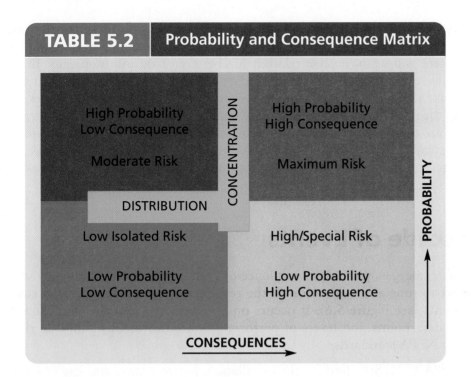

| TABLE 5.2 | Probability and Consequence Matrix |

High Probability
Low Consequence

Moderate Risk

CONCENTRATION

High Probability
High Consequence

Maximum Risk

PROBABILITY

DISTRIBUTION

Low Isolated Risk

Low Probability
Low Consequence

High/Special Risk

Low Probability
High Consequence

CONSEQUENCES

the number of resources placed throughout a community. Concentration is the number of resources needed in a given area in the community. Specific issues that determine concentration vary and include the number of events (calls) for service; the risk factors of the area; the availability and reliability that units will be available to respond; and the time of arrival of secondary responding units. A challenge for the fire chief and local government is balancing the distribution and concentration of resources in a community to achieve the best results. If the *distribution* (fire stations and companies) is too low in comparison to *concentration* (resources, staffing), the outcomes will not be successful.[4]

The main objective is to have a distribution of resources available to reach a majority of events, regardless of their significance, within the protected jurisdiction and meet the service level objectives established by their communities. Many factors make up the risk level that would indicate the need for a higher concentration of resources, such as the inability of occupants to take self-preserving actions, construction features, lack of built-in fire protection, hazardous structures, lack of adequate **fire flow**, and nature of the occupancy or its contents. Evaluation of such factors leads to determining the number of personnel required to conduct the critical tasks necessary to contain the event in an acceptable time frame.

fire flow
■ The flow rate of a water supply, expressed in gallons per minute (gpm), measured at 20 pounds per square inch (psi) residual pressure, that is available for firefighting.

An agency's level of service should be based on its ability to cope with the various types and extents of emergencies that it can reasonably expect after conducting a risk assessment. The risk assessment process starts with viewing the most common community risk, the potential fire problem, target hazards, the critical infrastructure, and a review of historic call data.

RISK ASSESSMENT MODEL

The risk assessment model incorporates the following: the various elements of risk to the relationships among the community as a whole, the frequency of events that occur, the severity of potential losses, and the usual distribution of risks. Overall, any community is likely to have a wide range of potential risks, and, there will be an inverse relationship between risk and frequency. In short, the daily events are usually the routine ones and result in minimal losses, whereas the significant events are less frequent and result in greater losses. Moving up the probability and consequence matrix toward the highest risk levels, the events are less frequent. If the community's risk management system is working, a catastrophic loss should be an extraordinary event. Community-based risk management involves trying to keep routine emergencies from becoming serious loss situations.

Cascade of Events

cascade of events
■ In every emergency there is a sequence of events, which are critical elements in respect to time and evaluation of the response system, known as the cascade of events; and it occurs on every emergency call.

In every emergency there is a sequence of events, which are critical elements in respect to time and evaluation of the response system, known as the **cascade of events** (see Figure 5.6). It occurs on every emergency call[5] (see Table 5.3). These data points and levels of performance are universally found in the following NFPA standards:

NFPA 1221, Standard for the Installation, Maintenance, and Use of Emergency Services Communications Systems

FIGURE 5.6 Cascade of Events *(Courtesy of Center for Public Safety Excellence)*

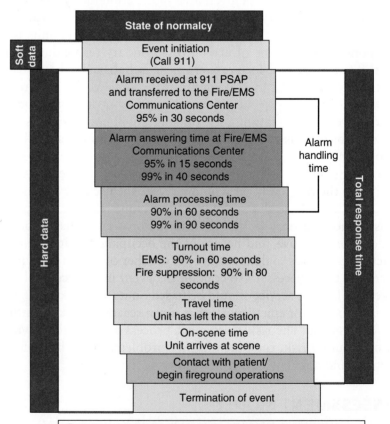

Response service level objectives will be based on risk factors, which translate into distribution and concentration of resources.

TABLE 5.3	Cascade of Events
Event initiation point	The point at which a human being or technologic sentinel (i.e., smoke alarm, infrared heat detector, etc.) becomes aware that conditions exist requiring activation of the emergency response system. These are factors that occur, resulting in activation of the emergency response system. Factors may delay the initiation or call for help by seconds, minutes, hours, or even days before assistance is requested. An example is the patient who ignores chest discomfort for days until it reaches a critical point at which time he or she makes the decision to seek assistance.
Alarm received and transmitted	The point at which a call is received and answered at the 911 Public Safety Answering Point (PSAP) and is transmitted to the Communications Center, if separate from the PSAP.
Alarm answering time	The point at which the alarm is transmitted from the PSAP and answered at the Fire/EMS Communications Center.
Alarm processing time	A process by which an alarm answered at the Communications Center is retransmitted to emergency response facilities (ERFs) or to emergency response units (ERUs) in the field.
Turnout time	The interval between the activation of station and/or company alerting devices and the time when the responding crew is aboard the apparatus and the apparatus is en route (wheels moving) to the call as noted by the mobile computer terminal or by voice notification by the officer to dispatch that the company is responding.
Travel time	The point at which the responding apparatus signals the dispatch center that it is responding to the alarm and ends when the responding unit notifies the dispatcher of its arrival on scene (via voice or mobile computer terminal notification).
On-scene time	The point at which the responding unit arrives on the scene of the emergency.
Initiation of action	The point at which operations to mitigate the event begin. This may include size-up, investigation, resource deployment, and/or patient contact/intervention.
Termination of incident	The point at which units have completed the assignment and are available to respond to another request for service.
Total response time	The time from the call being received at the 911 PSAP and the point at which crews arrive and intervention begins.

Source: CFAI, *Creating and Evaluating Standards of Response Coverage for Fire Departments*, 4th ed. 2003 (CD-ROM).

NFPA 1710, Standard for the Organization and Deployment of Fire Suppression Operations, Emergency Medical Operations, and Special Operations to the Public by Career Fire Departments

NFPA 1720, Standard for the Organization and Deployment of Fire Suppression Operations, Emergency Medical Operations, and Special Operations to the Public by Volunteer Fire Departments

These data points are incorporated as benchmark performance measures in the CFAI self-assessment process.

The total response performance continuum is shown in Table 5.3.

Each of the elements within the cascade of events provides specific data points, which should be measured and reported in respect to the agency's overall response and performance.

A 2010 study by the Fire Protection Research Foundation provides a quantitative evaluation of fire emergency and EMS mobilization times and identifies key factors affecting the performance of the fire service. The study supplies a statistical analysis of actual fire emergency and EMS alarm handling and turnout times based on data collected across a diverse representative population of North American fire service organizations. The results present measured data for validation and refinement of requirements provided by nationally recognized standards, and additionally indicate the most significant and variable factors (e.g., difference in daytime and nighttime events).

PEARSON
myfirekit

Visit MyFireKit Chapter 5 for the full report.

The Impact of Time on Outcomes

Firefighters encounter a wide variety of conditions at each fire. Some fires will be at an early stage, and others may have already spread throughout the building. This variation in conditions complicates attempts to compare fire department capabilities. A common reference point must be used so that the comparisons can be made under equal conditions. In the area of fire suppression, service level objectives are intended to prevent the **flashover point**. This is the temperature point at which the heat in an area or region is high enough to ignite all combustible material simultaneously.

flashover point
■ The temperature point at which the heat in an area or region is high enough to ignite all combustible material simultaneously. This particular point of a fire's growth dramatically increases the threat to life and property.

THE SIGNIFICANCE OF FLASHOVER

Flashover is the particular point of a fire's growth that dramatically increases the threat to life and property. As has been discussed, fire suppression tasks required at a typical fire scene can vary a great deal depending upon the building, the health and age of the occupants, and the conditions encountered. What fire companies must do, simultaneously and quickly, if they are to save lives and limit property damage is to arrive within a short period of time with adequate resources to do the job. Matching the arrival of resources within a specific time period is the objective of developing a comprehensive standard of cover based upon the anticipated risk to be addressed as well as to provide intervention prior to flashover (see Figure 5.7).

Flashover is a critical stage of fire growth, because it creates a quantum jump in the rate of combustion and necessitates a significantly greater amount of water to reduce the burning material below its ignition temperature. A fire that has reached flashover often indicates that it is too late to save anyone in the room of origin and that a greater number of firefighters are required to handle the larger hose streams needed to extinguish the fire. A post-flashover fire burns hotter and moves faster than a pre-flashover fire, compounding the search-and-rescue problems in the remainder of the structure, at the same time more firefighters are needed for fire attack (see Table 5.4).

FIGURE 5.7 Structure Fire *(Courtesy of Rick Black. Center for Public Safety Excellence)*

The fire service can reasonably predict staffing and equipment needs for different risk levels and fire stages. The correlation of staffing and equipment needs in accordance with the stage of growth is one of the key determinants in the development of response coverage (see Figure 5.8, page 156). The goal is to maintain and strategically locate enough firefighters and equipment so that a minimum acceptable response force can reach a reasonable number of fire scenes before flashover and provide much needed care in critical medical emergencies.

PEARSON
myfirekit

Visit MyFireKit Chapter 5 for information on The Stages of Fire Growth.

EVALUATING EMS CAPABILITIES

Survival of cardiac death is often time driven. The brain can be without oxygen for only a short period of time: four to six minutes. Therefore, rapid intervention is necessary to prevent brain death.

TABLE 5.4	Pre- and Post-Flashover	
PRE-FLASHOVER	**POST-FLASHOVER**
Limited to one room	May spread beyond one room
Requires smaller attack line	Requires larger and more attack lines
Search and rescue is easier	Compounds search and rescue
Initial assignment can handle	Requires additional companies

FIGURE 5.8 Time versus Products of Combustion

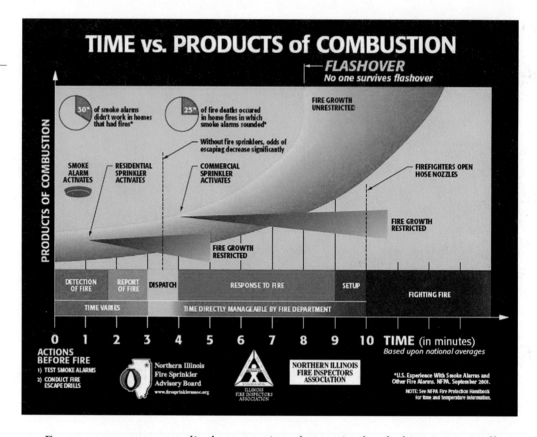

From an emergency medical perspective, the service level objective typically is to provide medical intervention within a six-minute time frame, because brain damage is very likely at six minutes without oxygen. However, in a cardiac arrest situation, survivability dramatically decreases beyond four minutes without appropriate intervention, which includes early recognition and bystander CPR. Those attending a major international meeting held in the Utstein Abbey just outside Stavanger, Norway, in 1991, sought to establish a set of common definitions and key data points for events associated with cardiac arrest. The resulting criteria, known as the Utstein Criteria, created this set of definitions and spelled out exactly how to report cardiac arrest data. The medical community hoped the criteria would serve as a tool to facilitate EMS leaders and researchers as they sought to understand the reasons for different survival rates among communities. The research recommends using the Utstein reporting criteria for outcomes research and capture of the following time stamps/points in the cascade of events in an EMS call that should be tracked: early access, early CPR, early defibrillation, and early advanced cardiac life support (ACLS).

Early defibrillation is often called the critical link in the chain of survival because it is the only way to treat most sudden cardiac arrests successfully (see Figure 5.9). When cardiac arrest occurs, the heart starts to beat chaotically (fibrillation) and cannot pump blood efficiently. Time is critical; if a normal heart rhythm is not restored within minutes, the person will die. In fact, for every minute without defibrillation, the odds of survival drop 7 to 10 percent. A sudden cardiac arrest victim who is not defibrillated within eight to ten minutes has virtually no chance of survival.[6]

The shortest possible response times create the highest probabilities of resuscitation. An important evaluation point lost on most agencies is the time that crews

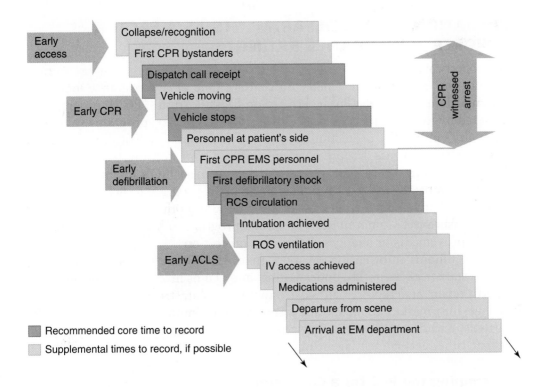

FIGURE 5.9 Events Associated with Cardiac Arrest

Early access
- Collapse/recognition
- First CPR bystanders

Early CPR
- Dispatch call receipt
- Vehicle moving
- Vehicle stops
- Personnel at patient's side

Early defibrillation
- First CPR EMS personnel
- First defibrillatory shock
- RCS circulation
- Intubation achieved
- ROS ventilation

Early ACLS
- IV access achieved
- Medications administered
- Departure from scene
- Arrival at EM department

CPR witnessed arrest

■ Recommended core time to record
□ Supplemental times to record, if possible

reach the patient's side. Instead, the clock often stops when the vehicle arrives or stops at the address. The key to a successful outcome is the point the patient is actually contacted. In high-rise communities or other larger complexes, the time period between arrival at the scene and arrival at the patient's side can be substantial, and this can most certainly affect the outcome due to delayed intervention.

For most departments today, EMS accounts for the majority of response calls. This trend will not only continue in the future but also likely increase as the population ages and more people use the EMS system as their primary source of medical care. This shift will require a new level of analytics to evaluate current deployment strategies and the impact on the overall response system.

Methods of Risk Management

With the development of the National Board of Fire Underwriters, the first process was implemented that began to assess risk versus response capability. Today's fire service leadership has several methods to assist in this evaluation. The ISO Fire Suppression Rating Schedule, the CFAI accreditation process, and NFPA Standards 1710 and 1720 on deployment provide various means to evaluate departmental performance. Progressive leaders will utilize all three to assist in the evaluation of their departments' capabilities (see Figure 5.10). Progressive departments are incorporating each method of evaluation, NFPA 1710 for career and NFPA 1720 for substantially volunteer, into their departmental strategic planning and standard of cover development process.

By doing so, these departments provide a comprehensive overview of existing performance and help to develop an effective plan to meet their future demands for service.

FIGURE 5.10 ISO Logo *(Courtesy of Insurance Services Office, ISO)*

EVALUATION OF FIRE DEPARTMENTS AS A FACTOR IN PROPERTY INSURANCE RATING

U.S. property insurance companies have funded initiatives aimed at fire mitigation since the middle of the nineteenth century. One of the insurance industry's most important tools is the Public Protection Classification (PPC™) program administered by the Insurance Services Office (ISO). ISO's PPC program evaluates a community's public fire protection capability and assigns it a protection-class rating from 1 to 10. Class 1 represents exemplary fire protection; Class 10 means that the area's fire suppression program does not meet ISO's minimum criteria. Insurance companies use information from the PPC program to help market, underwrite, and price homeowners, business owners, and other types of commercial property insurance. Communities rely on the program to help plan, budget, and justify improvements or changes in their fire protection.

ISO maintains information on more than 45,000 fire protection areas and provides statistical, actuarial, underwriting, and claims information and analyses; property repair and replacement cost estimations; catastrophe modeling; policy language; information about specific locations and communities; fraud identification tools; and data processing services. It serves the insurance industry, regulators, and other government agencies.

Determining the PPC for a Community

ISO's evaluation of a community's fire suppression system includes a review of the dispatch center, fire department, and water supply infrastructure. A community's strengths and/or weaknesses relative to specific criteria in each of those categories will determine the community's public protection classification. Communities can have different combinations of strengths and weaknesses yet still receive the same PPC. Therefore, the PPC number alone does not fully describe all the features and capabilities of an individual fire department. Generally, the classification numbers suggest the following:

- Classes 1 through 8 indicate a fire suppression system with a creditable dispatch center, fire department, and water supply.
- Class 8B recognizes a superior level of fire protection in an area lacking a creditable water supply system. Such an area would otherwise be Class 9.
- Class 9 indicates a fire suppression system that includes a creditable dispatch center and fire department but no creditable water supply.
- Class 10 indicates the area's fire suppression program does not meet minimum criteria for recognition.

For many jurisdictions, ISO publishes a "split class," such as 6/9. In such jurisdictions, all properties within 1,000 feet of a water supply (usually a fire hydrant) and within five road miles of a fire station are eligible for the first class (Class 6 in the example). Properties more than 1,000 feet from a water supply (usually a fire hydrant) but within five road miles of a fire station are eligible for Class 9. All properties more than five road miles from a fire station are Class 10.[7]

Proposed Revisions to the FSRS

The *Fire Suppression Rating Schedule* (FSRS) is the basis for the ISO's public fire protection classification activities nationwide. The FSRS focuses review on three specific areas of operations—the dispatch center, fire department, and water supply. Although it is essential to consider these criteria in the community planning process, do not mistake these evaluations as a measurement of the quality of services being provided. As noted in a Charleston, South Carolina (an ISO-graded Class 1 department), post-incident report, in which nine firefighters were lost, several performance issues contributed to the outcome of this particular incident; while the FSRS has an important place in the community risk assessment model, it should not be the only criteria utilized in the evaluation process. The FSRS schedule is currently under review and is scheduled to be updated in the near future.

PEARSON
myfirekit
Visit MyFireKit Chapter 5 for more details on the Proposed Revisions to the FSRS.

The Effect of PPC on Insurance Premiums

ISO provides insurance companies with public protection classifications and associated details, including fire station locations, response area boundaries, the location of hydrants, and other water supply details. But because insurance companies, not ISO, establish the premiums they charge to policyholders, it is difficult to generalize how an improvement (or deterioration) in PPC will affect individual policies, if at all.

However, ISO's studies have consistently shown that, on average, communities with superior fire protection have lower fire losses than do communities whose fire protection services are not as comprehensive. Consequently, PPC does play a role in the underwriting process for many insurance companies.

PEARSON
myfirekit
Visit MyFireKit Chapter 5 for more information on the ISO and Fire Suppression Rating Schedule.

THE COMMISSION ON FIRE ACCREDITATION INTERNATIONAL (CFAI)

The Commission on Fire Accreditation International (CFAI) is a nonprofit entity formed in 1986 through a memorandum of understanding (MOU) between the ICMA and the IAFC. In 2006, the corporate name was changed to the Center for Public Safety Excellence (CPSE). The CPSE oversees both the Commission on Fire Accreditation International (CFAI) and the Commission on Professional Credentialing (CPC) (see Figure 5.11).

The mission of the CFAI is to assist fire and emergency service agencies throughout the world in achieving excellence through self-assessment and accreditation in order to provide continuous quality improvement and enhancement of service delivery to their communities. The CFAI offers a comprehensive system of fire and emergency service evaluation by analyzing over 250 performance indicators and 90 criteria within 10 main categories, which research has shown helps to determine organizational effectiveness. In addition, it aids local government executives evaluate expenditures that are directly related to improved or expanded service delivery in the community and provides a nationally accepted set of criteria by which communities can judge the level and quality of fire, EMS, and other services they provide. The CFAI also offers a measurement tool for community use in gauging the effectiveness of the fire and emergency service agency. Decisions on accreditation, general organizational operation, and special programs and activities for the CFAI are made by a commission that also oversees revisions to the accreditation model and self-assessment process, training, education, research, and development, as well as other issues related to self- and peer assessment. Eleven CFAI commissioners represent a variety of fire service stakeholders (see Figure 5.12).

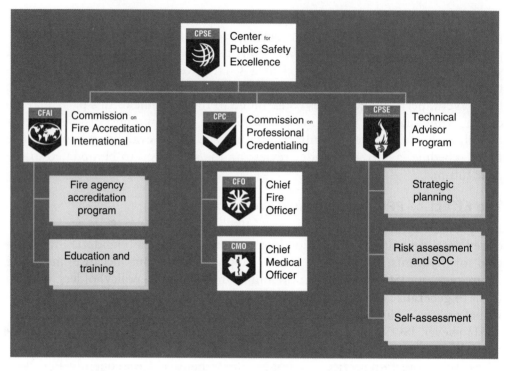

To be an effective leader today, it is critical to be able to evaluate all governmental functions. The CFAI process provides for periodic organizational evaluation to ensure effectiveness, help in managing change, and raise the level of professionalism within the organization and the profession. This is accomplished through the use of a self-assessment process for fire and emergency service agencies that is designed to answer three very basic questions:

1. Is the organization effective?
2. Are the mission, goals, and objectives of the organization being achieved?
3. What are the reasons for the success of the organization?

FIGURE 5.12 CFAI
Organizational
Chart *(Courtesy of
Center for Public Safety
Excellence)*

Program evaluation focuses on identifying the efficiency and significance of the organization's activities. It seeks to answer these questions:

- Is the organization producing benefits that justify the expenditures?
- Are there better ways to achieve the organizational objectives?
- Are performance measures used to quantify the efficiency and effectiveness of departmental programs?
- Are the goals and objectives being achieved with the appropriate allocation of resources?
- Will the achievement of the goals contribute to providing a quality and stated level of service for the area being protected?
- To what extent are the activities of the organization making an impact on the mission of the organization?

The accreditation model includes 10 categories used in the evaluation of departmental performance:

- Governance and Administration
- Assessment and Planning
- Goals and Objectives
- Financial Resources
- Programs
- Physical Resources
- Human Resources
- Training and Competency
- Essential Resources
- External Systems Relationships

PERFORMANCE INDICATORS

Within each category of the accreditation model, there are criterion and performance indicators. A criterion is a measure, a set of measures, or an index on which a judgment may be based to evaluate performance. Each criterion defines a major area within each category of the self-assessment. A performance indicator is the desired level of achievement toward a given objective, defining the ability to demonstrate doing a particular task as specified in the accreditation process. There are 253 performance indicators in the self-assessment process, 82 of which are core competencies.

PEARSON
myfirekit
Visit MyFireKit Chapter 5 for more information on performance indicators.

THE OBJECTIVES OF THE CFAI PROGRAM

The objectives of this program are to provide not only a quality improvement process to enhance the capability of a fire agency but also a mechanism for the community to recognize and understand its respective community risks and associated emergency protection needs. The system has been proven to improve the local fire agency's resources and emergency service delivery systems. Because this program is designed to improve the quality of life in the communities served by the fire and emergency service organizations and to provide recognition for quality service, it serves as a mechanism in the design of future plans for improvement.

The CFAI program defines a model accreditation system that is credible, realistic, usable, and achievable. It is designed to be used by fire service agencies, city administrators, city/county managers, and elected officials to evaluate community fire risks using state-of-the-art practices. Those communities that have successfully utilized the self-assessment process develop policies that reduce fire/EMS risks and gain results in improved delivery of services. This program, a voluntary system, reflects a unique combination of fire protection engineering standards, community values, and an inventory of commonly used acceptable practices. The project involves a "systems" model instead of a singular, stand-alone "pass or fail" document. System modeling is an organized collection of processes, services, and products that are highly integrated to accomplish an overall goal. The system has various inputs, which go through certain processes to produce certain outputs and together accomplish the desired goal(s) for the system. The program's ultimate goal is to supply an accreditation process that will improve the ability of communities to recognize and understand their respective fire and life safety risks, supply a balance of public and private involvement in reducing the risk, and improve the quality of life for citizens within the communities using this model.

The primary missions of today's fire service agency are to prevent fires from starting; avoid the loss of life and property when a fire does occur; and provide means to evaluate a variety of other locally related services, such as emergency medical response, specialized rescue service, and public education efforts, to satisfy the needs of its jurisdiction and citizens. In contrast, the traditional responsibility of a fire service agency has been to prevent fires and to suppress them if they should occur. As has been discussed previously in this text, modern-day career, volunteer, or combination fire agencies have expanded their services beyond basic prevention and fire suppression, due to changes in society's demands.

STANDARD OF COVER

In the research that was conducted over a 10-year period to develop the initial accreditation process, standard of cover is the prime concept that emerged as having had broad impact on the fire service. As noted previously in the chapter, this concept was born during World War II in Great Britain and has gained acceptance in many countries.

Standards of cover (SOC) can be defined as those written policies and procedures that establish the distribution and concentration of an organization's fixed and mobile resources. The accreditation process requires an agency to have adopted a system that defines risk categories within the response area of that jurisdiction and, based upon this risk, the anticipated workload. As such, a written policy statement should be developed that defines the service level objectives for emergencies such as fires, medical calls, and other anticipated emergencies. The original modeling developed by the CFAI was based upon building occupancy. However, because of the range of complex issues faced today by the fire service, it was necessary to expand the model to include critical infrastructure, key assets, national disaster modeling, integration of deployment strategies, and the use of geographic information systems (GIS).

RISK ASSESSMENT MODEL

Any risk assessment model must take into account that the relative magnitude or the degree of risk will be different in every community. In general, it is anticipated that in most communities the vast majority of the risk will fall into the low-risk and moderate-risk categories, with smaller percentages being distributed among the other probability quadrants. The majority of fire service concern should be directed toward the development of fire defense strategies for occupancies that fall into the high-probability and/or high-consequence category. The development of an organizational strategy to achieve the desired service level objectives is very important to the credibility of any fire organization.

Historically, the risk levels in a community were primarily based on structural occupancies and business types that existed in the community. However, many other risks must be assessed today as well. For example, an agency that has wildland firefighting responsibilities may define its risk on the basis of topography, geography, fuel cover, and weather conditions. An area with an urban–wildland interface may have a risk assessment that combines structural conditions with the existing or potential natural fuel load. Areas that have major transportation routes, a commercial airport, a rail line intersecting the jurisdiction, commercial ports, and/or key national assets contain risks that must be factored into any risk assessment process.

The purpose of risk assessment is not only to evaluate risks and hazards in the jurisdiction's response area but also to provide a basic methodology to evaluate existing response coverage. The process begins with the identification of community hazards and risks. A hazard is defined as a source of potential danger or an adverse condition. A risk is defined as the possibility of loss or injury or the exposure to the chance of loss; the combination of risk with the probability of an event equals the significance of the consequence (impact) of the event: risk + probability = impact. The initial attempt at determining risk was based upon assessing the building inventory in a community. Although this model has evolved over time to include more than the building industry, the building risk is an important element. It can be divided into the following four classifications:

- *High-hazard risk.* Schools, hospitals, nursing homes, explosive plants, refineries, high-rise buildings, and other high-life hazard or large fire potential occupancies demonstrate high-hazard risk. This type of risk frequently indicates a fire agency's need for multiple-alarm capability or ability to concentrate adequate resources to control loss when a fire occurs.
- *Medium-hazard risk.* Apartments, offices, mercantile, and industrial occupancies do not normally require extensive rescue or firefighting forces.
- *Low-hazard risk.* One-, two-, or three-family dwellings and scattered small businesses and industrial occupancies demonstrate low-hazard risk. It should be noted that approximately 84 percent of all deaths occur in residential dwellings. Even though major **property loss** is usually low, the potential for life loss is high.
- *Rural-operation risk.* Small commercial structures remote from other buildings, such as detached residential garages and outbuildings, demonstrate rural-operation risk. Areas may be classified as remote/isolated rural risks if they are isolated from any centers of population and contain few buildings, for example, rural land with no occupied structures or recreational areas. Topographic and geological conditions may make a rural area a risk that needs to be dealt with in a more comprehensive fashion. Some of the reasons might be wildland cover or exposure to urban–wildland interface.

An evaluation system currently under development by the CPSE, International Association of Fire Fighters (IAFF), IAFC, and other partner organizations

PEARSON
myfirekit
Visit MyFireKit Chapter 5 for more information on Integrated Risk Management.

property loss
■ Direct loss of property due to fire.

including the National Institute of Standards and Technology (NIST) and Worcester Polytechnic Institute (WPI) explores several factors that impact community risk and ultimately the needed resources to deploy to an emergency. Each category contains subsets of risks/hazards relevant in determining overall community risk and vulnerability (see Table 5.5).

Once the details of risks/hazards are determined for a community, then the community can design deployment of resources (or other activities, e.g., promoting smoke detectors, public education, disaster planning, building/fire code amendments, etc.) either to manage the known risks or to respond to and mitigate the emergency when an adverse risk event occurs. (Fires and medical emergencies are adverse risk events, as are natural or other disasters.)

Department leaders must provide sufficient information to the elected officials so the latter can determine (1) what resources to commit to risk management (prevention/preplanning/preparation), (2) what resources to commit to response/mitigation, and (3) what level of risk to accept. These concepts are built upon the existing basic infrastructure, the response capability, the current level of community preparedness, scene operations, the ability to assemble an effective fire force, response reliability, and community service level objectives.

The number and type of tasks needing simultaneous action will dictate the minimum number of firefighters required at different types of emergencies. The following tables are examples of the number of firefighters assigned to the initial tasks, which usually are performed simultaneously in a response to a single-family residential structure, a three-story multiple-unit apartment, and a high-rise structure. The tasks identified usually occur within the first 5 to 15 minutes of emergency operations (see Tables 5.6 through 5.8).

WHAT IS AN EFFECTIVE RESPONSE FORCE?

An **effective response force** is defined as the minimum amount of staffing and equipment that must reach a specific emergency zone location within a targeted travel time. This level of force should be able to handle the typical emergency medical incident or fire that is reported shortly after it starts and is within the maximum prescribed travel time for the type of medical emergency or risk level of the structure. Considering that the fire department cannot hold fire or other risks to zero or successfully resuscitate every patient, its response objective should find a balance among effectiveness, efficiency, and reliability, which will keep community risk at a reasonable level. At the same time, the department needs to yield the maximum life and property savings and provide for the safety of the responding firefighters, paramedics, police officers, and other first responders.

effective response force
■ The minimum amount of staffing and equipment that must reach a specific emergency zone location within a maximum prescribed travel or driving time, which is capable of initial fire suppression, EMS, and/or mitigation.

RESPONSE RELIABILITY

Response reliability is defined as the probability that the required amount of staffing and apparatus will be available when a fire or emergency call is received. Ideally, the response reliability of the fire department would be 100 percent if every piece of its apparatus were available every time an emergency call was received. Realistically, there are times when a call is received for a particular company, but the company is already on another call. This situation requires a substitute (second-due) company to be assigned from another station to answer that call. As the

TABLE 5.5 Propagation of Community Risk

PRIOR EVENTS	COMMUNITY DESCRIPTION	PREVENTION		EVENTS		COMMUNITY OUTCOME
			Notification	Response		
				Initial	Final	
Fire	Demographic Profile	Prevention	Fire	Initial Assessment	Intervention	Community Outcomes
EMS	Public Assets	✓Engineering ✓Education ✓Enforcement	EMS	Emergency Response	Emergency Response	
Special Operations	Response Capacity		Special Operations			

TABLE 5.6 | Initial Response to a Low-Hazard Risk

One-, two-, or three-family dwellings and scattered small businesses and industrial occupancies

TASK	NUMBER FIREFIGHTERS ASSIGNED	EXAMPLE OF COMPANY ASSIGNMENTS
Attack line	2	1st Engine
Rapid intervention team	2	Truck/engine
Search and rescue	2	Truck
Ventilation	2	Truck
Backup line	2	2nd Engine
Safety officer	1	Assigned
Pump operator	1	1st Engine
Aerial operator (optional depending on the incident)	1	Assigned
Water supply	1	2nd Engine/WT
Command officer	1	Battalion chief
Command aid (optional depending on the incident)	1	Assigned
Investigator	1	
Total personnel	17	

Source: Fresno Fire Department, Standard of Cover, 2009.

number of emergency calls per day increases, so does the probability that a needed piece of apparatus will already be busy when a call is received. Consequently, the response reliability of the fire department for that company decreases, which will have an impact on department travel times to emergencies. The size of the area that a station covers, the number of calls, the types of calls, and the population density all affect response reliability. The more densely populated, the more likely a second-due call will occur. An analysis of current response data can reveal variations in the response reliability among stations. The optimal way to track response reliability would be to analyze the total call volume for a particular **fire management area**, which is a geographic area of a jurisdiction that is classified according to one or more risk categories. Then track the number of double and triple calls to assess what the true response reliability is for that given area and the companies assigned to respond in the area.

Even though the community establishes the department's service level goals, these goals are often influenced by federal and state legislation, such as two-in/two-out regulations, federal and state OSHA requirements, ISO grading schedule,

fire management area
■ A geographic area of a jurisdiction that is classified according to one or more risk categories. The size and classification of a fire analysis area are usually based upon either a specific area or in some cases a building or complex.

TABLE 5.7 — Initial Response to a Medium-Hazard Risk

Apartments, offices, mercantile, and industrial occupancies not normally requiring extensive rescue or firefighting forces

TASK	NUMBER FIREFIGHTERS ASSIGNED	EXAMPLE OF COMPANY ASSIGNMENTS
Attack line	4	1st Engine
Rapid intervention team	2	Truck/engine
Search and rescue	2	Truck
Ventilation	2	Truck
Backup line	4	2nd/3rd Engine
Safety officer	1	Assigned
Pump operator	2	1st/2nd Engine
Aerial operator (optional depending on the incident)	1	Assigned
Water supply	1	4th Engine
Command officer	1	Battalion chief
Command aid (optional depending on the incident)	1	Assigned
Investigator	1	
Total personnel	22	

Source: Fresno Fire Department, Standard of Cover, 2009.

national standards such as those developed by the NFPA, and best practices found in the CFAI agency accreditation process. The service level goals identified for the community are based upon the events the fire department is called to respond to and the service provided by the fire department. These service goals are the benchmark of performance in respect to travel times, but do not necessarily measure other aspects of performance.

As the magnitude of emergencies ranges from small to catastrophic, the requirements for resources consequently vary greatly. A high-risk area could necessitate a timely deployment of more fire companies for several reasons; for example, more resources are required for the possible rescue of persons trapped within a high-risk building with a high-occupant load as compared to a low-risk building with a low-occupant load, or to control fires in large, heavily loaded structures than in small buildings with limited contents. Therefore, creating a level of service consists of the analysis made regarding the distribution and concentration of resources needed in relation to the potential demand placed upon them by the level of community risk.

TABLE 5.8 Initial Response to a High-Hazard Risk

Schools, hospitals, nursing homes, explosive plants, refineries, high-rise buildings, and other high-life hazard or large fire potential occupancies

TASK	NUMBER FIREFIGHTERS ASSIGNED	EXAMPLES OF COMPANY ASSIGNMENTS
Attack line	2	1st Engine
Rapid intervention team	2	2nd Engine
Search and rescue	2	Truck
Ventilation	4	Truck
Backup line	2	2nd Engine
Safety officer	1	Assigned
Pump operator	1	1st Engine
Aerial operator (optional depending on the incident)	1	Assigned
Water supply	1	4th Engine
Command officer	1	Battalion chief
Command aid/PIO liaison (optional depending on the incident)	2	Assigned
Staging officer	1	3rd Engine
Lobby control	1	3rd Engine
Base	1	5th Engine
Stairwell support	1	5th Engine
Investigators	1	
Total personnel	24	

Source: Fresno Fire Department, Standard of Cover, 2009.

EXAMPLES OF COMMUNITY SERVICE LEVEL GOALS

Table 5.9 is an example of how a department can show service level goals related to travel times of the initial response to an emergency. The times shown represent the benchmark this department is trying to achieve in each service and response category.

The formulation of such benchmark performance measures and their incorporation into organizational documents such as the standard of cover (SOC), strategic plan, and period report to the elected officials are important in keeping everyone focused on the achievement of service level goals. The establishment of service level benchmarks, which are communicated to the community, elected

TABLE 5.9	Fire Department Travel Time Service Level Benchmarks	
	FIRST-ARRIVING UNIT	**BALANCE OF FIRST-ALARM ASSIGNMENT OR SPECIALIZED UNITS = FULL EFFECTIVE RESPONSE FORCE**
Suppression	4 Minutes/90 Percent Travel time of the first unit to an emergency once notified of the event to arrival at the incident.	8 Minutes/90 Percent Travel time of the balance of a first-alarm assignment (typically three engines, two trucks, and a battalion chief) to an emergency once notified of the event to arrival at the incident.
EMS	4 Minutes/90 Percent Travel time of the first unit to an emergency once notified of the event to arrival at the incident.	8 Minutes/90 Percent Travel time of the balance of a first-alarm assignment (typically a truck response in cases of patient extrication) to an emergency once notified of the event to arrival at the incident.
Specialized Services		
Hazardous materials	4 Minutes/90 Percent Travel time of the first unit to an emergency once notified of the event to arrival at the incident. Initial response companies provide first-responder operational HazMat mitigation and are Level B personnel protective qualified.	15 Minutes/90 Percent Travel time of the HazMat team, which includes fully equipped HazMat response vehicle and seven qualified HazMat specialists.
Urban search and rescue	4 Minutes/90 Percent Travel time of the first unit to an emergency once notified of the event to arrival at the incident. Initial response companies provide first-responder to USAR incidents and are qualified to the RS1 level.	15 Minutes/90 Percent Travel time of the USAR team, which may include one of several USAR vehicles and six technical rescue specialists.
Aircraft Rescue Firefighting		
ARFF units	3 Minutes/100 Percent Within three minutes from the time of the alarm, at least one required aircraft rescue and firefighting vehicle must reach the midpoint of the farthest runway serving air carrier aircraft from its assigned post or reach any other specified point of comparable distance on the movement area that is available to air carriers, and begin application of extinguishing agent.	4 Minutes/90 Percent Within four minutes from the time of alarm, all other required vehicles must reach the point from their assigned posts and begin application of an extinguishing agent. 8 Minutes/90 Percent Travel time to an in-flight emergency for off-site resources is two engines, one truck, and one battalion chief.

(continued)

	FIRST-ARRIVING UNIT	BALANCE OF FIRST-ALARM ASSIGNMENT OR SPECIALIZED UNITS = FULL EFFECTIVE RESPONSE FORCE
Suburban		
Suppression	5 Minutes/90 Percent Travel time of the first unit to an emergency once notified of the event to arrival at the incident.	10 Minutes/90 Percent Travel time of the balance of a first-alarm assignment (typically three engines, one truck, and a battalion chief) to an emergency once notified of the event to arrival at the incident.
EMS	5 Minutes/90 Percent Travel time of the first unit to an emergency once notified of the event to arrival at the incident.	10 Minutes/90 Percent Travel time of the balance of a first-alarm assignment (typically a truck response in cases of patient extrication) to an emergency once notified of the event to arrival at the incident.
Rural		
Suppression	10 Minutes/90 Percent Travel time of the first unit to an emergency once notified of the event to arrival at the incident.	14 Minutes/90 Percent Travel time of the balance of a first-alarm assignment (typically three engines, one truck, and a battalion chief) to an emergency once notified of the event to arrival at the incident.
EMS	10 Minutes/90 Percent Travel time of the first unit to an emergency once notified of the event to arrival at the incident.	14 Minutes/90 Percent Travel time of the balance of a first-alarm assignment (typically a truck response in cases of patient extrication) to an emergency once notified of the event to arrival at the incident.
Specialized Services		
Hazardous materials	As listed above for suburban and rural.	25 Minutes/90 Percent HazMat team travel time includes fully equipped vehicle and seven qualified HazMat specialists.
Urban search and rescue	As listed above for suburban and rural.	25 Minutes/90 Percent USAR team travel time may include one of several vehicles and six technical rescue specialists.

Source: Fresno Fire Department, Standard of Cover, 2009

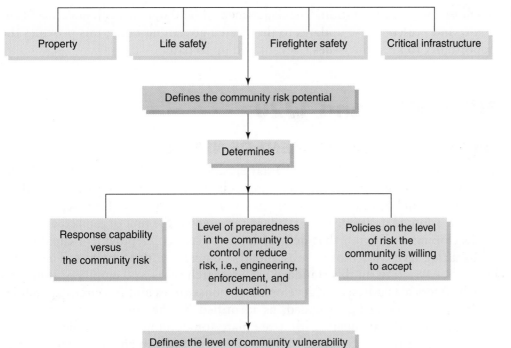

FIGURE 5.13
Community
Risk/Hazard Factors

| Property | Life safety | Firefighter safety | Critical infrastructure |

Defines the community risk potential

Determines

| Response capability versus the community risk | Level of preparedness in the community to control or reduce risk, i.e., engineering, enforcement, and education | Policies on the level of risk the community is willing to accept |

Defines the level of community vulnerability

officials, and throughout the organization, provides the visible target the organization is trying to achieve.

The information compiled regarding the factors of community risks/hazards will help determine the needed management and response capability. Resources committed for risk management, response, and mitigation of risk events that occur will determine the overall community vulnerability when an adverse event occurs (see Figure 5.13).

INTEGRATION, REPORTING, AND POLICY DECISIONS

With the completion of an SOC document, an agency will have an in-depth understanding of how its current level of resources matches the community risk factors. The SOC is often used as the basis for the creation of a strategic planning document, which helps to provide a clear, comprehensive picture of what has been found and what recommendations will be necessary for future planning and implementation. The document, with the use of graphs and mapping-based displays, should foster informed policy discussion. It can do this by presenting the following key points: the existing baseline of current performance, historical performance, identification of community risk factors, current service level objectives, critical task analysis for anticipated events, company distribution and concentration analysis, company reliability analysis, proposed service level objective statements with associated cost-benefit analysis, and recommendations to accomplish the objectives.

In public presentations, care must be taken to inform elected officials and the public as to the current levels of services and the proposed service level objectives. Ultimately, the final deployment plan will be determined based upon the desire of the residents served, policy makers, community expectations, values, and economics.

Once this final deployment has been determined, the development of a strategic plan will help to prioritize the changes noted in the SOC with recommended timelines for implementation. The department should update both documents annually and present these updates to the elected governing body; this provides the opportunity to keep the issues relevant and up to date.

NFPA 1710 and 1720

ORIGIN AND DEVELOPMENT OF NFPA 1710

The development of the NFPA 1710 standard, known as the Standard for the Organization and Deployment of Fire Suppression Operations, Emergency Medical Operations, and Special Operations to the Public by Career Fire Departments, was first adopted in 2000. This standard represented the first organized approach to developing a criterion that defined levels of service, deployment capabilities, and staffing levels for those "substantially" career fire departments. The NFPA committee working on this standard used research work and empirical studies in North America as a basis for developing response times and resource capabilities for those services being provided, as identified by the fire department. NFPA 1710 provides departments with a template for developing an implementation plan in respect to the standard.[8] NFPA 1710 sets forth in concise terms the recommended resource requirements for fires, emergencies, and other incidents. It expects the emergency response organization to evaluate its performance and report it to the authority having jurisdiction (AHJ). The *scope* and *purpose* help to define what the standard does and what it covers. In both cases, the standard defines the minimum acceptable requirements, while still allowing more stringent or more comprehensive ones if a community so decides.

With respect to scope, these minimum requirements are related to how fire, EMS, and special operations are organized and deployed in departments that are substantially career. They address these organizations' objectives as well as their functions. Not surprisingly, NFPA 1710 emphasizes three key areas of a successful operation: service delivery, capabilities, and resources. The standard sets forth the minimum criteria related to the effectiveness and the efficiency of public entities that provide fire suppression, emergency medical service, and special operations. Both effectiveness and efficiency are specifically related to protecting the public and fire department responders.

When the NFPA 1710 standard was adopted, many in local government were concerned that it might create a number of legal implications. The one issue that seemed to generate the most concern during the standard's development was whether jurisdictions could be held liable for failing to comply with the standard. The courts have traditionally been reluctant to hold cities, towns, and fire departments liable for the consequences of their discretionary decisions related to fire department resource allocations.

The standard's organizational statement sets forth the minimum information required concerning what the organization does, how it is structured, and the staffing required to achieve its objectives. Service delivery objectives found in the standard are specific requirements for deployment, staffing, response times, and the necessary support systems. These support systems include safety and health, communications, incident management, training, communications, and pre-incident planning. A system is a functionally related group of components. These are areas in which a set of needs or requirements work closely together and are interrelated to achieve a key result. The NFPA 1710 standard addresses five of these systems.

- *Safety and health.* Each organization must have an occupational safety and health program meeting the requirements of NFPA 1500, Standard on Fire Department Occupational Safety and Health Program.
- *Incident management.* Each organization must have in place an incident management system designed to handle expected incidents. The system must be in accordance with NFPA 1561, Standard on Emergency Services Incident Management System.
- *Training.* Each organization must ensure members are trained to execute all responsibilities consistent with its organizational statement. This training must be accomplished using a programmatic approach that includes a policy.
- *Communications.* Each organization must have a communications system characterized by reliability; promptness; and standard operating procedures, terminology, and protocols. Departments must also comply with all the requirements set forth in NFPA 1221, Standard for the Installation, Maintenance, and Use of Emergency Services Communications Systems.
- *Pre-incident planning.* Safe and effective operations are grounded in identifying key and high-hazard targets. The standard requires departments to develop operational requirements to gather information regarding these locations.

There are several time components defined in the NFPA 1710 standard relating to emergency response system performance. All components must be measured and documented by departments in an annual report provided to their governing body. Figure 5.14 (page 174) illustrates the NFPA 1710 **total response time** elements that have been developed to measure fire suppression incidents and EMS response.

The purpose of department evaluation and reporting is to measure and document its compliance with the NFPA 1710 standard. According to the standard, a department must perform an annual evaluation of service based on actual response data. Using these data, the department prepares and submits an annual written report to the governmental authority noting compliance or noncompliance with the standard.

The report must explain any deficiencies, the consequences of the deficiencies, and the offer of improvements on how the department plans to become compliant.

total response time
■ The time from the call being received at the 911 public safety answering point and the point at which crews arrive and intervention begins.

Visit MyFireKit Chapter 5 for more details on NFPA 1710.

NFPA 1720

In 2001, the first edition of NFPA 1720 was known as the Standard for the Organization and Deployment of Fire Suppression Operations, Emergency Medical Operations, and Special Operations to the Public by Volunteer Fire Departments. This standard was the first organized approach to defining levels of service, deployment capabilities, and staffing levels for substantially volunteer fire departments.[9] Approximately three out of every four fire departments in the United States are volunteer; therefore, this standard, as well as its related practices (accreditation, certification, etc.), has a profound effect on the direction of the volunteer fire service. This standard was revised in 2004 and 2010 and is scheduled for revision again in 2015.

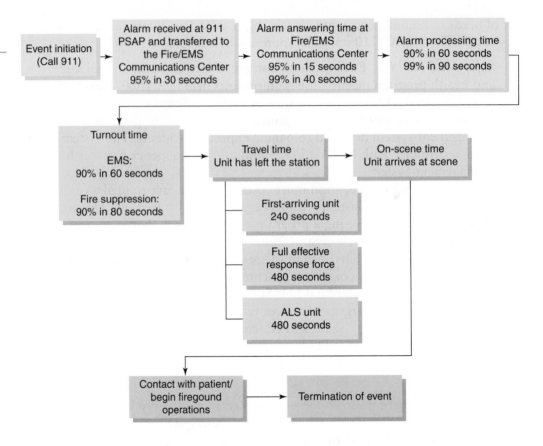

FIGURE 5.14 Total Response Time

The scope of the standard includes minimum requirements relating to the organization and deployment of fire suppression operations, emergency medical operations, and special operations to the public by substantially all volunteer fire departments. The requirements address functions and outcomes of fire department emergency service delivery, response capabilities, and resources. NFPA 1720 also contains minimum requirements for managing resources and systems, such as health and safety, incident management, training, communications, and pre-incident planning. This standard addresses the strategic and system issues involving the organization, operation, and deployment of a fire department and does not address tactical operations at a specific emergency incident. The standard's focus is on the organization, operation, and deployment of resources in the following key areas (see Tables 5.10 through 5.12):

- Fire Suppression Organization
- Community Risk Management
- Hazardous Materials
- Staffing and Deployment
- Reporting Requirements
- Annual Evaluation
- Suppression Operations
- Initial Attack
- Intercommunity Organization
- Emergency Medical Services

TABLE 5.10	Staffing and Response Time			
DEMAND ZONE[a]	DEMOGRAPHICS	MINIMUM STAFF TO RESPOND[b]	RESPONSE TIME (MINUTES)[c]	MEETS OBJECTIVE (%)
Urban area	>1,000 people/mi^2	15	9	90
Suburban area	500–1,000 people/mile^2	10	10	80
Rural area	<500 people/mile^2	6	14	80
Remote area	Travel distance ≥ 8 miles	4	Directly dependent on travel time	90
Special risks	Determined by AHJ	Determined by AHJ based on risk	Determined by AHJ	90

[a] A jurisdiction can have more than one demand zone.
[b] Minimum staffing includes members responding from the AHJ's department and automatic aid.
[c] Response time begins upon completion of the dispatch notification and ends at the time interval shown in the table.

Source: National Fire Protection Association, *NFPA 1720: Standard for the Organization and Deployment of Fire Suppression Operations, Emergency Medical Operations, and Special Operations to the Public by Volunteer Fire Departments* (Quincy, MA: National Fire Protection Association, 2010).

In addition, the standard has set requirements for support systems including:

- Safety and Health
- Incident Management
- Training
- Communications
- Pre-Incident Planning

Elements of NFPA1720 are shown in Table 5.10.

Table 5.11 (page 176) shall be used by the authority having jurisdiction (AHJ) to determine staffing and response time objectives for structural firefighting, based on a low-hazard occupancy such as a 2,000-foot, two-story, single-family home without a basement and exposures as well as the percentage of accomplishment of those objectives for reporting purposes. Table 5.12 (page 177) depicts the difference of time when the PSAP is located in another location from the communications center, and the 911 call is transferred to a communications center that dispatches the emergency units.

Conclusion

The challenge for today's fire service leader is to possess the ability to analyze a community's risk factor(s) and propose the appropriate deployment level for the community. Fire service leaders have realized over the course of the past 25 years

TABLE 5.11	Alarm Time Where the PSAP Is the Communications Center				
EMERGENCY EVENT	ALARM INITIATED	ALARM SOUNDS AT COMMUNICATIONS CENTER	STARTS ALARM PROCESSING	DISPATCH STARTS	ERUs RESPOND
→ Detection →	Transmission →	Answered →	Alarm Processing →		
→ Time Varies →	Not Specified →	95% ≤ 15 sec → 99% ≤ 40 sec	90% ≤ 60 sec → 99% ≤ 90 sec	IAW → SOP/SOG	

Alarm sounds are audible, visual, or both.
Alarm processing begins when the alarm is answered by the telecommunicator, including interrogation of the caller, and ends at the beginning of ERF/ERS notification.
IAW = In accordance with.

Source: National Fire Protection Association, *NFPA 1720: Standard for the Organization and Deployment of Fire Suppression Operations, Emergency Medical Operations and Special Operations to the Public by Volunteer Fire Departments* (Quincy, MA: National Fire Protection Association, 2010).

that, to be effective, a deployment resource model must be built on sufficient, accurate, and complete data no matter what methodology or model is chosen. In the fire service one of the difficulties is the selection of the right tool from the toolbox to address the potential risk faced by a community. Sometimes the tools that are chosen determine the way the problem or the situation is addressed. There is an old saying that if you give a child a hammer, everything in the house looks like a nail. In today's fire service, the "one size fits all" approach, or one solution for every fire department in respect to the response delivery system, is simply not feasible. The fact is the risk analysis should be problem driven instead of method driven, which is to say today's leadership must start with the identification of the problem and then seek appropriate methods to address the issue, rather than start with a methodology and look for a problem to apply it to!

In this chapter, several methods have been outlined from which to choose to begin risk analysis of a community, which will assist in the development of a comprehensive deployment package. The consistent underlying theme is that the development of a community risk management model must take into account all the factors of the community risk level(s) and the response needed regardless of the system chosen. The fire service has witnessed a convergence of thought on this topic in the past decade. If one were to overlay the ISO grading schedule, CFAI accreditation model, and NFPA 1710 and/or 1720, one would begin to see threads of commonality. As the fire service becomes more adept and proficient at risk analysis in the United States, it will see more of this convergence of thought and application in respect to how it analyzes risk in this country and how to deploy resources to minimize the effects of emergencies in local jurisdictions. Most progressive chief officers are using the models applicable to their organization. It is not uncommon to find an evaluation of performance based upon the accreditation process, ISO, and appropriate national standard documents. No matter what evaluation processes are chosen for utilization, they must be based upon good data, performance measurement, and national research as the foundation for future planning and educating elected officials and the residents served.

PEARSON
myfirekit

Visit MyFireKit Chapter 5 for Perspectives of Industry Leaders.

TABLE 5.12 | Alarm Time Where the PSAP Is Other than the Communications Center

EMERGENCY EVENT	ALARM INITIATED	ALARM SOUNDS AT PSAP	ALARM TRANSFERRED TO CC	ALARM SOUNDS AT CC	STARTS ALARM PROCESSING	DISPATCH STARTS	ERUs RESPOND
→ Detection →	Transmission →	Answered →	Transfer →	Answered →	Alarm Processing →		
→ Time Varies →	Not Specified →	95% ≤ 30 sec →	95% ≤ 15 sec →	90% ≤ 60 sec →	IAW SOP/SOG →		
			99% ≤ 40 sec	99% ≤ 90 sec			

Alarm sounds are audible, visual, or both.

Alarm processing begins when the alarm is answered by the telecommunicator, including interrogation of the caller, and ends at the beginning of ERF/ERS notification.

CC = Communications Center.

IAW = In accordance with.

Source: National Fire Protection Association, *NFPA 1720: Standard for the Organization and Deployment of Fire Suppression Operations, Emergency Medical Operations, and Special Operations to the Public by Volunteer Fire Departments* (Quincy, MA: National Fire Protection Association, 2010).

Review Questions

1. Explain how history has evolved to create the current standard of cover practice the fire service has embraced.
2. Describe the three key considerations made in a community risk assessment.
3. Compare and contrast the difference between the Insurance Services Office's (ISO) Fire Suppression Rating Schedule, NFPA Standards 1710 and 1720, and the Commission on Fire Accreditation International (CFAI) bases for assessing fire risks and community response capabilities.
4. Describe how the probability and consequence matrix works. Give an example for each quadrant.
5. Explain what standard of cover is and how it is applied in a community.
6. Describe the response performance continuum; discuss each of the components.
7. Describe the CFAI self-assessment process.
8. Compare and contrast the difference in requirements of NFPA 1710 and NFPA 1720.
9. Explain the four classifications used for building risk assessment and how they impact the standard of cover.
10. The propagation of community risk takes into account five main focus areas. Explain each one and why it is important.

PEARSON
myfirekit™

For additional review and practice tests, visit www.bradybooks.com and click on MyBradyKit to access book-specific resources for this text!

Register for MyFireKit by following directions on the MyFireKit student access card provided with this text. If there is no card, go to www.bradybooks.com and follow the MyBradyKit link to buy access from there.

References

1. The Fire Brigades Union, *Integrated Risk Management Planning: The National Document* (Birmingham, England: FBU National IRMP Department, Folium Group Ltd., 2004).
2. Commission on Fire Accreditation International, *Creating and Evaluating Standards of Response Coverage for Fire Departments,* 4th ed., 2003 (CD-ROM).
3. National Fire Academy, "Analytical Approaches to Public Fire Protection," *Community Risk Management* (2004), pp. 6–9.
4. Commission on Fire Accreditation International, *Self-Assessment Manual,* 6th ed. (Chantilly VA: Author, 2000).
5. Commission on Fire Accreditation International, *Creating and Evaluating Standards of Response Coverage for Fire Departments,* 4th ed. (Chantilly, VA: Author, 2003).
6. University of Washington School of Medicine, *2009 Survive Cardiac Arrest.* Retrieved

September 30, 2010, from http://blacklistinc. com/survive/utstein.php

7. Dennis Gage, *Evaluation of Fire Departments as a Factor in Property Insurance Rating* © ISO Properties, Inc., 2005.

8. National Fire Protection Association, *NFPA 1710: Standard for the Organization and Deployment of Fire Suppression Operations, Emergency Medical Operations, and Special Operations to the Public by Career Fire Departments* (Quincy, MA: National Fire Protection Association, 2010).

9. National Fire Protection Association, *NFPA 1720: Standard for the Organization and Deployment of Fire Suppression Operations, Emergency Medical Operations, and Special Operations to the Public by Volunteer Fire Departments* (Quincy, MA: National Fire Protection Association, 2010).

6 CHAPTER

Managing in a Changing Environment: Measuring System Performance

(Courtesy of Rick Black, Center for Public Safety Excellence)

OBJECTIVES

After completing this chapter, you should be able to:

- Define the term *performance measurement*.
- Describe the difference between qualitative and quantitative data.
- Describe the three types of performance measures: workload, efficiency, and effectiveness.
- Describe the seven pitfalls of performance measurement systems.
- Describe what Vision 20/20 is.
- Define the terms *baseline* and *benchmark*.
- Describe how benchmarking can be used to evaluate and improve organizational performance.

PEARSON

myfirekit™

For additional review and practice tests, visit www.bradybooks.com and click on MyBradyKit to access book-specific resources for this text!

Introduction

Performance measurement today is vastly different from what it was when many of us entered the fire service. With the development of the fire service accreditation model in the late 1980s, significant analysis and research were conducted on what makes a department credible. This effort was made to determine which criterion and performance indicators would ultimately be included in the accreditation model. Departments that were viewed at the time as being progressive were analyzed against international counterparts, and it was soon found that the level of performance measure being utilized in North America was substantially different from that in use internationally. In fact, it is safe to say that, at that time, and unfortunately for many departments still today, the product of measurement for the fire organization had nothing to do with performance and everything to do with reporting statistics of activity.

Early in the discussion on accreditation and performance, Chief Ron Coleman stated that, when he had entered the fire service, some of the initial performance reported to the local governing bodies was how many feet of hose were laid the previous year, the number of and total feet of ladders that were put up on fires, and how many times the aerial ladder had been raised (see Figure 6.1).

During that time, statistical reporting often focused on meaningless information that had no relationship to performance outcomes. As the fire service has evolved and progressed in the last 15 to 20 years, the focus on measuring not only activity but also multiple activities of performance and services that occur daily in organizations, has become very important.

Early efforts to measure performance by fire departments and local government centered on determining what type of measurements would provide meaningful data. As this process has continued to evolve, local government is now

FIGURE 6.1 Engine at Fire Scene *(Courtesy of Tony Escobedo)*

asking what to measure in terms of what the citizens may want to know, which has led to a number of efforts nationally to develop a common set of data points for comparative analysis between organizations. As local government has become more experienced in setting targets and measuring performance, it also has begun to utilize these national statistics in conjunction with information from its citizens to establish performance standards, measure specific performance, assess citizen satisfaction levels, and set funding priorities.

One of the challenges the fire service has faced to date has been the lack of standardized data points, which can be used at the local level to help analyze and assess departmental performance. Standardized data points are specific points of measurement that can be shared, evaluated, and integrated across a business function to measure performance. An example of standardized data points for overall response time is the cascade of events (see Chapter 5). Despite the significant efforts of several organizations, national associations, and organizations such as the Commission on Fire Accreditation International, there still is a lack of standardized data elements required to be reported and analyzed across the fire service profession for its multiple areas of service. In part, this is due to the lack of federal direction, specifically in the United States, in respect to what performance should be measured and what data are to be collected. As previously noted, the United States has not been able to collect valid data from each and every fire service provider across the country. In fact, many departments still do not report to the federal government the types of incidents responded to or the causes and origins of the fires responded to, let alone performance data of other services they provide.

If the fire service is to use performance measures and to gauge national risk and departmental effectiveness, future legislation will need to be passed mandating that fire departments report to both the federal and state levels of government. Specific data points that are well quantified will provide for effective analysis of the nation's fire problem. With this said, progressive departments in the fire service have already taken significant steps forward to improve the performance measurements being utilized today.

As the refinement of these data points continues to evolve, it will enable cause and effect to be pinpointed, which, in turn, will allow the profession to develop more targeted strategies to address the problems communities are facing in respect to fire, life safety, and risk reduction. The days are gone in which fire department officials can stand before their elected officials or at a news conference and use scare tactics such as the threat of people dying in fires to try to obtain or maintain organizational resources. Such statements made by members of the fire service simply do not work in most jurisdictions today.

What community leaders and citizens have come to expect is facts supported with analysis and the use of valid business analytics. If the fire service fails to provide this documentation, the respect for the department gained from a political standpoint and from the customers will evaporate, and the department's ability to move forward its agenda for the safety of its community will diminish. Fortunately, in most cases, the data are available to help frame the discussion, but unfortunately, many departments are not collecting the data or conducting effective analyses to help do this.

Many fire service leaders have finally realized that their customer base has elevated its level of understanding and need for information, maybe even more so than the profession has. The quality improvement movement has touched almost

FIGURE 6.2
Firefighters at Structure
Fire   *(Courtesy of Rick
Black, Center for Public
Safety Excellence)*

every profession; as such, fire service customers (residents) must be tuned in to measuring their own performance in the workplace. Subsequently, they demand the same of the local fire/EMS provider. Over the course of the next decade, fire service leaders will continue to see customers become much more informed about how government works, not only focusing on the basic performance analytics such as the number of calls responded to, fire loss, and call breakdown but also looking at the level of outcomes the department is having with the money being invested. Whether it is involved in fire protection, police services, public works, or public utilities, every department in local government, if it is not analyzing how well it is providing service for the resources being expended, is at risk of losing the support of the community and its elected officials.

Today, the fire service has to prove the services it provides are a value-added proposition for the community. Some performances will be viewed from a **qualitative** standpoint (relating to or expressed in terms of quality), often based upon the perception of the community residents formed by what they see the fire department doing every day (see Figure 6.2).

The public admires the perception of the fire service and the work it does. However, so often it is the **quantitative** analyses of the numerical data that allow leaders to continue to move their organizations forward with the elected officials and within the department's strategic planning process.

In the fire service, the concept of business analytics will become much more important in the future. Business analytics is the extensive use of quantitative and qualitative analysis, both of which are explanatory and predictive in nature (see Table 6.1). Quantitative data are the tangible results that can be measured based on the outcomes the organization is trying to achieve. Qualitative data are the intangible and subjective elements, which can be equally important depending on what is being evaluated.

Qualitative data are not measured data; however, measurements can be designed to determine the degree of qualitative behaviors. Therefore, they can be

qualitative
■ Relating to or expressed in terms of quality; qualitative research is based on individual, often subjective, analysis.

quantitative
■ Relating to or expressed in terms of quantity; quantitative research is based on numerical data.

TABLE 6.1	The Difference Between Qualitative and Quantitative Data
QUALITATIVE DATA	**QUANTITATIVE DATA**
Overview	Overview
• Deals with descriptions	• Deals with numbers
• Data that can be observed but not measured	• Data that can be measured
• Colors, textures, smells, tastes, appearance, beauty, etc.	• Length, height, area, volume, weight, speed, time, temperature, humidity, sound levels, cost, members, ages, etc.
• Qualitative ⟶ Quality	• Quantitative ⟶ Quantity

measured. Social scientists do this regularly to establish trends in human behavior. Although human behavior is generally observable and described qualitatively, quantitative measurements of these qualitative behaviors are designed to provide comparisons. This is important to developing valid and reliable research results, particularly in the fire service because many of the causes that necessitate an emergency response have a behavioral basis.

The acquisition of business intelligence helps to create a fact-based management process to drive effective decision making. As has been mentioned, the future of the fire service will be dictated by the degree of business intelligence its leaders possess. So much of what the fire service has done in the past has been based on the same practices and traditions utilized for over one hundred years: the placement of stations, the way it responds, and the intuitive processes that so often drive decision making on a day-to-day basis. Although some of these are still valid, if the fire service cannot show its value added from an analytical perspective to its elected officials and community, then its leaders often cannot make the case for the resources needed to enhance or expand its services. A challenge to the fire service is the need to expand the ability to provide analytical assessment on a wide range of topics, each of which depends upon good data. The most important factor in being prepared for business analytics is sufficient high-quality data. The difficulty in sharing data quality is integrating and reconciling data across various systems and deciding what subsets of data to make available for analysis. Simply stated, if one agency is to be compared to another to assess the fire problem throughout the United States, then the data have to be collected, analyzed, and displayed in the same manner to identify the trends from a national or local perspective and ensure they are accurate.

The simplest definition of analytics is "the science of analysis," or how an entity or business arrives at an optimal or realistic decision based on existing data. Instead, managers often choose to make decisions based on past experiences, their own intuitive sense, or other qualitative aspects of decision making. Yet, unless data were involved in the process, it would not be considered analytics.

Common applications of analytics include the study of business data using statistical analysis in order to discover and understand historical patterns with an eye to predicting and improving department performance in the future. Although some people will use this term to denote the use of mathematics in business processes, others hold that the field of analytics includes the use of operations research, statistics, and probability. In the fire service, to limit the field of analytics to only statistics and mathematics would be a disservice because much of its modeling includes visual modeling through the use of such tools as geographic information systems.

The innovations and trends of business analytics spanning the areas of process, new technologies, user interface design, and system integration, are driven by trying to create *business value* (efficiency and effectiveness) or *value added* (for the customer). Business value is measured as progress toward bridging the gap between the needs of the user and the accessibility and usability of analytic tools. In an effort to make analytics more relevant and tangible to the user, solutions are focusing on specific applications and tailoring the results and interfaces toward the audience, so they are comprehensible and provide human-level insight.[1] Therefore, an important component of measurement is not only what to measure and how to measure a specific aspect of the department's operation but also the means to do it. So often a department will purchase software for a single application and find the data cannot be interfaced with other databases, making overall performance measurement time consuming and labor intensive. An important part of providing good business analysis is to have a plan in place for effective computer and software system integration.

Local government is presently in a period of significant transition and increased scrutiny by the public. Most in local government have been asked to maintain existing service levels with far fewer resources. Local government will be required to provide more efficient services for the citizens, more information, and better technology. To do so will necessitate advanced analytics to analyze large amounts of information quickly, efficiently, and effectively, so the fire service may elevate a wide range of organizational objectives to ensure the organization is meeting the changing demands of the customer. In times such as this, local government needs to be proactive and timely in responding to the changes the public is demanding. Real-time information, sharing situational analysis, promoting awareness, and accurate decision making in dealing with the volumes of data available become extremely important. What is done at the local governmental level in respect to utilizing business analytics may ultimately dictate in many cases the future path of the department.

PEARSON
myfirekit

Visit MyFireKit Chapter 6 for information on Visualizing and Mining Data.

For example, a properly designed strategic plan with calculated outcomes (as discussed in Chapter 4), linked to an operational plan with measurable goals, and action plans, which are reflected in a budget designed around programs, can be evaluated for effectiveness via statistical analysis to provide for locally relevant data. This supplies elected officials and the public with information to use in evaluating the fire department's effectiveness and cost benefit.

As the fire service continues to perfect its performance measures in each organization, one must keep in mind that performance measures should be used to improve organizational performance through better planning, budgeting, managing, and communicating. With this said, one still has to be open to looking past what has always been done no matter how much data have been collected and analyzed. A good example of this happened in my own department while we were going through an update of the strategic planning process. Like many organizations, we were evaluating options to reduce services due to the economic issues the department faced at the local level. As the information and business analytics were analyzed to help structure the decision, we were doing so in the context of what had always been done in the past. The issue at hand was the placement of truck companies throughout the city and the jurisdictions the department serves.

The business analytics being utilized was the travel time performance for these truck companies in respect to assembling an effective response force to provide service to the residents based upon the adopted service level objectives at 8 minutes, 90 percent of the time. This is a valid performance measurement based on a national standards and the accreditation model. The discussion involved the reduction of one or two truck companies and what trucks would be placed out of service, but the organization never asked the question whether the trucks were in the right place to begin with. As the dialogue continued, the assessment and evaluation led the department to a new realization and a totally different question. If the department approached the problem from an alternative vantage point, would it come up with a different answer? It was easy to provide the business analytics on what was done in the past and to say whether one or two companies were placed out of service, which companies were placed out of service based upon coverage, number of calls, assess valuation protected, and effective fire force. However, when the problem was viewed from another vantage point, the answer was quite different. The question was reframed to this: If no trucks were in service today and four, five, or six trucks were located to cover the city, where would they be located to better serve the community based on the 8 minutes 90 percent service level objective? Posing this question revealed that the distribution originally formulated was actually not the best, even when zero reductions of existing resources were factored into the assessment. The point to be made, even with good business analytics, is that unless the problem is viewed from a different perspective, the business analytic may tell a story, but possibly not the complete story. It is imperative to allow analysis of *all* the existing options, because when this occurs, one may find a completely different answer.

Measuring System Performance

Accountability has never been more important in local government than it is today. The commonly used term *transparency in government* speaks to the ability of government to be able to articulate how it is performing. Citizens have the right to expect local resources will be managed carefully, services will be provided efficiently, and programs will live up to what has been promised. Responsible elected leaders and managers in local government seek every opportunity to reassure their citizens that these expectations are being met. Performance measurement becomes the documentation of departmental and program accomplishments and is a key element of organizational accountability; however, it is more than that.

Local governments serious about managing and improving performance use performance measures to gauge their programs' effectiveness and to identify improvement strategies. For these organizations, performance measurement is more than a public information device; it is an essential leadership and management tool.

The performance measures of a department or program will reflect the array of services provided and, ideally, much more. A good set of measures also reveals the quality of these services, the efficiency with which they are delivered, and the effectiveness of the services in achieving their intended purposes. Unfortunately, many sets of performance measures in local government fall short of this ideal, often providing little more than raw counts of clients served and services rendered. Such limited sets of performance measures are of only modest managerial and policy value.

PEARSON
myfirekit™
Visit MyFireKit Chapter 6 for more information on Performance Measures.

Seeking More Useful Performance Measures

Three types of performance measures that may be found in the performance reports of local governments are at the forefront of efforts to improve performance management and accountability:

- Workload (output) measures
- Efficiency measures
- Effectiveness (outcome) measures

WORKLOAD MEASURES

Workload measures—also called output measures—provide information on the amount of work performed or capable of being performed or how many units of service were provided. Such measures indicate, for instance, how many calls were received, how many claims were processed, how many clients were served, how many training programs were offered, and how many safety inspections were conducted. They tell nothing about the efficiency (cost) with which these services were provided and nothing about the quality or effectiveness of the services. Although they might inform us of the demand for a given service, *in their raw form* workload measures offer little else of managerial value.

workload
- The amount of work performed or capable of being performed.

EFFICIENCY MEASURES

Ideal measures of efficiency relate outputs produced to resources consumed. Local governments that report, for example, unit costs for processing claims or claims processed per staff-hour have combined workload information (number of claims) with resource consumption information (costs of staff hours) to create a measure of efficiency. Other measures included in this category address efficiency somewhat less directly than by output-to-input ratios, focusing instead on turnaround time or other aspects of *process efficiency*. The result in either case is a measure reflecting the stewardship of public resources to achieve a public purpose (see Figure 6.3).

Collecting measures of this type provides important operational feedback, which challenges managers to consider whether their level of efficiency is satisfactory and, if not, to contemplate options for improving it.

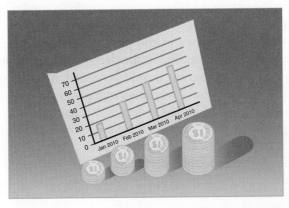

FIGURE 6.3 Value-Added Tool

EFFECTIVENESS MEASURES

Measures of effectiveness, or outcomes, gauge the extent to which objectives are being met. Examples of outcomes include accident and injury rates, loss ratios, and subrogation success rates. Typically, this category also includes measures of service quality. Like efficiency measures, effectiveness measures provide feedback that challenges managers to assess the adequacy of their services and to consider ways to improve service quality and effectiveness.[2]

Performance measurements provide a means of defining program service levels. Whether measuring fire suppression, fire education, arson investigation, or any other fire service program, performance measures can supply clarity of mission. Additionally, performance measurement systems provide a rational methodology to report program accomplishments to managers, customers, and policy makers.

Managers use operational performance measures to plan and control programs at the operational level, whereas strategic performance measures provide guidance to both managers and policy makers who have to make decisions from a more global (big picture) perspective. Performance measures provide a means to clarify and improve the services of programs in understandable terms to citizens, customers, fire managers, and firefighters. These terms are typically formulated as inputs, outputs, and outcomes. Program costs can be calculated by evaluating the efficiency, effectiveness, and equity of the program. Although costs are not always quantitative—that is, measured in dollars and cents—there is often a tendency to consider only the financial costs. Equity and effectiveness costs typically must be measured in qualitative terms, which are much harder to measure and justify because they are based mostly on a set of values or assumptions about what is in the best interest of the public. Despite the difficulty in costing qualitative measures, it is extremely important for managers to give it their best effort.

Leading-edge organizations use performance measurement to gain insight into, and make judgments about, the effectiveness and efficiency of their programs, processes, and people (see Figure 6.4).

These best-in-class organizations choose what indicators to use to measure their progress in meeting strategic goals and objectives, gather and analyze performance data, and then use the data to drive improvements and successfully translate strategy into action. Fire service officials can utilize the information collected about the programs offered to evaluate outcome performance for customers and how well these programs and services are meeting the strategic objectives of the organization and the community. They can then use evaluations based on predetermined performance measures to support requests for additional resources or reallocation of existing resources. Data can also help in analyzing how efficiently current resources are being utilized. Leadership can use the same data to help identify both strengths and weaknesses in the program, thus supporting decisions to modify a program or sometimes aiding in the decision to end a program. Although an evaluation may suggest a program should be ended, a good performance measurement system could provide early warnings of a program weakness, which could lead to changes to improve the service(s) before it is too late.

The levels of public services provided by any jurisdiction are often political-based issues that require political/policy decisions. The strongest, most comprehensive, and most understandable performance measurement systems do not change this fact, nor should they. Political leaders and policy makers—including

FIGURE 6.4 Aerial Operations at Night *(Courtesy of Rick Black, Center for Public Safety Excellence)*

city council members; fire district board members; and state, federal, tribal, and private responders—are elected and appointed to make decisions about the allocation of resources. Fire service leaders are responsible to develop performance measurement systems that can meet their communities' objectives in the best way possible. In this sense, managers and leaders in the fire service are public safety policy entrepreneurs who are constantly searching for opportunities to implement creative and innovative fire service programs. These programs must show that they simultaneously meet the needs of their customers and provide for the overall public safety concerns of the community before it is reasonable to expect they will be funded.

The data provided by a good performance measurement system can be an effective tool in influencing political decisions. However, performance measures neither make decisions nor replace people. Their intent is to provide a systematic management approach that provides better data and evaluation opportunities, which the fire service leadership then uses to make important programmatic decisions. For example, a well-planned public education program for community youth based on sound research and analysis and supported by a clear mission statement for the program can make the difference between gaining and not gaining community support. Support can be translated into budget dollars and staff to implement the program, and leadership can use the performance measurement results to substantiate how well the program is meeting its objectives.

It is optimal to have good performance measures for every fire department program integrated into the strategic planning process of the fire department. A holistic approach to strategic planning provides a set of programs complementary to the mission of the organization, which identifies the most appropriate level of service for each program. Performance measurement systems succeed when the organizations' performance measures are related to—that is, are in alignment with—overall organizational goals. Performance measures are also needed because they address the community's strategic plan in a more comprehensive way. Yet, fire departments should have a strategic plan that provides direction to and guidance for the development of fire service programs that correspond to the objectives of the community. For example, if the community is concerned about its youth, then youth education programs are very important. On the other hand, if the community is primarily a resort town or retirement village, other services may take precedence.

Performance Measurement Dimensions

Several dimensions that are important to the development of an effective performance measurement system include inputs, outputs, outcomes, efficiency, and effectiveness (quality).

INPUTS

inputs
■ Resources such as people, raw materials, energy, information, and finance that are put into a system (such as an economy, a manufacturing plant, a computer system, a department) to obtain desired output(s). Inputs are classified under "costs" in accounting.

Inputs are the resources such as people, raw materials, energy, information, and finance that are put into a system (such as an economy, a manufacturing plant, a computer system, a department) to obtain desired output(s). Inputs consist of dollars, staffing (additional personnel and time allocation of existing personnel), equipment, supplies, and other tangible goods or commodities. Inputs can also include demand characteristics of a program based on target populations. Cost analysis requires a review of all inputs, direct or indirect. Indirect inputs tend to cause the greatest amount of anxiety among program managers and elected officials as they cope with sensitive resource allocation decisions. Yet indirect costs (sometimes caused by unintended consequences) are important as well and, if overlooked, can cause a program to fail due to lack of adequate resources or other organizational support. Accurate inventory and analysis of inputs are critical so that realistic cost projections and cost comparisons can be completed. A thorough understanding of program inputs is essential to the development of meaningful performance measures; often many of the simplest items are overlooked and not accounted for in a program. An accurate description and accounting of program inputs also provide an opportunity for program managers and other stakeholders to have meaningful dialogue about the allocation of these scarce resources. Program managers who are struggling to stretch already overextended staffs can do an even better job of time management when inputs are accurately and honestly identified. Inputs in fire service programs are articulated in such things as the following: firefighters per thousand population, the number of mechanics per apparatus serviced, the number of inspectors per inspectable occupancies, the number and type of apparatus required to be dispatched to handle a specific type of incident adequately, and the number of personnel assigned to a piece of apparatus.

OUTPUTS

outputs
■ Services provided or products produced by a program.

Outputs are the services provided or products produced by the program.[3] Outputs, which are generally internal in nature, refer to how much activity is generated within a program. Examples of activities are the number of complaints answered, the number of responses to an event, and the number of personnel required to complete a job. Outputs reflect how busy an agency is. In the most simplistic terms, efficiency is measured by dividing the number of outputs by the number of inputs.[4] Understanding outputs is required before improvements can be made in the way something is being done or to determine the right way to do something. However, outputs are not indicators of whether the activity is the right thing to do but only of how efficiently dollars are being spent, staff time is being used, and supplies are being allocated to accomplish the stated activities. The fire service, much like other organizations, tends to rely heavily on outputs as a measure for defending a program or requesting supplemental resources.

Output reports written to describe the amount of activity within a program are ubiquitous; examples of measuring outputs are the number of fires dispatched, the number of EMS incidents dispatched, the number of inspections conducted, and the number of public education events held. Each of these examples has a common theme; that is, the number of something is tallied. In other words, it is easy to categorize and count what was done, but that is not enough; critical questions must also be asked. For example, when working with public education programs, should the program manager count the number of participants or the number of educational offerings (i.e., presentations)? This question is important because the resulting numbers can be drastically different. A single focus on outputs can misrepresent the true performance of a program. For example, the number of apparatus dispatched to an incident is an output measure. Most departments count an incident as one event regardless of the number of trucks dispatched. However, some fire departments report each apparatus dispatched on an incident as an event. Note how this can skew the output and distort a comparison of departments.

OUTCOMES

Outcomes are measures of effectiveness or the quality of a program, service, or product. Outcomes describe results and whether the program goals are being met.[5] An outcome measure is "the end result that is anticipated or desired (such as the community having clean streets or reduced incidents of crimes or fires)."[6] However, outcome measures should be developed with the user in mind. Managers need operational outcome measures useful to them in planning and controlling their programs at the operational level. Strategic outcome measures are very important to those persons responsible for developing, guiding, and evaluating the performance outcomes in relation to the overall mission of the organization and the community.

Mathematicians and budget analysts try to describe outcomes in terms that can be explained by some numerical formula using ratios, ordinal scales, or some other technique that reduces the subjectivity in the analysis of outcome measures. Although the objective is to eliminate subjectivity completely, we must address those intangibles, which are not easily quantifiable, such as the "public good," "quality of life," and "political considerations," using precise mathematical formulas. Elected officials are typically most interested in outcome measures. Because politicians want to know what the "bottom line" is for a given service or program, they prefer to have the information in simple, easy-to-understand language that can be delivered in 30-second sound bites. However, this admirable

outcomes
■ Measures of effectiveness or the quality of a program, service, or product. The way the customer responds to the product or service. A description of the intended result, effect, or consequence that will occur from carrying out a program or activity. A long-term, ultimate measure of success or strategic effectiveness. A measure of the extent to which a service has achieved its goals or objectives and, as defined, met the needs of its beneficiaries. The result of the organization's taking inputs and transforming them into products or services.

objective is not always a realistic expectation. The fire service manager is often caught in the dilemma to provide sound analytic work in an informational format that satisfies the elected officials.

Outcome measures, like outputs, can also be used to measure efficiency. Simply divide number of outcomes by number of inputs. However, the real strength of good outcome measures lies with the story they tell regarding the quality of the service or the effectiveness of the program. Outcome measures go beyond measuring mere activity. Whereas outputs can measure whether something is being done, outcome measures can determine whether the right thing is being done. Although this may look like a subtle difference, it is an extremely important distinction. Outcome measures not only can help to bring clarity to what a program is supposed to accomplish but also, when clearly articulated, can be used to steer training and education programs internally as well as future resource allocation.

Additionally, outcome measures can be the focus of reports that communicate how well the department is meeting its program goals internally and externally. Inside the organization, reports based on outcome measures give managers and their subordinates important responses to how well they are doing at meeting their stated goals. Outside the organization, the same reports provide stakeholders and policy makers with feedback on how effective their decisions were and how well resources are being used to meet customer needs.

These reports also communicate how effectively the same resources are meeting the overall mission of the organization and community. Whereas output performance measures calculate the number of apparatus dispatched to a fire, outcome performance measures evaluate how effective firefighters were after their arrival on the scene of an incident. If it takes 15 minutes to hook up to the available water supply and the house burns down, the output measures were not affected because only the resources dispatched and their arrival times were measured. However, the outcome was a disaster. An example of an operational performance outcome measure is the number of fires kept to the room of origin after the arrival of the fire department. A fire department may want to define this outcome measure further by describing what fire department resources must arrive before it is reasonable to measure the outcome. This requires clearly stated goals and objectives that accurately reflect the fire department's expectations, which should then be provided to the members of the department in the form of operating procedures. As important is determining how the data will be acquired to measure the outcome.

One example of a strategic outcome performance measure is the reduction in dollars of fire loss in the community. This may be the result of an active fire prevention program or the implementation of a sprinkler ordinance. These programs, although very different from each other, contribute to the organization's overall strategic mission. The number of youths taught how to call the fire department through 911 is an output measure by itself. However, if the number of youths who call 911 to report real emergencies increases, it is a positive outcome directly correlated to the fire department's educational program. This example demonstrates how an outcome measure may be simple to state, but a veritable challenge may be collecting accurate and meaningful data. For example, using the 911 training, how do the current data compare with the number of calls received before the educational program was implemented?

The same approach can be taken in training programs, apparatus maintenance programs, or facilities maintenance programs. In the classroom, inputs include the classroom, supplies, teacher(s), resources necessary to get the fire-fighters to class, and other requirements to conduct the class. Outputs are the number of firefighters who attended or the content of the material presented by the class facilitator. However, without outcome measures, the department cannot verify whether anyone learned anything. One way to measure outcomes is through testing in the classroom, yet many will argue classroom tests do not represent the real world. Thus, other ways of more accurately measuring the real outcome of the class are needed. Examples include the use of preceptor programs (such as those used in paramedic programs), peer review programs in the field, and critiques after incidents.

EFFICIENCY

Efficiency is a measure that compares the cost of something, in terms of resources used, in relation to the production of something, in terms of service, products, energy expended, or some other input. Efficiency is number of outputs (or sometimes outcomes) divided by number of inputs, thus providing a unit-cost ratio.[7] It is important to note that, if an organization uses outputs instead of outcomes to measure efficiency, a lower unit-cost ratio may achieve the desired result at the expense of the quality (i.e., outcome) of the service. Efficiency is an easily understood concept that has an important impact on decision makers in all sectors in society. Efforts to improve efficiency in both the public and private sectors sometimes manifest themselves in downsizing, layoffs, and mergers to reduce costs and improve margins of profit. Decision makers in the public sector typically look for ways to reduce taxes and government spending. This often results in efforts to privatize public services or contract services with the private sector, create public–private partnerships, or reduce staffing without reducing the levels of service.

Reductions in costs of inputs without reductions in outputs improve efficiency as do reductions in costs without commensurate reductions in service (usually measured in outputs). It is important not to confuse efficiency with effectiveness. Efficiency addresses only the cost to do something. Efficiency measures are important for several reasons. First, when resources are scarce, decisions must be made to allocate those resources based on some set of criteria, which is used to set priorities. Cost is certainly one criterion for making a decision, and for many policy makers it is their most important criterion. Second, managers must frequently be able to demonstrate they are getting the maximum outputs and outcomes from their already limited inputs before decision makers are willing to give them more resources.

Third, efficiency measures provide the persons doing the work feedback on how well they are utilizing their resources. Continuous improvement efforts in many organizations are focused on being more efficient. Fourth, in a free-market economy, the more efficient a private firm is, the more competitive it can be, thus improving its profits. The public sector is also challenged to be more competitive (usually defined as doing the job cheaper, i.e., more efficiently) through competitive bidding with the private sector. Finally, efficiency measures are important to the public sector because public managers are responsible for eliminating wasteful spending of tax dollars, and citizens are holding them accountable for doing their job. Public managers use efficiency measures to report their achievements to everyone in the system, from those who do the work, to those who receive the service, and ultimately to those who pay their taxes.

efficiency
■ A measure that compares the cost of something, in terms of resources used, in relation to the production of something, in terms of service, products, energy expended, or some other input. Capacity to produce desired results with a minimum expenditure of time, energy, money, or materials. Comparison of what is actually produced or performed with what can be achieved with the same consumption of resources (money, time, labor, etc.). It is an important factor in determination of productivity.

Because the fire service is generally part of the public domain, fire departments are not immune from the challenges presented by both elected officials and the public to be more efficient. Fire departments cannot ignore these challenges but must meet them realistically and responsibly. They can most often achieve operational efficiencies through implementation of sound management practices and resourceful leadership. Creativity and innovation can be critical elements in enhancing the development of sound performance measurement systems.

EFFECTIVENESS

Effectiveness is a measure of achieving the desired result and does not necessarily mean at the lowest cost. Effectiveness measures address whether the right thing is being accomplished, with consideration given to the quality of the service provided or product produced. Effectiveness measures also rely on clearly stated program goals and objectives. Otherwise, most discussions about what a program is doing and whether it should be doing it usually regress into debates over the very subjective and personal values of those involved in the discussion. Therefore, effectiveness measures should be written by program managers after such discussion has taken place. After coming to a clear understanding of what the program is intended to accomplish, preferably based on a consensus of the stakeholders, program goals and objectives can then be decided. Some organizations use other terms in place of the word *effectiveness*.

Effectiveness is one of the most important concepts in performance measurement, because it focuses on the quality of the service or program and is a clearer reflection of the purpose or scope of the program than is the measure of efficiency. Effectiveness raises issues of customer expectations. Even though most people want to achieve their expectations at the lowest possible cost, they usually are not willing to affect outcomes severely just to achieve reduced costs. Effectiveness is also an important concept that must be included in any discussion regarding outsourcing, privatizing, downsizing, or reducing levels of service. The performance measures used to evaluate the effectiveness of the program in meeting its strategic organizational objectives must be clear. Sometimes it may make more sense to eliminate a program altogether rather than to cut the costs (by reducing the inputs) below the level that provides for a minimum level of effective service delivery.

Fire protection is surprisingly difficult to measure well. Because the main purpose of fire protection is to reduce loss of life and property, it is difficult to measure or even estimate what tragedies have been averted as a direct result of education and prevention programs. Two key measurement strategies follow: (1) measuring losses that occur, how they change over time in light of outside explanatory factors such as socioeconomic conditions and environment, and how the losses compare to those in similar municipalities; and (2) measuring intermediate fire protection efforts known to contribute to the desired goals.[8] Building inspections without the staff or resources to conduct adequate follow-up to assure compliance may not be an effective use of resources. Offering emergency medical service without the ability to support continuing education for EMTs and paramedics reduces the effectiveness of the EMS program. Lack of personal protective clothing hinders the fire department's ability to deliver effective fire suppression. Severely cutting or eliminating training in a fire department may save the department money, but the outcome may be an even greater reduction in the effectiveness of firefighters' capability during emergencies.

Fire departments are often required to make budget cuts with the caveat that the cuts must not reduce service levels. Although it is often possible to accomplish these two conflicting objectives, it is only for the short term. The long-term consequences of such decisions are detrimental. This is why performance measurements are essential tools for fire service managers and leaders to understand the efficiency and effectiveness of existing efforts. When current efforts are neither efficient nor effective, then discontinuing a program's resources may not have serious consequences. However, whether expanding or reducing a program or service, it is imperative that efficiency and effectiveness are measured, understood, and clearly communicated to all stakeholders.

Other performance measures may be expressed in terms such as the following:

Quality. The degree to which a product or service meets customer requirements and expectations.

Timeliness. Measures whether a unit of work was done correctly and on time. Criteria must be established to define what constitutes timeliness for a given unit of work. These are usually based on customer requirements.

Productivity. The value added by the process divided by the value of the labor and capital consumed.

Safety. Measures the overall health of the organization and the working environment of its employees.

Performance measures quantitatively indicate something important about products, services, and the processes that produce them. They are a tool to help personnel understand, manage, and improve what organizations are doing. Effective performance measures can let personnel in the department know:

- How well they are doing
- Whether they are meeting their goals
- Whether their customers are satisfied
- Whether their processes are in statistical control
- Whether and where improvements are necessary

For fire service leaders, these performance measures provide the information necessary to make intelligent decisions about what is occurring in their organizations.

System Performance Measurements

Performance measurement is the "heart and soul" of the performance-based management process. Flowing from the organizational mission and the strategic planning process, it provides the data that the fire service will collect, analyze, report, and ultimately use to make sound business decisions. It directs the business function by justifying budgetary expenditures, documenting progress toward established objectives, identifying areas of both strength and weakness, providing an ongoing assessment of the current "organizational climate," and driving business improvement. In a nutshell, performance measurement supports organizational existence.

Performance measurement systems succeed when the organization's strategy and performance measures are in alignment and when leaders convey the organization's mission, vision, values, and strategic direction to employees and external stakeholders. The performance measures give life to the mission, vision, and strategy by providing a focus that not only lets each employee know how he or she contributes to the success of the department but also provides its stakeholders with a means to measure community expectations.

Integration places performance measures where they are the most effective: integrated with the organization's strategic plan. This makes it possible for the measures to be effective agents for change. If the measures quantify results of the activity, one need only compare the measured data with the desired goals to know whether actions are needed. In other words, the measures should carry the message.[9]

Change might be inevitable, but all too often it occurs like an unguided missile seeking an elusive target at unpredictable speeds. For most activities, it is far better to manage change with a plan—one that includes clear goals and useful indications of progress toward a desired objective. Participants in any activity need to know what outcome is expected, how their work contributes to the overall goal, how well things are progressing, and what to do if results are not occurring as they should. This approach places performance measures where they are the most effective: integrated with the activity.

Inappropriate measures are often the result of random selection methods. For example, brainstorming exercises can get people thinking about what is possible and provide long lists of what *could* be measured. Unfortunately, such efforts by themselves do not provide reliable lists of what *should* be measured. Unless the measures are firmly connected to results from a defined process, it is difficult to know what corrective actions to take as well as to be able to predict with confidence what effects those changes will have.

In order to be able to identify effective corrective actions to improve products and services, it is vital to measure the results of all key processes. In this way, specific processes that need to change can be identified when progress is not satisfactory.

The performance process model provides a good overview of how measurement can be integrated throughout the agency (see Figure 6.5).

Noted performance measurement expert Mark Graham Brown points out the following:

- Measurement reduces emotionalism and encourages constructive problem solving. Measurement provides concrete data on which to make sound business decisions, thus reducing the urge to manage by "gut feeling" or intuition.
- Measurement increases one's influence. Measurement identifies areas needing attention and enables positive influence in that area. Also, employees "perform to the measurement," an example of how measurement influences employee performance.
- Improvement is impossible without measurement. If you don't know where you are, then you can't know where you're going, and you certainly can't get to where you want to be. It's akin to traveling in unknown territory without a compass or a map. You're totally lost.[10]

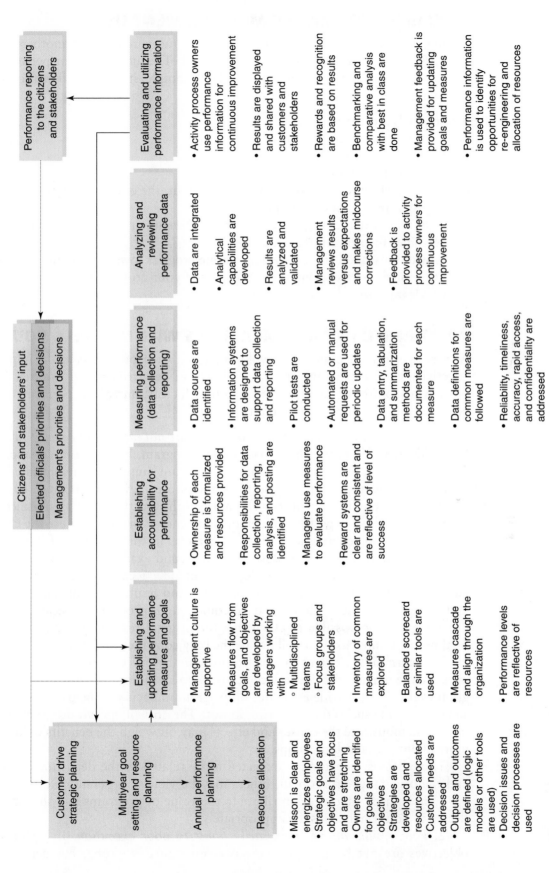

FIGURE 6.5 Performance Measurement Process Model

HOW IS PERFORMANCE MEASUREMENT USED?

Asking, What are the benefits of performance measurement? is another way of asking how performance measurement is used. The answer is that performance measurement has many beneficial uses. For example, it can be utilized to do the following:

- Set goals and standards.
- Detect and correct problems.
- Manage, describe, and improve processes.
- Document accomplishments.
- Gain insight into, and make judgments about, the effectiveness and efficiency of programs, processes, and people.
- Determine whether organizations are fulfilling their vision and meeting their customer-focused strategic goals.
- Provide measurable results to demonstrate progress toward goals and objectives.
- Determine the effectiveness of your part of your group, department, division, or organization.[11]

WHAT PERFORMANCE MEASURES WILL NOT TELL YOU

At times performance measures will not paint a clear picture as to the impact or effectiveness of a program or service. This can be attributed to external forces outside the control of the agency such as time factors, unrealistic performance expectations, or use as a numeric quota.

1. *As such, the cause and effect of outcomes are not easily established.* Outcomes can, and often do, reveal the impact of the program, but without collaborating data, it is difficult to demonstrate that the program caused the outcome(s). Realistically, the outcomes of public sector services are inevitably affected by many events outside public control. For example, in the weatherization assistance program, it is not always possible to demonstrate energy savings because the changes introduced to homes may also result in changes in the behavior of inhabitants, which confounds the analysis. Assume, as a second example, that the goal of fire research is to encourage the development of new technologies, which will be adopted by the profession and allied industries and result in a reduction of fire loss and civilian casualties. The desired outcome may not occur for decades, and although it may be possible to claim that the original research contributed to the final product, most likely it will not be the only contributing factor.

To determine the extent to which a program has affected the outcomes and to measure the impact, the fire service needs to do an in-depth analysis. Special program evaluations provide estimates of program impacts and help determine why some programs succeed and others do not. The cost of special program evaluations to demonstrate the causes and effects may outweigh the benefits of knowing more about causal relationships.

Though most benefits are expected to be related to the department's efforts and the original program plan, others may be viewed as serendipitous impacts. Such unplanned outcomes contribute to the value of programs and should be reflected in performance results appropriately.

2. *Poor results do not necessarily point to poor execution.* If performance objectives are not being met, obviously something is wrong; but performance

information itself does not always provide the reason. Instead, it raises a flag requiring investigation. Possibilities include performance expectations that were unrealistic or realigned work priorities. Every organization should be able to explain performance results and to define and address the contributing factors that led to the end result.

3. *Numerical quotas do not fix defective processes.* A danger also exists when performance objectives become numerical quotas. The setting of numerical goals and quotas does nothing to accomplish improvements in the process. Instead, identifying the challenges and changing the processes are needed to improve performance and achieve desired outcomes.

4. *Measurements only approximate the actual system.* Performance measurement provides a valuable tool for management and continuous improvement. However, people might try to "game" the system in a way that will make their programs look good. Additionally, accurate data may not be available. These are among the reasons why a department may need to recognize that the measured system is not the same as the actual system.

5. *Performance measures do not ensure compliance with laws and regulations.* Performance measures help form the basis for sound performance-based management; however, they do not provide information on adherence to laws and regulations or the effectiveness of internal controls. Bypassing internal controls or noncompliance with laws and regulations may expedite operations and thus result in a "favorable performance" statistic, which does not necessarily indicate good performance. For example, a building could be constructed more quickly if safety controls and funding limitations were ignored. Because compliance and internal controls often have a direct effect on performance, care should be taken to supplement performance measurement with other oversight activities to ensure that controls are in place and working as intended and that activities are adhering to laws and regulations.

Note: Performance measures *can* be constructed in such a way that ensures compliance with laws and regulations. However, it should not be assumed that they automatically ensure compliance.[12]

PERFORMANCE MEASUREMENT ROADBLOCKS

During the course of the performance measurement implementation phase and the maintenance phase, your performance measurement system may run into a few obstacles or roadblocks to success. Some may be laced with good intentions; some may be deliberate attempts at derailment. Whatever the case, be aware of them and consider the following:

- *Stay focused on the objective.* Establishing and implementing a performance measurement system is an in-depth and continuous process. As a result, it is very easy for personnel to get caught up in the process of developing and perfecting the process. Unfortunately, when this preoccupation occurs, the original intent of improving performance "takes a back seat" while people totally engross themselves in charts, graphs, and meetings to design, redesign, and re-redesign the system. Beware of this cycle of "analysis paralysis"; do not let the design process take over the overall project to improve performance.

- *Do it right the first time.* Performance measurement systems take time to design, implement, and perfect. It can be a difficult process. As a result, a strong urge may surface to take the easy way out, to get it over quickly, to measure the trivial or the obvious, or to set goals or targets that are easily attainable. Resist this urge. If you and your organization are sincerely committed to the performance measurement process, then make a sincere effort to follow through with it. Look at it as a positive process, and take a long-term view. The investment of time and planning at the front end will save thousands of hours in the end and will also result in a better process.

■ *The mission is not impossible.* The factor to deal with here is the complete opposite of the preceding one. Rather than take the "low road" of easiness, the person, group, or organization decides to take the "high road" of impossibility. Adorned with good intentions and lofty aspirations, it establishes immeasurable objectives and sets unreachable goals. Then, when failure sets in, the entire performance measurement system gets "scrapped." Instead, when establishing a performance measurement system, be realistic. Do not set yourself up for failure. Take small steps at first, and then let your system grow with your organization.

■ *Anticipate resistance!* It is inevitable that resistance to the performance measurement process will surface, usually in the development phase. That is human nature. It is a fact: many people do not like change. Performance measurement may expose weak areas in employee performance and carries with it an "accountability factor." The only way to deal with resistance is by involving employees in the performance measurement process from start to finish; this encourages employee buy-in.

■ *Do not use it as a hammer.* Some organizations may use their performance measurement system as a punitive measurement system, a way to catch employees doing something wrong and to punish them. If such is the case, can you imagine how the employees feel? Certainly they are not very committed or very loyal. Playing "gotcha" decreases employee performance, which in turn decreases organizational performance. It is not conducive to the development of a high-performance organization and should be avoided at all costs.

■ *Counter "it won't work here!"* This cry is the mantra of skepticism regarding performance measurement doubters. It goes along with such statements as "It will never happen!" There are two reasons why such an attitude exists: (1) past organizational experience shows a trend of unfinished or failed projects, or (2) the person is a natural resister to change. In the first case, organizational leadership must champion the cause and show unending commitment to the performance measurement undertaking. In the second case, the employee may need to be educated on the purpose of and expected outcomes for the department.

■ *Recognize the rearview mirror approach.* Some business entities focus only on bottom-line results. For example, each morning a restaurant owner may look only at the previous day's profits as a performance indicator. Unfortunately, that shows only how the restaurant did in the past. It does not predict how the restaurant is going to do in the future. Many in the performance measurement arena call this "driving through the rearview mirror." You are not focused on where you are going, only on where you have been. And, if you are driving through the rearview mirror, you will not see the large hole in the road until you have driven right into it. So, when setting up a performance measurement system, remember to measure the critical things that impact the results, not just the results themselves.

■ *Avoid tunnel vision.* Although tunnel vision, or data tunnel vision, comes well after the performance measurement system has been developed and

FIGURE 6.6 Tunnel

implemented, fire officials need to keep it in mind while they are doing the developing and implementing. After measurement data have been collected, analyzed, and reported, it is possible for fire officials to focus on only a particular piece or area of measurement data on which to base their decisions, completely forgetting about the consequences of doing so. For example, if a department concentrates only on the travel time to emergencies, then such an action could lead to compromising employee and public safety, jeopardizing the original mission, which was to get there in the first place. When working with performance measurement, one should look at the big picture first and then focus on the pieces that fit into that picture (see Figure 6.6).

■ *The system of performance measurement belongs to everyone.* Organizational dynamics play a big part in this scenario. One person or a group of people may consider the performance measurement process or system to be a part of "their turf" and will not want to relinquish control to anyone else. This situation precludes the total organizational involvement necessary for establishing and implementing the system. The result: a failed system and several employees saying, "See. I told you so!" When implementing changes and designing the system, give control over to those held responsible for performance and improvement.[13]

MAJOR PITFALLS OF MEASUREMENT SYSTEMS

It is rare to find an organization that has no problems with its performance measurement system. Some may need only simple fixes whereas others may require major overhauls. However, learning from and avoiding others' mistakes are wise things to do. Here are some of the errors other organizations have made. Avoid these pitfalls of measurement systems.

1. *Amassing too much data.* It results in "information overload." Managers and employees will either ignore the data or use them ineffectively.

The most common mistake organizations make is measuring too many variables. The next most common mistake is measuring too few.

Mark Graham Brown, *Keeping Score* (1996)[14]

2. *Focusing on the short term.* Most organizations collect only financial and operational data. They forget to focus on the longer-term measures—the very ones on which the Malcolm Baldrige National Quality Award focuses—of customer satisfaction, employee satisfaction, product/service quality, and public responsibility.

3. *Failing to base business decisions on the data.* A lot of managers make decisions based on intuition and past experience rather than the data being reported to them. If the data are valid, they should be used appropriately.

4. *"Dumbing" the data.* Sometimes data can be summarized so much that they become meaningless. If business decisions are going to be based on the data, then the data need to be reported clearly and understandably.

5. *Measuring too little.* Making business decisions with too little data is just as problematic as basing them on too much data. Some organizations tend to measure too few key variables to get the "whole picture" of the health of their organizations.

6. *Collecting inconsistent, conflicting, and unnecessary data.* All data should lead to some ultimate measure of success for the department. An example of conflicting measures would be measuring reduction of inspectors per inspectable occupancies while, at the same time, measuring staff satisfaction with their workload.

7. *Driving the wrong performance.* Exceptional performance in one area could be disastrous in another. Mark Graham Brown tells a poignant anecdote about "chicken efficiency" in which the manager of a fast-food chicken restaurant scores a perfect 100 percent on his chicken efficiency measure (the ratio of how many pieces of chicken sold to the amount thrown away) by waiting until the chicken is ordered before cooking it. However, the end results of his actions were dissatisfied customers (from waiting too long) and lack of repeat business. Thus, the "chicken efficiency" was driving the wrong performance . . . and driving the customers away!

8. *Encouraging competition and discouraging teamwork.* Comparing performance results of organizational unit to organizational unit, or one employee to another, sometimes creates fierce competition to be "number one," with the result of destroying a sense of teamwork. Remember to compare to stated performance goals.

9. *Establishing unrealistic and/or unreasonable measures.* Measures must fit into the organization's budgetary and personnel constraints, be cost-effective, and be achievable. Nothing can demoralize an employee or the department quicker than establishing goals that can never be reached.

10. *Failing to link measures.* Measures should be linked to the organization's strategic plan and should cascade down into the organization (horizontal and vertical linkage). Measures without linkage are like a boat without water: they are useless and are not going anywhere.

11. *Measuring progress too often or not often enough.* There has to be a balance here. Measuring progress too often results in unnecessary effort and excessive costs, with little or no added value. On the other hand, measuring progress not often enough engenders a situation in which nothing is known about potential problems until it is too late to take appropriate action.

12. *Ignoring the customer.* Management often wants to measure only an organization's internal components and processes, which it can "command and control." However, in reality, it is the customer who drives any organization's

performance. Most of the best-in-class organizations place customer satisfaction above all else.

13. *Asking the wrong questions or looking in the wrong places.* Sometimes chief officers ask who is to blame instead of asking what went wrong. They look for the answers in the people instead of the process. A faulty process makes employees look faulty.

14. *Confusing the purpose of the performance measurement system.* The purpose of a performance measurement system is not merely to collect data but rather to collect data upon which to make critical organizational decisions that will, in turn, drive improvement. Knowing that one is 10 pounds overweight is just knowledge of a fact. Taking improvement actions based on that knowledge is where "the truth" lies.[15]

PEARSON
myfirekit
Visit MyFireKit
Chapter 6 for more
information on Per-
formance Measures.

Vision 20/20 Model Performance Measures for Fire Protection

An excellent example of this in the U.S. fire service today is a project known as Vision 20/20 (see Figure 6.7). What started out as a discussion of several key leaders in the profession has grown into a focused effort to move forward with strategies to reduce fire losses in the United States.

PEARSON
myfirekit
Visit MyFireKit
Chapter 6 for more
information on
Vision 20/20.

MODEL PERFORMANCE MEASURES FOR FIRE PREVENTION PROGRAMS

The purpose of the Vision 20/20 project is to outline potential model performance measures for local, state, and national fire prevention program managers. The goal is to begin reporting fire prevention efforts in a consistent enough fashion to allow for legitimate program comparison and the establishment of both baseline performance measures and benchmark standards. In other words, program managers can compare departmental results to their own history, and to other jurisdictions, to begin formulating management decisions based on evidence. The establishment of consistent and accepted performance measures will allow for demonstrated evidence that prevention programs are producing a desired result.

FIGURE 6.7 Vision 20/20 Logo *(Courtesy of Jim Crawford, Project Manager, Vision 20/20)*

To the US Community of Fire Safety Professionals and Advocates:

We all know the devastation from fire that occurs every single day and we are all dedicated to making our nation safer from that devastation. The goal of Vision 20/20 is a simple one—to marshal forces for the development and support of a national strategic agenda for fire loss prevention. In the spring of 2008, the Vision 20/20 National Forum brought some of the brightest minds in fire safety together in Washington DC to help determine how best to achieve this goal. The Forum was an important first step in the continuing Vision 20/20 initiative.

Objectives of the Vision 20/20 initiative include:

- Provide a forum for sustained, collaborative planning to reduce fire loss in the United States
- Involve agencies and organizations with expertise and commitment to fire loss reduction in this collaborative effort
- Focus on actions that are needed to bridge the gap between recommended solutions and the current status of fire prevention activity
- Communicate recommendations and actions clearly with all levels of the fire safety community
- Build on the success and momentum of existing efforts

The report of National Strategies for Fire Loss Prevention presented here is the end-result of the Vision 20/20 National Forum. It is derived from the collective knowledge of experts from across the nation of the fire problem and from experience in the field. These strategies were determined to have a direct impact on the loss of life and property from fire in both the short- and long-term. This report is also an active, growing and evolving document that will serve as a blueprint for continued refinement by task groups working to earn commitment for the recommended actions.

We are grateful for funding from the DHS Assistance to Firefighters Prevention and Safety Grant program and to our sponsor, the Institution of Fire Engineers, USA Branch, for making this project possible. A number of professionals in the fire prevention community have been talking about this effort for some time and, without the leadership of IFE-USA, it would not have been possible. The funding provided by DHS is a key component and the returns on this investment in saving lives will be one that will resonate for years to come.

We appreciate the support of the many volunteers who have stepped up to help including the members of the Steering Committee, the Web Forum Satellite hosts, additional funding support, and participants in the Web and National Forums who took valuable time out of their busy schedules to contribute to this landmark opportunity.

Thank you for your commitment to reducing fire loss in the United States. Let us move forward as a nation, focusing on strategies that have been recommended by our leaders in fire prevention and capitalizing on the strengths of our diversity and shared mission.

Sincerely,

Jim Crawford   Bill Kehoe
Vision 20/20 Project Manager   IFE-USE Branch Treasurer

Source: Courtesy of Jim Crawford, Vision 20/20, 2009.

However, a word of caution is necessary. The number of potential variables and the complexity involved in establishing model performance measures for fire prevention programs have often proved problematic.

Generally, the results that are expected from fire prevention (and related life safety) programs include documenting educational gain (people actually learning something as opposed to just sitting in class); documenting risk reduction where increased safety behaviors or decreased hazard-producing behaviors can be acknowledged (e.g., hazards noted and abated during fire code compliance inspections); and finally documenting reductions in losses. Losses in this context mean reductions in deaths, injuries, and both direct and indirect economic losses.

However, often extraneous factors can affect the data received and their perception. So although prevention programs exist to increase safety knowledge and reduce risks and losses, decision makers should be cautioned about abandoning efforts (as some have) due to inadequate data or study. For example, numbers for incident rates may rise, fall, or remain constant because of prevention efforts, random chance, data entry errors, or (in the case of wildland fires) the weather.

For these reasons, most of the model measures have been described as changes, and the act of beginning to evaluate performance of prevention programs in common terms is the goal. With field experience over time, as evaluation methods (and the measures ultimately recommended) increase in sophistication, program managers can begin to compare the results of their efforts in more scientific terms.

For comparative purposes, Vision 20/20 groups these results-oriented performance measures into common evaluation terms used and taught at the National Fire Academy, the National Fire Protection Association, and other organizations. They include (for post-program evaluation) process, impact, and outcome evaluation measures, which are recommended for comparative analysis of the results that prevention programs achieve.

Why Evaluate Fire Safety Programs?

Vision 20/20 outlines a series of strategies and action items, which will be discussed in Chapter 10, to reduce the nation's fire loss. To determine the impact, performance measures were crafted and select terminology developed to help in evaluating fire safety programs:

- Understand how to improve service
- Learn whether programs have any unexpected benefits or problems
- Monitor whether program is having desired results
- Monitor progress toward program's goals
- Produce data on which to base future programs
- Demonstrate effectiveness to target population, to public, to others who want to conduct similar programs, to those who want to fund future programs

Terminology

The difference between the terms listed here is actually driven by the stated goals of the particular program being evaluated. But generally, they are defined in the following fashion relative to fire prevention programs.

- **Outcome evaluation.** The mechanism of determining how well a program achieves its ultimate goals (such as reduced losses).
- **Impact evaluation.** The mechanism of measuring changes in the target population that the program is intended to produce. These measures could be considered advanced indicators of successful outcomes (such as reduced risk).
- **Process evaluation.** The mechanism of testing whether a program is attaining its goals, such as reaching target populations with quantifiable numbers expected. Process evaluation measures often include measures for workload (e.g., number of inspections done per inspector) and milestones in achievement of process objectives.

Benchmarking

It has been stated that competitive athletic activities would be a terrible waste of time if nobody kept score. No one really wants to watch a lot of physical effort between two groups of individuals unless the excitement of finding out who did the best job of playing the game can be enjoyed. In the fire service the opposing sides are often the demand for services and the delivery of those services versus the efficiency and effectiveness of those services. Keeping score in these cases is a matter of measuring the level of fire service effort combating a community's fire problem, EMS, inspections, or any service a department delivers. In this context, the use of two different terms in keeping score is important. A foundational step in improving your quality of service is the development of two scorecards: one is the *baseline*; the other is the *benchmark*.

A **baseline** is defined as a database from which something can be judged. Baselines consist of the current level of performance at which a department, process, or function is operating. Only the agency that is delivering the activities and programs can create a baseline because it is a measurement internal to the organization. For example, a vital baseline for a fire service agency is its experience in responding to alarms. Over a period of years, the department should have developed a database clearly identifying how fast the agency can receive, process, dispatch, and respond to an alarm. These data are the real-life experiences of the organization and constitute a baseline for the agency. In fact, when fire officials stop and think of all the services provided or programs managed, there are literally hundreds of important data elements that need to be measured. They need to begin by choosing to measure the one that will promote value-added service to the customer(s).

A **benchmark** is defined as a standard from which something can be judged. Searching for the best practices for a specific service or action will help define superior performance. Therefore, a benchmark is the best performance that can be found by a department or others performing similar services or functions. A benchmark is a standard against which the organization can assess itself to see whether its baseline is achieving an acceptable level of performance in the context of fire service standards. In the alarm processing area, a benchmark to compara-

baseline
■ A database from which something can be judged. Baselines consist of the current level of performance at which a department, process, or function is operating. Only the agency that is delivering the activities and programs can create a baseline because it is a measurement internal to the organization.

benchmark
■ A standard from which something can be judged. Searching for the best practices for a specific service or action will help define superior performance. Therefore, a benchmark is the best performance that can be found by a department or others performing similar services or functions. A benchmark is a standard against which the organization can assess itself to see whether its baseline is achieving an acceptable level of performance in the context of fire service standards.

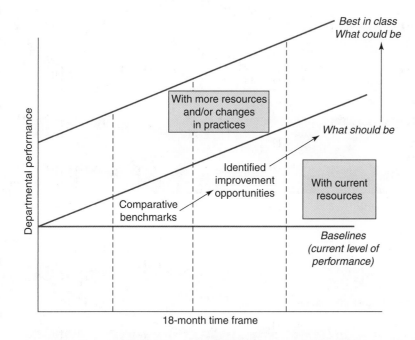

FIGURE 6.8
Performance Levels

18-month time frame

ble fire organizations might be the ability to receive, process, and dispatch apparatus and personnel in 60 seconds. Many departments today are using fire service standards such as NFPA 1710, 1720, 1221, and others as benchmarks to achieve.

Now, in comparing two concepts, an organization may have a baseline of 90 seconds for alarm processing, whereas the benchmark is 60 seconds. Therefore, the agency has identified an opportunity for improvement to reduce alarm processing time by 30 seconds (see Figure 6.8).

Another prime example of the use of baselines and benchmarks would be the travel time of apparatus. A benchmark goal might be that the organization wants to achieve a travel time of less than four minutes to 90 percent of its incidents. The baseline may indicate the organization is able to achieve this goal in only 75 percent of its incidents. The difference between the baseline and the benchmark can become a performance improvement goal for the organization.

Another way to use a benchmark is in the development of the department's standard of cover document. For example, if a department has adopted service level objectives, fire officials can show the performance of the organization in relation to the benchmark, what the department is striving to accomplish. When done using fractile measurement, it clearly demonstrates the opportunity for improvement. Examples of this are illustrated in Figures 6.9 through 6.11.

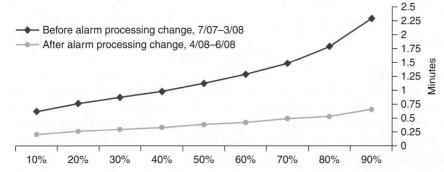

FIGURE 6.9 Alarm Processing Time—Emergency Calls

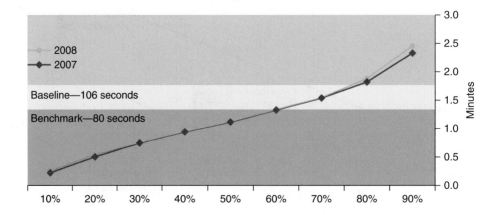

FIGURE 6.10 Turnout Times—Fire Calls

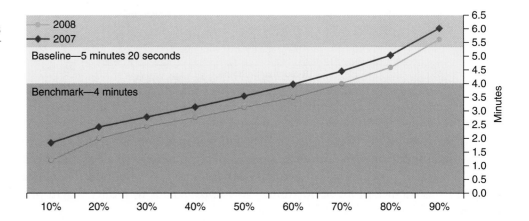

FIGURE 6.11 Travel Time—First-in Fire Calls

One of the most important aspects of quality improvement is measurement. Companies that have recognized the importance of quality realize they cannot afford to go by their own perceptions about what their customers want or by how well they think they are providing a service. The process of a department's benchmarking its service(s) against others allows it to evaluate its performance as compared to that of others.

The basic idea behind benchmarking is for two or more organizations of similar size and makeup to work together to build a common set of performance measurements. Then they exchange information on their own performance and make comparisons on what has and has not worked successfully (see Figure 6.12).

As more organizations are added to this process, a performance curve begins to appear, with some organizations at the bottom and some at the top. By comparing the key practices and processes of the best performers with those of the worst, before long some important lessons begin to emerge. Effective benchmarking usually results in a series of "aha!" moments for underperformers, as they discover how their counterparts have successfully resolved similar problems in their work processes. This will help the organization take appropriate action to improve its performance. However, it is not generally a matter of only the worst performers profiting from the process. Usually, even the best performers can learn from others, as they inquire about the processes other organizations use.

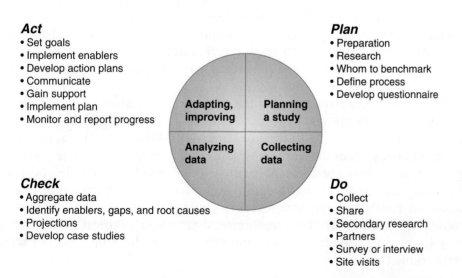

FIGURE 6.12
Benchmarking Process
Flowchart

The fire service can use benchmarking as a tool to improve organizational efficiency and effectiveness in the following ways:

- Identifying customer needs
- Selecting processes that are key to accomplishing organizational success
- Collecting quantifiable data with the ability to compare the internal baseline against benchmarking statistics
- Providing insights into new ways of doing business and delivering service
- Assisting in analyzing both internal and external data collection
- Identifying those items that enable consistent superior performance
- Assisting in learning the best practices being used by others who have achieved superior performance
- Assisting in developing goals, objectives, strategies, tactics, and organizational action plans aimed at making our departments the best in the market they serve

It is a natural element of the plan-do-check-act (PDCA) component of the quality improvement process (see Figure 6.13).

Act
- Set goals
- Implement enablers
- Develop action plans
- Communicate
- Gain support
- Implement plan
- Monitor and report progress

Plan
- Preparation
- Research
- Whom to benchmark
- Define process
- Develop questionnaire

Adapting, improving | Planning a study

Analyzing data | Collecting data

Check
- Aggregate data
- Identify enablers, gaps, and root causes
- Projections
- Develop case studies

Do
- Collect
- Share
- Secondary research
- Partners
- Survey or interview
- Site visits

FIGURE 6.13
Benchmarking for
Competitiveness

Benchmarking does not consist of visiting another organization to see how it operates, searching for a cure-all for an organization's insufficiencies, merely copying what others are doing, or a stand-alone exercise. It certainly is not quick and easy. According to Osborne and Gaebler (1992) in *Reinventing Government*, "If you don't measure results you can't determine success from failure."[16] Benchmarking has become one of the power tools government uses to improve its performance. Benefits of benchmarking include the following:

PEARSON
myfirekit

Visit MyFireKit
Chapter 6 for more
information on
Benchmarking.

- Is a great way of improving effectiveness and efficiency
- Helps to accelerate the change process
- Assists in setting and achieving higher performance goals
- Provides the bridge to overcome the not-invented-here (NIH) traditional aspects found in many fire departments today
- Provides a means to meet customer expectations better

COMPARATIVE SURVEY AND ANALYSIS

The process of collecting and analyzing information to be used as performance measurements and to compare the department with other organizations is a key component of system improvement.

These surveys often prove useful in providing broad organizational comparisons of one agency to another. Total quality management espouses the principle that customers, or citizens in the case of municipalities, usually desire the best service or product at the least cost. Although best service is always a customer concern, customers are often willing to pay a premium for superior service when it can be shown what it means to the customer.

Conclusion

As leaders focus their organizations on moving forward in the future, it is imperative that they do so with a clear understanding of both what their organizations are facing today and the challenges and the opportunities that will be encountered in the future. During this recent economic downturn not only in the United States but worldwide, it should have become clear to every fire service leader that performance measurements of the services and programs offered in their departments are critical, not only to their future success but also to their survival. Performance measurement is about measuring results in terms of effectiveness, efficiency, workload, and quality.

This chapter has provided multiple examples of measuring system performance, and hopefully fire service leaders can examine these concepts and begin to implement those applicable to their organizations. Of course, this is often easier said than done. One of the barriers organizations often have to overcome when they begin to measure performance is their poor ability to compile and analyze the existing data and to identify data that are not being collected. For many organizations today, this requires a re-engineering of their current data systems. Although this in itself can be a monumental task, it is a critical element in their ability to measure the effectiveness and efficiency of the programs and services offered to the community.

Another trend that will continue in the fire service is benchmarking performance. The process of benchmarking includes measuring and comparing performance against similar programs and services through the use of comparative analysis with other departments, established national standards, norms in the region, past performance, and/or internal established benchmarks. This allows leaders to set the target that their organizations should continually strive to meet. Performance measurement and benchmarking help leaders answer critical questions such as the following: what are the overall goals, what specific outcomes are they aiming for, how will they know whether these goals are being achieved, how well should they be doing, how are they doing versus their peer group, how can they improve, and how will they know when they are doing better.

The bottom line with performance measurement and benchmarking is that they help the organization improve its level of accountability to the governing body, the community leadership, and the residents by enhancing the department's ability to make more informed decisions. When the department can do this, it provides a more effective level of service and will use taxpayer dollars as efficiently as possible. As importantly, the department will be able to articulate the performance of the programs and services provided to all of its internal and external stakeholders. This is a critical element for organizations today in the competitive environment of local government.

PEARSON
myfirekit

Visit MyFireKit Chapter 6 for Perspectives of Industry Leaders.

Review Questions

1. Describe the difference between qualitative and quantitative data. Provide an example of how both may be used.
2. Define *workload*, *efficiency*, and *effectiveness*. Provide a specific example of how your organization or local fire service provider uses these concepts to evaluate performance.
3. Describe the seven pitfalls of performance measurement systems.
4. Define the system performance measures that begin the cascade of events. Provide an analysis of performance for your department or local fire service provider based upon the key data points noted in the models.
5. Define the terminology used for evaluation in the Vision 20/20 model performance measures.
6. Complete an analysis of your department or local fire provider using the outcome measurement example for code compliance, effectiveness, and fire investigation programs.
7. Define the terms *baseline* and *benchmark*. Cite examples of how your department or local fire service provider uses them.
8. Assemble a report on comparative survey data with 10 departments similar to yours. You are in Department A.

PEARSON

myfirekit™

For additional review and practice tests, visit www.bradybooks.com and click on MyBradyKit to access book-specific resources for this text!

Register for MyFireKit by following directions on the MyFireKit student access card provided with this text. If there is no card, go to www.bradybooks.com and follow the MyBradyKit link to buy access from there.

References

1. Ron Kohavi, Neal J. Rothleder, and Evangelos Simoudis, "Emerging Trends in Business Analytics," *Communications of the ACM,* 45, no. 8 (August 2002), 6.
2. David N. Ammons and Benjamin B. Canada, *Tracking Performance in Local Government Risk Management, A Catalog of Performance Measures,* Institute of Government, University of North Carolina at Chapel Hill (Fairfax, VA: Public Entity Risk Institute, 2000), pp. 2–3.
3. ICMA, Arizona Leadership Academy, *Performance Modern Dimensions,* 1997.
4. Ibid.
5. Ibid.
6. Ibid.
7. Ibid.
8. H. P. Hatry, L. H. Blair, D. M. Fisk, J. M. Greiner, J. R. Hall Jr., and P. S. Schaenman, *How Effective Are Your Community Services? Procedures for Measuring Their Quality,* 2nd ed. (Washington, DC: ICMA the Urban Institute Press, 1992), p. 93.
9. Will Artley (Oak Ridge Institute for Science and Education) and Suzanne Stroh (University

of California), *The Performance-Based Management Handbook, Volume 2, Establishing an Integrated Performance Measurement System,* prepared by the Training Resources and Data Exchange Performance-Based Management Special Interest Group, September 2001, p. 1.

10. Ibid., p. 5.
11. Ibid., pp. 5–6.
12. Ibid., pp. 6–7.
13. Ibid., pp. 7–8.
14. Mark Graham Brown, *Keeping Score: Using the Right Metrics to Drive World-Class Performance* (New York: Productivity Inc., 1996).
15. Will Artley (Oak Ridge Institute for Science and Education) and Suzanne Stroh (University of California), *The Performance-Based Management Handbook, Volume 2, Establishing an Integrated Performance Measurement System,* prepared by the Training Resources and Data Exchange Performance-Based Management Special Interest Group, September 2001, pp. 9–10.
16. David Osborne and Ted Gaebler, *Reinventing Government: How the Entrepreneurial Spirit Is Transforming the Public Sector* (Reading, MA: Addison-Wesley, 1992).

7 CHAPTER

Fiscal Management in Difficult Times

Adopted Budget
FISCAL YEAR 2009-2010

(www.fresno.gov)

OBJECTIVES

After completing this chapter, you should be able to:

- Describe the history of budgeting in the United States.
- Describe the various budget methods used at the federal, state, and local levels.
- Explain what GASB is.
- Define what GAAP is.
- Describe the impact of health care and pension costs on local government.
- Describe the current state of facilities and equipment in the U.S. fire service.

PEARSON

For additional review and practice tests, visit www.bradybooks.com and click on MyBradyKit to access book-specific resources for this text!

Introduction

Since 2000 the state of the economy in the United States as well as worldwide has been unprecedented, and it is almost impossible to predict what may occur in the future. During the last five years, the severity and the impact of the economic downturn have affected all levels of government to a degree not experienced since the 1930s. The news seemed to get worse each day, and the daily reports of fire service cutbacks became widespread from the largest to the smallest departments, each with staff to pay, ongoing pension costs, and many with outstanding capital bond payments and lease payments. The choices for many departments were bad choices. The departments that were able to minimize the impact of this crisis had a good understanding of budgeting and the importance of forecasting trends in revenues and expenses. If this economic meltdown has done nothing else for the fire service, it has highlighted the importance of chief executives understanding all facets of the budget.

Depending upon the conditions at any given time in history, budgets have emphasized financial control, managerial improvements, planning, or all three. In reviewing U.S. history, it is difficult to imagine government without budgeting. However, unlike many other American institutions and practices, the concept of budgeting, as it is known today, was not introduced to the United States by the early colonists. Rather, it developed in the latter part of the 19th century.

The Origins and Historical Context of Budgeting

At the turn of the 19th century, the United States was "the only industrialized nation without a budget system."[1] Congress raised and voted on the money needed to operate the national government in a more or less haphazard manner. The same was true of states and local governments, with the exception that more state and local variations existed due to their different governing agencies. In part, this legislative dominance was due to the notion of separation of powers and the fear of a strong executive branch, resulting from the American colonial experience. As a consequence, budgeting—to the extent it can be said to have existed in the sense known today—was the preserve of the legislative branch. This pattern dominated all levels of government in the United States. However, it was the cities that sparked reform. Patronage-based political machines of the 19th century may have fostered the development of budgeting as it is now known. In the last decade of the 19th century, budgeting was defined as "a valuation of receipts and expenditures or a public balance sheet, and as a legislative act establishing and authorizing certain kinds and amounts of expenditures and taxation."[2]

The idea of a budget as a control mechanism had been developing since the 1830s, but it gained momentum after the Civil War with the growth of cities and the expansion of municipal services. By the end of the 1890s, three basic forms of municipal budgeting existed. Some cities simply used a tax levy, an approach disliked by reformers due to the lack of control through inattention to the expenditure side of budgeting, coupled with dominance by the city council. A second approach was a tax levy accompanied by detailed appropriations. Missing, of course, were particulars regarding revenue estimates. Still other cities used a tax levy but preceded it with detailed estimates of receipts and expenditures, a practice that found favor with business and middle-class reformers. However, city councils were not legally bound to adhere to these estimates. Can you imagine an approach being used today that does not provide detailed revenue and expense forecasts and transparency to the public?

Extremely important and influential in promoting municipal finance reform was the New York Bureau of Municipal Research, created in 1907, which highlighted poor fiscal procedures that resulted in inefficiency. This agency worked to inject uniformity and responsibility into governmental finance. This bureau's approach utilized the business corporation as the model for bureaucratic organization and encouraged using scientific management concepts to promote planning, specialization, quantitative measurement, and standardization.

The ensuing public administration movement with its scientific management ethic succeeded in creating widespread dissatisfaction with the budgetary methods of the time. As mentioned earlier, budgetary practices at most levels of government in the United States were dominated by the legislative branch. As such, departmental budget estimates at the local level, in most states, and at the federal level were submitted directly to the legislative body. Seldom were supporting data included with estimates, and spending requests were not related to revenue projections or overall spending. At that time, a lack of standardization existed in accounting, and departments bargained with legislative appropriations committees directly or with local city council members. Little central oversight of departmental spending existed.

THE TRANSITION

PEARSON
myfirekit

Visit MyFireKit Chapter 7 for more information on budget reforms.

The early work of the 20th century led to a number of reforms and introduced several budget methodologies, which have provided the transition to the practices and procedures that are in use today. Through the past hundred years, numerous pieces of legislative reform have been enacted at all levels of government to promote ethical fiduciary responsibility.

Budget Reform in the 1990s

performance budgeting ■ Type of budget categorized by function or activity; each activity is funded based on projected performance; similar to program budgets, performance budgets, or outcome-based budgets.

In the 1990s, considerable attention had refocused upon **performance budgeting**, a type of budget categorized by function or activity. The concepts involved did not suddenly spring upon the scene, but rather became the object of renewed attention and interest as fiscal constraints persisted in many governmental entities. Several factors converged to shift attention to performance budgeting, wherein each activity is funded based on projected performance. One influence was the 1992 publication of *Reinventing Government: How the Entrepreneurial Spirit Is Transforming the Public Sector*, by David Osborne and Ted Gaebler.[3] They noted, among other things, something that was widely known—Americans are cynical about their government.

As one remedy, Osborne and Gaebler proposed a *results-oriented budget system* (although the term was not unique to them). The idea was to hold governments accountable for results rather than focus upon inputs, as traditional budgets and management did. Cost savings and entrepreneurial spirit would be rewarded. A long-term view would be facilitated in terms of strategy, costs, and planning for programs. Gaebler, a former city manager turned consultant, and Osborne, a writer and consultant, were exposed to a wide variety of governments; from this experience, they concentrated on examples they found of entrepreneurial management in government. A local government example they cited as a "performance leader" was Sunnyvale, California, with its focus on outcomes rather than inputs.

Interestingly, as is often the case, efforts to improve performance and increase the economy in the public sector had been underway before *Reinventing Government* was published. At the federal level, for example, the Chief Financial Officers Act of 1990 (P.L. 101-576) required the development and reporting of systematic measures of performance for 23 of the larger federal agencies. The Governmental Accounting Standards Board (GASB) had examined the use of *service efforts and accomplishments* (SEA) reporting for state and local government entities. Beginning in the early 1980s, it had encouraged governments to report not only financial data in budgets and financial reports but also information about service quality and outcomes.

FEDERAL REFORM EFFORTS IN THE 1990s

One thing that popular ideas, such as reinventing government, do is get the attention of the media and political leaders. Drawing on this attention, Congress passed the Government Performance and Results Act (GPRA) of 1993 (P.L. 103-62). This act required federal agencies to prepare strategic plans by 1997, to prepare annual performance plans starting with fiscal year 1999, and to submit an annual program performance report to the president and Congress comparing actual performance with their plans beginning in the year 2000.

STATE REFORM EFFORTS IN THE 1990s

State interest in the "new" performance budgeting of the 1990s is evident in a number of ways. The National Governors Association (NGA), for example, published *An Action Agenda to Redesign State Government* in 1993, which called for creating performance-based state government with measurable goals, such as benchmarks and performance measures (see Chapter 6), in order to move to performance budgeting. A year later, the National Conference of State Legislatures (NCSL) published a study entitled *The Performance Budget Revisited: A Report on State Budget Reform*. Together, these two studies generated interest in performance budgeting and measurement in a number of states. The rediscovery of, or renewed interest in, performance budgeting at the state and national levels appeared to be drawing on a continuing theme in American budgeting—the need to make resource allocation decisions based on more than the inputs used to carry out public programs. The national and state interest in the new performance budgeting draws upon a common source: the cynicism and loss of confidence in government in the United States.

LOCAL REFORM EFFORTS IN THE 1990s

Some local governments have responded by becoming more entrepreneurial in their approaches, such as Sunnyvale, California. As a council-manager city of about 130,000 people located south of San Francisco, Sunnyvale at the time was unique in its application of performance measurement and budgeting at the local government level. Its general plan looks 5 to 20 years into the future. The plan comprises seven elements and 20 sub-elements, which set goals and policies for the city. Its resource allocation plan is a 10-year budget to implement the general plan. Each year the annual budget is a performance budget targeting specific service objectives and **productivity** measures linked to the larger plan. Productivity is a measure of the performance of a worker or an operation system relative to resource utilization. Therefore, Sunnyvale's budget is a service-oriented document rather than the traditional line-item, input-oriented budget. In many ways, this city appears to be the embodiment of the contemporary interest in performance budgeting.[4]

productivity
■ A measure of the performance of a worker or an operation system relative to resource utilization: output divided by input.

PROSPECTS FOR BUDGETING IN THE 21ST CENTURY

Budgeting in the United States has experienced at least five phases: starting with control at the turn of the twentieth century, moving to management in the New Deal and post–World War II period, to planning in the 1960s, to prioritization in the 1970s and 1980s, to accountability in the 1990s. Budget reform appears to be alive and well in the United States. The federal government and many state governments, as well as local governments, are continuing to experiment with program and performance information. Professional organizations, such as the Government Finance Officers Association, continue to nurture change and advancement in budget presentation and financial reporting. Local governments are a focal point once more for budget innovation and change (see Table 7.1).

Over the course of the last century, many budget reform attempts have been made, with some more successful than others. Many of these reforms were driven by the need to have more control of the budget process to ensure ethical and appropriate activity, whereas others were promoted by a specific individual holding an office of influence.

Reform is often seen after the failure of a given budget process, prompting more regulatory oversight and reporting. Although many of these reforms are focused at the federal level, their repercussions always find their way into the

TABLE 7.1	Budget Reform Stages	
PERIOD	**BUDGET IDEA**	**EMPHASIS**
Early 1900s	Line-item budget Executive budget	Control
1950s	Performance budget	Management Economy and efficiency
1960s	Planning programming budgeting system	Planning Evaluation Effectiveness
1970s and 1980s	Zero base budgeting Tasked base budgeting Strategic planning	Planning Prioritization Budget deduction
1990s	New performance budget	Accountability Efficiency and economy management

operations of local government. The economic crisis experienced in the United States the past several years in the auto, housing, and financial sectors will undoubtedly promote a new generation of legislation reporting requirements and standards. This, too, will be felt at all levels of government.

Fiscal Management

As budgets are prepared at the local level, two factors have a great deal of impact on what happens at the local level—the Governmental Accounting Standards Board (GASB) and the generally accepted accounting principles (GAAP). Today, in local governments, there has been significant restructuring in respect to the general accounting practices and the regulations and guidelines applied at the local level by these two entities.

GASB

The **GASB** is the independent, private-sector organization that establishes the financial accounting and reporting standards for U.S. state and local governments. The GASB was established in 1984 and funded by publication sales, contributions from state and local governments, and voluntary assessment fees from municipal bond issues. The GASB is recognized by governments, the accounting industry, and the capital markets as the official source of generally accepted accounting principles for state and local governments.

> Accounting and financial reporting standards designed for the government environment are essential because governments are fundamentally different from for-profit businesses. Furthermore, the information needs of the users of government financial statements are different from the needs of the users of private company financial statements.
>
> The GASB is not a government entity; instead, it is an operating component of the FAF, which is a private sector not-for-profit entity. Funding for the GASB comes in part from sales of its own publications and in part from the municipal bond community. Its standards are not federal laws or regulations and the organization does not have enforcement authority. However, compliance with GASB's standards is enforced through the laws of some individual states and through the audit process, when auditors render opinions on the fairness of financial statement presentation in conformity with GAAP.[5]

On June 30, 1999, the GASB published comprehensive changes in state and local government financial reporting. GASB Statement No. 34, *Basic Financial Statements—and Management's Discussion and Analysis—for State and Local Governments,* provides a new focus of reporting public finance in the United States. Under this new standard, anyone with an interest in public finance—citizens, the media, bond raters, creditors, legislators, and others—will have more and easier-to-understand information available about his or her governments. Among the major innovations of Statement No. 34, governments are required to do the following:

- Report on the overall state of the government's financial health, not just its individual funds.
- Provide the most complete information available about the cost of delivering services to their citizens.

GASB
Governmental Accounting Standards Board is the independent, private-sector organization that establishes the financial accounting and reporting standards for U.S. state and local governments. The GASB was established in 1984 and is funded by publication sales, contributions from state and local governments, and voluntary assessment fees from municipal bond issues.

- Include for the first time information about the government's infrastructure assets, such as bridges, roads, and storm sewers.
- Prepare an introductory narrative section analyzing the government's financial performance.[6]

In June 2004, the Governmental Accounting Standards Board released GASB Statement No. 45, *Accounting and Financial Reporting by Employers for Postemployment Benefits Other than Pensions,* for retired public employees. GASB Statement No. 45 was intended to bring governmental accounting standards more in line with private company standards. This new accounting rule increases the amount of quality information included in government financial reports, particularly concerning the long-term costs of retiree health care and other retiree benefits.

It requires state and local governments to quantify the unfunded liabilities associated with retiree health benefits. Assessment results must be reported in governmental audits and updated regularly. Government financial statements then list an actuarially determined amount known as an annual required contribution.

In regard to health care, this contribution includes the normal costs—the amount to be set aside to fund future retiree health benefits earned in the current year—and unfunded liability costs—the amount needed to pay off existing unfunded retiree health liabilities over a period of no more than 30 years.[7]

GAAP

GAAP
■ Generally accepted accounting principles; a widely accepted set of rules, conventions, standards, and procedures for reporting financial information, as established by the Financial Accounting Standards Board.

GAAP is a widely accepted set of rules, conventions, standards, and procedures for reporting financial information, as established by the Financial Accounting Standards Board. GAAP can be principle-based on specific technical requirements, and most industries in the United States are expected to follow GAAP principles. Although different organizations contribute to GAAP, the Financial Accounting Standards Board (FASB) is its main contributor.

Generally accepted accounting principles are varied but based on a few basic principles that must be upheld by all GAAP rules. These principles include consistency, relevance, reliability, and comparability.

- Consistency means all information should be gathered and presented the same across all periods.
- Relevance means simply that the information presented in financial statements (and other public statements) should be appropriate and assist a person evaluating the statements to make educated guesses regarding the future financial state of the organization.
- Reliability means simply that the information presented in financial statements is reliable and verifiable by an independent party. Basically, an organization must confirm that if an independent auditor were to base reports off of the same information that the auditor would come up with the same results. Following the generally acceptable accounting principles also means the organization is representing a clear picture of what really happened (and is happening) with the organization.
- Comparability is one of the most important GAAP categories and one of the main reasons having something similar to GAAP is necessary. By ensuring comparability, an organization's financial statements and other documentation can be compared to similar businesses within its industry.

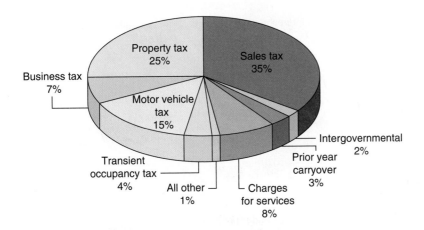

FIGURE 7.1
Municipal Revenue
Sources

Generally accepted accounting principles ensure that all organizations are on a level playing field; and the information they present is consistent, relevant, reliable, and comparable.[8]

LOCAL FISCAL MANAGEMENT

Regardless of the type of budget process an organization uses, there must always be a balance between the revenues brought in annually and the expenses articulated in the form of a budget document. Revenue streams will be different among cities, counties, fire districts, and volunteer fire agencies. Each can derive its revenue from different sources. The examples in Figures 7.1 and 7.2 show the breakdown of revenue and expenditures for a municipal fire department.

Cities often have a greater number of multiple revenue streams, such as property tax, sales tax, local impact fees, and fees for service, which help to create a larger pool of revenue from which to draw. However, at the same time, most cities have a much more diverse service delivery than that of special districts and/or volunteer fire departments. For special districts, the revenue is typically generated from property tax. In certain cases this is referred to as ad valorem tax, depending on the state. In many instances, capital expenditures are done with bond proceeds that are approved by the voters for specific capital projects. To pay the bonds, revenue is largely generated primarily from taxes on properties or parcels within the jurisdiction (see Figures 7.3 and 7.4).

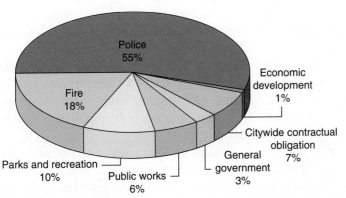

FIGURE 7.2
Municipal Expenditures

FIGURE 7.3 District
Revenues

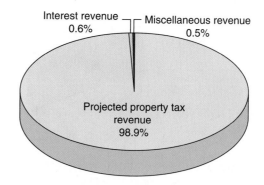

FIGURE 7.4 District
Expenditures

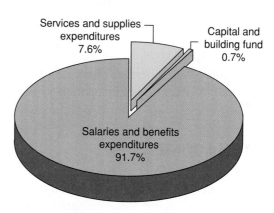

Volunteer organizations often derive their money from a variety of sources including general fund support from local governments. A common revenue stream is a specific tax for the volunteer department based upon a percentage of the property tax. In many cases, volunteer departments are very effective in raising their own money to support their efforts. Two factors are of importance in basic fiscal management: the revenue–expense ratio of an organization and understanding the various revenue streams that fund the budget.

Alternative Funding Mechanisms

Stimulated by local budget pressures, fire and emergency medical service departments in the United States are using a wide array of fund-raising approaches beyond traditional methods.

Each agency providing fire-related or emergency medical services must consider the range of opportunities and the local and state constraints in shaping its funding policy and determining its budget. Funding influences the amount of emergency resources available and the scope of prevention activities, which in turn influence life or death issues. The funding issue is not one to be left solely to accountants and finance officers; as a vital public policy issue, it can literally mean life or death in the community.

The choice of funding approaches also raises fundamental questions about governance and equity. Should only the users of local government services pay the burden of the services provided to them? Should fire protection and EMS be treated like insurance, in which the fees charged are a function of the risks presented? Should fire protection and EMS be considered services provided to people in need, with the costs spread evenly across society, with no "user charges"?

LOCAL GOVERNMENT FUNDING MECHANISMS

The wide variety of local funding approaches found in practice—and it is truly a wide variety—can be grouped into several major categories.

Taxes

These include not only general property taxes, local income taxes, and general sales taxes used to fund most local services but also transient taxes and other taxes earmarked specifically for fire and EMS services.

Borrowing

In addition to the familiar bonds used for purchasing capital equipment and facilities are *certificates of participation* (COPs). They operate much like home mortgages and are used to purchase equipment and facilities when the local jurisdiction is not allowed to use the more conventional debt instruments.

Leasing

Another way to avoid large capital outlay, especially for apparatus, is a leasing arrangement that often has a right to purchase at the end.

Benefit Assessment Charges

Administered somewhat like property taxes, these charges factor in not only size and type of property but also "benefits" from being close to fire stations, having reduced insurance, having special services available, and so on. These charges are a way to get around property tax limitations and can also improve the equity of charges for fire protection.

Fees

These range from small revenue producers such as fees for permits for new construction, special events, and operating hazardous functions, to fees for inspections and violations of codes, to fees for special services for which charges were not made in the past, such as pumping water out of a basement or rescuing a boater or hiker, to fees for virtually anything a fire department does. Perhaps the most lucrative new category of fees is for emergency medical service transport and emergency medical care.

Strategic Alliances

Fire departments are forming alliances within other agencies to provide the following: all the fire protection and EMS care in neighboring jurisdictions; selected services for parts of jurisdictions; and special services such as training, hazardous materials responses, and heavy rescue either under an annual contract or with a fee per usage or per student. These arrangements are mutually beneficial to the departments or agencies involved.

Cost Sharing and Consolidation

Many fire and EMS departments are joining together to pay for new facilities or services to reduce the burden on each department, especially where the facilities or services are not frequently used. Also, some fire and EMS departments are uniting to form single departments to save costs and improve service delivery.

Fines and Citations

Fire and EMS departments are charging fees for negligent fire, actions inconsistent with the law, and failure to comply with codes.

Sales of Assets and Services

Some fire and EMS agencies sell used equipment or services to produce much needed revenue.

Subscriptions

Most commonly used for emergency medical services, this is essentially a form of insurance in which a household pays a fixed fee per year, such as $35, and then does not have to pay anything additional for emergency medical service or transport it uses during the year. A major variation is where the subscription prevents any out-of-pocket expenditures beyond the subscription fee, but the local agency is free to charge the subscriber for fees that can be recovered from medical insurance or homeowners insurance. With either method, those not subscribing pay the full amount charged.

Impact Development Fees

In many communities new developments can be required to pay for the impact they have on capital purchases such as new fire stations and their full complement of equipment. This is accomplished by imposing an impact fee on each home or commercial property built. This revenue provides a portion of the money needed to build new facilities required as a result of new development. Thereafter, the provision of services is paid the same as for existing development.[9]

FUND-RAISING

In addition to using many of the sources just described, volunteer or combination fire and emergency services agencies have raised funds from the following:

- Food sales (including open grills, formal dinners, bake sales, pancake breakfasts, and barbecues)
- Entertainment events (such as dances, amusement park outings, carnivals, rodeos)
- Sports events (including turkey shoots, donkey baseball, softball, fishing, golf)
- Gambling (where legal, including casino nights, bingo, horse races)
- Raffles
- Door-to-door solicitation
- Direct-mail solicitation
- Public service announcements soliciting funds
- Selling space in annual reports
- Donations of services or money from industry

- Sale of honorary memberships
- Training and community education
- Sales of goods (such as logo-embossed clothing, calendars, antiques, beverage insulators)
- Sales of services (car washes, pet baths)
- Shared profits with private vendors of commercial services (such as photographs) or commercial goods[10]

Cost Impacts on the U.S. Fire Service Revenue and Budgets

A significant issue that has challenged the fire service since its inception has been attaining adequate resources to provide all of its services effectively. This is not meant to be an indictment of local governments in respect to the importance they place on the fire service. There are many causes of inadequate resources for the smallest of communities, which have a very limited resource base, to communities that have grown so rapidly they are continually behind the curve in maintaining performance to service demands. Whereas larger departments experience a more defined revenue stream, smaller departments often struggle to maintain even the basic level of fire protection with the available revenue stream. Whereas the majority of the U.S. population is served by departments that cover population densities over 25,000, the majority of departments and firefighters—primarily volunteer—are serving over 20,000 communities across the United States. The following table provides an overview of the number of firefighters, volunteer and career, and the populations (communities) they protect (see Table 7.2).

For many small communities, the challenges to obtain adequate revenue can be significant.

- Most of the revenues for all-volunteer or mostly volunteer fire departments comes from taxes, either a special fire district tax or some other tax, including an average of 64 to 68 percent of revenue covered for communities of less than 5,000 population.
- Other governmental payments, including reimbursements on a per-call basis, other local government payments, and state government payments, contributed an average of 11 to 13 percent of revenues for communities of less than 5,000 population.
- Fund-raising contributed an average of 19 percent of revenues for communities of less than 2,500 population.
- Used vehicles accounted for an average of 40 percent of apparatus purchased by or donated to departments protecting communities with less than 2,500 population.
- Converted vehicles accounted for an average of 14 percent of apparatus used by departments protecting communities of less than 2,500 population.[11]

If you work in a substantially career department, apply the above percentages to your own budget or operation. This provides some context to what many of the smaller departments face. For substantially career departments, the impact of increased benefit expenses has taxed many communities to maintain the level of

TABLE 7.2	Number of Career, Volunteer, and Total Firefighters by Size of Community		
POPULATION PROTECTED	CAREER FIREFIGHTERS	VOLUNTEER FIREFIGHTERS	TOTAL FIREFIGHTERS
1,000,000 or more	30,700	800	31,500
500,000–999,999	31,780	4,150	35,850
250,000–499,999	21,200	5,450	26,650
100,000–249,999	45,800	4,500	50,300
50,000–99,999	43,450	7,150	50,600
25,000–49,999	44,850	28,000	72,850
10,000–24,999	48,160	83,900	132,050
5,000–9,999	14,400	119,100	133,500
2,500–4,999	6,100	155,750	161,850
Under 2,500	7,750	398,350	406,100
Total	294,200	807,160	1,101,250

Source: U.S. Fire Administration, *Four Years Later—A Second Needs Assessment of the U.S. Fire Service*, A Cooperative Study Authorized by U.S. Public Law 108-767, Title XXXVI, FA-303/October 2006, p. iv.

needed resources to community growth. The percentage of benefit cost to base salary has continued to climb upward to over 40 percent for some organizations. What this translates to is an employee whose base salary is $50,000; the benefit cost for this employee is $20,000. The rise in benefit cost has had a tremendous financial impact on many jurisdictions throughout the United States, and the future projection is this cost will only continue to rise. Two of the main contributing factors are health care and pension costs; and many financial experts are predicting current practices and levels of benefits cannot be sustained in the future.

HEALTH CARE COST

By several measures, health care spending continues to rise at a rapid rate, forcing businesses and families to cut back on operating and household expenses, respectively. In 2008, total national health expenditures were expected to rise 6.9 percent—two times the rate of inflation. Total health care spending was $2.4 trillion in 2007, or $7,900 per person, which represented 17 percent of the gross domestic product (GDP).[12]

Health insurance premiums have increased rapidly in the recent past, growing a cumulative 78 percent between 2001 and 2007 and far outpacing cumulative wage growth of 19 percent over the same period.[13] These figures demonstrate the growing burden of health insurance costs on employers and employees, illustrate overall trends in health benefit costs, but do not show how this growing burden is affecting employers and employees in different settings.

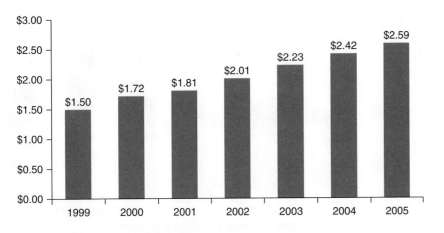

FIGURE 7.5 Mean Health Insurance Costs per Worker Hour for Employees with Access to Coverage, 1999–2005 *(Source: Kaiser Family Foundation calculations based on data from the National Compensation Survey, 1999–2005, conducted by the Bureau of Labor Statistics)*

Employer Costs for Health Insurance

Among workers with access to health benefits, average employer costs for health insurance per employee hour rose from $1.60 to $2.59 during the 1999 to 2005 period (Figure 7.5). This almost 62 percent increase in average costs per hour is much larger than the 23 percent increase in average employer payroll costs per hour for these workers during that same time frame.

Employer costs for health insurance viewed as a percent of payroll also showed significant variation. See Figure 7.6 for the average amount workers contribute to health insurance.

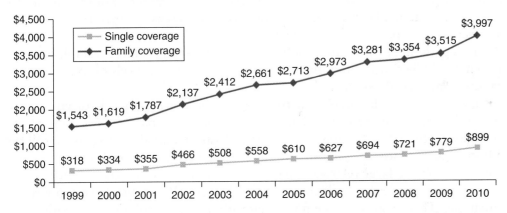

FIGURE 7.6 Average Amount Workers Contribute to Health Insurance, 1999–2010 *(Source: Kaiser/HRET Survey of Employer-Sponsored Health Benefits, 1999–2010)*

See Figure 7.7 for the percent of workers with health costs greater than 10 percent of payroll.

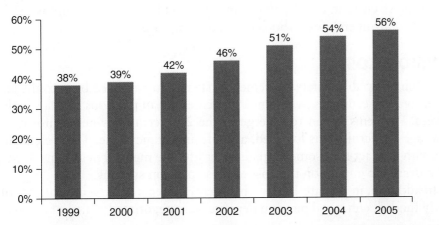

FIGURE 7.7 Percentage of Workers with Health Costs Greater than 10 Percent of Payroll for Employees with Access to Coverage, 1999–2005 *(Source: Kaiser Family Foundation calculations based on data from the National Compensation Survey, 1999–2005, conducted by the Bureau of Labor Statistics)*

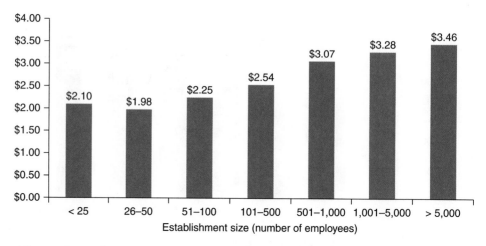

FIGURE 7.8 Average Hourly Health Insurance Costs for Employees with Access to Coverage, by Establishment Size, 2005 *(Source: Kaiser Family Foundation calculations based on data from the National Compensation Survey, 1999–2005, conducted by the Bureau of Labor Statistics)*

These charts have important implications for policy options that would require employers to devote a minimum percentage of payroll to health benefits. The wide variation in employer costs as a percentage of payroll means there is no typical or common percentage of payroll for health care that relates to most jobs. This suggests it would be quite difficult to develop a percentage of payroll requirement, which would either maintain the status quo for employers who do offer coverage or put employers who offer and do not offer on an equal basis with respect to health benefits.[14] The steady increase over time throughout the distribution suggests that a static percentage of payroll requirement would become less connected to actual health insurance costs as they increase over time and become a larger share of payroll (see Figure 7.8).

This pattern may indicate benefits are more generous in larger establishments, which would be consistent with the overall higher average levels of total compensation in large establishments.[15]

For service and laborer/cleaner/helper occupations, the levels of costs were relatively, but comparatively, high as a percentage of payroll. Looking at the change over the 1999 to 2005 period, health costs increased relative to payroll for all occupations. However, the size of this change was largest for those with health costs, which were large relative to payroll, such as in transportation, machine operators, laborers/cleaners/helpers, service, and clerical/administrative support occupations. There appears to be no agreement on a single solution to health care's high price tag. Many approaches may be used to control costs. What is known is if the rate of escalation in health care spending and health insurance premiums continues at current trends, the rising costs will affect the employers' bottom lines, the consumers' pocketbooks, and the employees.

PENSION COST

The origins of public safety retirement systems can be traced back to the late 19th and early 20th centuries, when most big cities began pensions for their public employees. Problems began to emerge in the 20th century when uniformed forces grew, age of recruits was lowered, and life expectancy rose. These elements, coupled with the fact that some systems offer lifetime medical health insurance, have had a dramatic impact on the cost of these pension systems.

In addition, in the late 1990s when many systems had the appearance of being overly funded, enhanced pension benefits were negotiated in lieu of normal salary increases. For example, in California, many public safety pensions now provide 3 percent at 50, with a cap of 90 percent of base salary. However, with cash-outs of sick leave and vacation, the retirement salary for some can exceed what the employee was being paid to come to work. We have all been witness to what has transpired since that time. Circumstances such as the dot-com bubble bursting, September 11, and most recently the housing foreclosure crisis and subsequent economic meltdown have severely impacted these funds negatively.

One of the first cities to experience this pension benefit crunch was the City of San Diego. San Diego's situation may be extreme, but many other cities also feel the burden of soaring pension costs. In San Diego's case in the 1990s, the city increased benefits, underfunded the system, and then experienced the turbulence of the financial market collapse, which has led to a substantial burden on the city to fund the pension obligation. A research brief on the financial condition of American cities by the National League of Cities in October 2010 found that leading the list of factors that finance officers say have increased over the seven years were employee benefit cost (83 percent) and pension cost (80 percent). The economic downturn and stock market nosedive, which cut income from pension-fund investments, forced cities to cover these gaps. Many resorted to either heavy borrowing through the use of pubic safety obligation bonds or contributing a greater percentage of their general fund to cover the debt.

Governments on the Rocks

Unfortunately, this is not a new phenomenon. In the past other local governments have faced big financial problems:

- Vallejo, California, in 2008, filed a case seeking bankruptcy protection and the adjustment of its departments under Chapter 9 of the United States bankruptcy code.
- Orange County, California, in 1994, declared bankruptcy when risky investments in the bond market triggered $1.7 billion in losses. The debt remaining today, which the county is paying off, stands at about half that account.
- Bridgeport, Connecticut, in 1991, filed for bankruptcy because of budget deficits, deteriorating roads and facilities, and a declining tax base. A bankruptcy judge dismissed the claim, saying the city had enough money to pay its bills.
- New York City, in 1975, flirted with bankruptcy when it ran out of money to pay its bills. President Ford initially rebuffed a plea for federal backing for new loans, promoting this front-page headline in the New York Daily News: FORD TO CITY: DROP DEAD. The federal government ultimately guaranteed loans to the city.[16]

The impact of pension cost and other post-employment benefits (OPEB) is quickly becoming a significant financial issue for state and local governments. Today, state and local government workers account for approximately 12 percent of the nation's workforce. The passage of tax initiatives, such as California's Proposition 13, and other such movements in many other states have severely impacted the dollars available to provide services and to address the rising cost of related benefits such as pension funding and other postretirement benefits.

Meanwhile, demographics and actuarial assumptions have changed, and there has been increased attention on the ability of state and local governments to pay the accrued costs of benefits for the expanding number of government employees.

In the United States, the unfunded pension liability of state and local governments could approach $700 billion to $1 trillion over the next 10 years. The cost of unfunded health benefits promised to retirees could push the number even higher.

There is little doubt public pension funds, along with private pension funds and 401K/individual retirement accounts (IRAs), shared the beating most investors suffered in the fall of 2008 market collapse. According to the Center for Retirement Research (CRR) at Boston College, the value of all retirement accounts fell $4 trillion between October 9, 2007, and October 9, 2008. Of this amount, about $1 trillion was felt by public pension plans. Another $1 trillion was lost in private pension plans, and individuals lost the other $2 trillion in their 401Ks and IRAs. Even by mid-year 2009, equities, as measured by the broadly based Wilshire 5000 index, remain 40 percent below the peak they reached on October 9, 2007.[17]

Defined benefit plans promise a pension benefit to participants, based on a formula usually considering length of service and final pay. Because the benefit is fixed, the burden of the investment losses falls to the public employer, which contributes dollars, or "funding," each year based on calculations of what will be needed to pay the benefit. In some cases, employees also contribute from their own pay, which may be an agency policy or part of a negotiated contract.

According to CRR, the funding level in 2007 for a sampling of 120 state and local plans was at 87 percent of the promised benefit. After the market collapse, CRR estimates the funding ratio would have declined to 65 percent. But, because of smoothing, the actual impact would probably fall between 59 percent and 75 percent funding, depending on future market performance. The U.S. Government Accountability Office considers 80 percent funding acceptable for public plans.[18]

Pension Plans Roll with the Financial Punches

The bad news is pension funds have suffered some deep losses in the recent market turmoil. As a result, both the employee and the employer should be prepared for an increase in required contributions to keep the plans solvent (see Figure 7.9).

The aggregate market value of state and local government pension funds declined from $3.2 trillion in 2007 to about $2.3 trillion in October 2008, a 28 percent loss, according to an issue brief published in December by the National Association of State Retirement Administrators (NASRA), based in Baton Rouge, Louisiana. However, in the near term, only a portion of those losses will be reported because of accounting procedures, commonly called the smoothing process, used by most pension funds to report gains and losses over long-term investments.

Although the smoothing process may delay the consequences of pension funds' recent investment losses, fund participants should begin preparing for those consequences. Most likely, state and local governments and their employees will have to increase their contributions to the plans in the next two to three years. The amount of the increase will depend on a specific pension fund's condition before the decline and other factors.

It is hoped local and state economies will recover enough over the next two years to bear the extra costs. It is possible, during this time frame, that markets will begin to reverse (and mitigate the losses).[19]

This is not a new problem. Historically, pension systems at the state and local levels have at various times been underfunded for most of the last 50 years. The average funding ratio has both grown and declined over time (see Table 7.3).

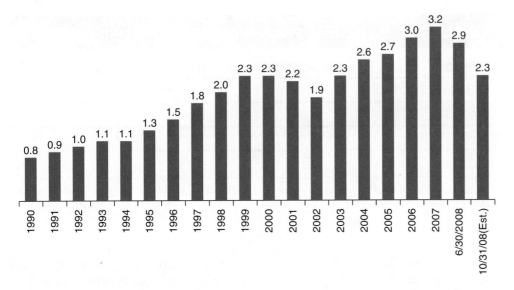

FIGURE 7.9
Aggregate Market Value of State and Local Government Pension Funds, 1990–2008, in $Trillions *(Source: U.S. Federal Reserve, Quarterly Flow of Funds Report, 2008)*

Today, many factors drive this:

- Historically, extraordinary personnel growth plus political pressure contributed to the rise of pension liabilities.
- The up market for investments in the late 1990s and from 2003 to 2005 has helped investor return and narrowed the underfunding gap.
- There are implicit obstacles to solving pension liabilities.
- There is political pressure to increase pension benefits when current salaries are limited by restricted revenues.
- State and local legislatures listen and respond to employee unions and increase benefits without providing corresponding sources of funding.
- The ever-increasing demand for infrastructure improvements and expanded public safety services has more than strained state and local budgets.
- Pension obligation bonds (POB) have masked the real systemic problem that needs to be addressed and have been a Band-Aid and short-term fix for significant budget loopholes and the consistent current underfunding of pension obligations.
- Defined benefit (DB) plans, as compared to defined contribution (DC) plans, are difficult to maintain financially as benefits promised cannot be easily provided, especially given the revenue restraints facing states and cities.
- The transition to a DC plan is less volatile and more predictable from a cost perspective and, if funded appropriately, should provide an acceptable level of retirement.
- The transition to a DC plan from a DB plan is costly and complicated.
- Expectations of government employees and unions are high and not easily changed, and efforts to increase employees' contributions are not well received.
- Many state constitutions protect pension benefits from being changed retroactively and some prospectively.
- In the absence of state constitutional provisions, certain states have adopted legislation to prohibit diminishing or impairing public employee pension rights.[20]

TABLE 7.3	Funding Percentage of Total Pension Liability

PERIOD	FUNDING PERCENTAGE OF TOTAL PENSION LIABILITY
Mid-1970s	50%
1990	80%
2000	100%
2003	77%

Source: Standard & Poor's Research, *Managing State Pension Liabilities: A Growing Credit Concern*, January 2005. Retrieved October 6, 2010, from http://www.nasra.org/resources/sandp0501.pdf

From a financial management perspective, one can anticipate an increased focus on total compensation analysis in negotiations of labor contracts as well as financial forecasting to determine future financial exposure. Although some departments have used this methodology in the past, more often than not, they have used an approach of looking at the cost over the period of a particular contract. However, as contracts are negotiated over several periods of time, the cumulative financial impacts must be fully understood. Over time these individual actions can overlay on one another and cause unintentional financial consequences in respect to pension, benefits, and other employee costs.

When doing total compensation review, it is important to know and understand the key terms used:

- *Payroll costs* include employer payments for wages, salary, overtime, vacation, holiday, sick days, bonus, and other cash compensation to employees, excluding severance payments and unemployment benefits.
- *Health costs* include all employer payments for health coverage, excluding employee contributions to premiums or out-of-pocket medical cost sharing.
- *Non-health fringe benefits* include employer payments for life and short-term disability insurance, defined benefit and defined contribution plans, workers' compensation, Social Security, and Medicare.
- *Total compensation* is defined as the sum of payroll and all fringe benefit costs including health.
- *Establishment of organization size for comparative purposes* is the number of employees at a department or agency.

PEARSON
myfirekit™

Visit MyFireKit Chapter 7 for Comparative Cost—Top 25 Cities.

An example of this type of analysis is a report that the Bureau of Labor Statistics publishes by industry group. The analysis breaks down the total compensation by wages and salaries, paid leaves, supplemental pay, insurance cost, retirement and savings, and legally required benefits (see Table 7.4).

The State of Fire Service Infrastructure

While health and pension issues are of long-term concern, so too is the matter of fire service infrastructure, including facilities, apparatus, and equipment. If you have the opportunity to travel in your state, region, or nationally, you can

TABLE 7.4 State and Local Government, by Major Occupational and Industry Group

Employer costs per hour worked for employee compensation and costs as a percent of total compensation: state and local government workers, by major occupational and industry group, September 2008.

| | OCCUPATIONAL GROUP (1) | | | | | | | | INDUSTRY GROUP | |
| | ALL WORKERS | | MANAGEMENT, PROFESSIONAL, AND RELATED | | SALES AND OFFICE | | SERVICE | | SERVICE-PROVIDING (2) | |
COMPENSATION COMPONENT	COST	PERCENT	COST	PERCENT	COST	PERCENT	COST	PERCENT	COST	PERCENT
Total Compensation	$39.18	100.0	$48.21	100.0	$26.69	100.0	$28.56	100.0	$39.24	100.0
Wages and Salaries	25.77	65.8	32.80	68.0	16.49	61.8	17.30	60.6	23.82	65.8
Total Benefits	13.41	34.2	15.41	32.0	10.21	38.2	11.26	39.4	13.42	34.2
Paid Leave	3.25	8.3	3.85	8.0	2.42	9.1	2.56	9.0	3.25	8.3
Vacation	1.15	2.9	1.19	2.5	1.08	4.0	1.11	3.9	1.15	2.9
Holiday	1.07	2.7	1.28	2.7	.77	2.9	.83	2.9	1.07	2.7
Sick	.81	2.1	1.07	2.2	.47	1.8	.50	1.8	.81	2.1
Personal	.22	.6	.32	.7	.10	.4	.12	.4	.22	.6
Supplemental Pay	.35	.9	.27	.6	.21	.8	.56	2.0	.35	.9
Overtime and premium (3)	.18	.4	.08	.2	.12	.4	.36	1.2	.17	.4
Shift differentials	.05	.1	.03	.1	.02	.1	.09	.3	.05	.1
Nonproduction bonuses	.13	.3	.16	.3	.08	.3	.12	.4	.13	.3
Insurance	4.39	11.2	4.89	10.1	3.97	14.9	3.55	12.4	4.40	11.2
Life	.10	.2	.14	.3	.05	.2	.05	.2	.10	.2
Health	4.21	10.8	4.65	9.7	3.86	14.4	3.45	12.1	4.22	10.8
Short-term disability	.03	.1	.03	.1	.02	.1	.02	.1	.03	.1
Long-term disability	.06	.1	.07	.2	.04	.2	.03	.1	.06	.1
Retirement and Savings	3.09	7.9	3.69	7.7	1.92	7.2	2.71	9.5	3.09	7.9
Defined benefit	2.75	7.0	3.27	6.8	1.68	6.3	2.47	8.6	2.75	7.0
Defined contribution	.34	.9	.42	.9	.23	.9	.25	.9	.34	.9

(continued)

TABLE 7.4 State and Local Government, by Major Occupational and Industry Group (*continued*)

Legally Required Benefits										
	2.33	5.9	2.71	5.6	1.68	6.3	1.87	6.6	2.33	5.9
Social Security and Medicare	1.82	4.6	2.24	4.7	1.30	4.9	1.26	4.4	1.82	4.6
Social Security (4)	1.40	3.6	1.72	3.6	1.03	3.9	.97	3.4	1.40	3.6
Medicare	.41	1.1	.52	1.1	.27	1.0	.29	1.0	.42	1.1
Federal Unemployment Insurance	(5)	(6)	(5)	(6)	(5)	(6)	(5)	(6)	(5)	(6)
State Unemployment Insurance	.06	.2	.06	.1	.06	.2	.07	.2	.06	.2
Workers' Comp	.45	1.1	.40	.8	.33	.12	.55	1.9	.44	1.1

(1) This table presents data for the three major occupational groups in state and local government: management, professional, and related occupations, including teachers; sales and office occupations, including clerical workers; and service occupations, including police officers and firefighters.
(2) Service-providing industries, which include health and educational services, employ a large part of the state and local government workforce.
(3) Includes premium pay for work in addition to the regular work schedule (such as overtime, weekends, and holidays).
(4) Comprises the Old-Age, Survivors, and Disability Insurance (OASDI) program.
(5) Cost per hour worked is $0.01 or less.
(6) Less than .05 percent.

Source: Bureau of Labor Statistics, *Economic News Release*, Table 3. Retrieved October 31, 2010, from www.bls.gov/news.release/ecec.t03.htm

witness everything from the Taj Mahal fire station with top-of-the-line equipment to a pole barn building with a 40-year-old apparatus and hand-me-down equipment. This snapshot of fire service infrastructure is systematic of several issues: lack of investment, lack of effective strategic planning, lack of adequate replacement and maintenance schedules, and ineffective deployment modeling in the past.

The ability to provide adequate service depends directly on having the proper number of facilities located appropriately and quality equipment to safely do the job. For many departments, these are not the circumstances. As leadership into the 21st century is considered, one of the most critical and fundamental concerns is having adequate resources to do the job effectively. This requires excellent fiduciary stewardship, which is based upon an effective strategic planning process for many fire departments. The historic challenge has been insufficient resources, with no game plan to correct the situation, resulting in many organizations operating with too few stations, outdated facilities, and either insufficient equipment or equipment past its useful life. In 2006, the U.S. Fire Administration, in cooperation with NFPA, released a report entitled, *Four Years Later—A Second Needs Assessment of the U.S. Fire Service,* which examined facilities, apparatus, and equipment.

PEARSON
myfirekit

Visit MyFireKit Chapter 7, for Station Placement Methodology.

FACILITIES

One of the analyses provided in the 2006 report was the number of the fire stations that were over 40 years old, had backup power, and had a vehicle exhaust system (see Figure 7.10 and Table 7.5). The latter two would indicate a degree of reinvestment into the facility.

The results of the survey were broken down further to include specific characteristics based upon community size (see Table 7.6). This survey identifies a critical need to focus on the adequate maintenance and renewal of fixed-site infrastructure.

FIGURE 7.10 Fire Station (*Courtesy of Fresno Fire Department*)

TABLE 7.5	Total Number of Fire Stations with Indicated Characteristics in Communities of This Population Size		
POPULATION PROTECTED	OVER 40 YEARS OLD	NO BACKUP POWER	NOT EQUIPPED FOR EXHAUST EMISSION CONTROL
1,000,000 or more	485	578	10
500,000–999,999	462	466	566
250,000–499,999	340	375	573
100,000–249,999	724	667	877
50,000–99,999	713	568	829
25,000–49,999	1,102	1,037	1,633
10,000–24,999	2,281	2,056	3,571
5,000–9,999	2,157	2,721	4,376
2,500–4,999	2,490	4,141	6,029
Under 2,500	6,524	13,388	16,481
Total	17,279	25,999	34,944
Percent of U.S. Total	36%	54%	72%

Source: U.S. Fire Administration, *Four Years Later—A Second Needs Assessment of the U.S. Fire Service*, A Cooperative Study Authorized by U.S. Public Law 108-767, Title XXXVI, FA-303/October 2006, p. 59.

PLANNING FOR FACILITY RENEWAL

Two critical issues facing many organizations today are the current state of their facility infrastructure and the lack of a comprehensive maintenance plan. Today, many agencies find themselves competing for capital dollars to fix existing structures, versus the building of new facilities such as pools, libraries, and community centers. This is often a difficult choice for many communities. The development of a capital renewal plan to keep existing infrastructure in top condition is a critical element to the provision of services as is offering employees a healthy working environment.

WHY DOES FACILITIES MAINTENANCE MATTER?

As America's public safety facilities age, we face the growing challenge of maintaining facilities at a level that enables firefighters to meet the demands of the 21st century as well as current health and safety codes and standards. Although the construction of new facilities supports these tasks, many older buildings have developed modularly over time (see Figure 7.11). A 1920s-era fire station may have received an addition in 1950, which in turn received an addition in 1970, and yet another in 1990. An initial well-thought-out plan of its layout and future use was never done.

TABLE 7.6	Number of Fire Stations and Selected Characteristics by Community Size			
POPULATION OF COMMUNITY	**AVERAGE NUMBER OF STATIONS**	**PERCENT OF STATIONS OVER 40 YEARS OLD**	**PERCENT OF STATIONS HAVING BACKUP POWER**	**PERCENT OF STATIONS EQUIPPED FOR EXHAUST CONTROL**
1,000,000 or more	75.83	42.6	49.2	99.1
500,000–999,999	35.89	33.9	65.8	58.5
250,000–499,999	20.20	31.2	65.6	47.5
100,000–249,999	10.83	30.8	71.6	62.7
50,000–99,999	5.49	29.0	76.9	66.3
25,000–49,999	3.39	30.3	71.5	55.1
10,000–24,999	2.07	37.5	66.2	41.3
5,000–9,999	1.50	38.2	51.8	22.5
2,500–4,999	1.39	36.8	38.8	10.9
Under 2,500	1.30	36.5	25.1	7.8
Total	1.78	36.5	38.9	17.5

Source: U.S. Fire Administration, *Four Years Later—A Second Needs Assessment of the U.S. Fire Service*, A Cooperative Study Authorized by U.S. Public Law 108-767, Title XXXVI, FA-303/October 2006, p. 73.

FIGURE 7.11 Fire Station Under Renovation *(Courtesy of Steve Derenia)*

The task of caring for these old buildings, many of which are historically or architecturally significant, at a level that supports contemporary response practices is a significant challenge. At the same time, maintaining the workings of new, more technologically advanced facilities also demands considerable expertise and commitment.

PAY ME NOW OR PAY ME LATER

You may remember the TV commercial for car oil filters, with the tag line, "You can pay me now or pay me later." The ad's underlying message was clear: if you spend a few dollars now to change your car's filter, you could avoid more expensive repairs in the future. This is "preventive maintenance" in its simplest form—spending a little money now to perform regular inspections and maintenance in order to minimize future big-ticket costs. The same philosophy holds true if the department desires to prolong the functional lifetime of its buildings and equipment.

Because routine and unexpected maintenance demands are bound to arise, every organization must proactively develop and implement a plan for dealing with these inevitabilities. Thus, an organization must be ready to meet the challenges of effective facilities maintenance. It is simply too big of a job to be addressed in a haphazard fashion. After all, the consequences may affect the department's ability to respond, the health of its personnel, the ability for the building to support day-to-day operations, and the organization's long-range fiscal outlook.

A sound facilities maintenance plan serves as evidence that facilities are, and will be, cared for appropriately. On the other hand, neglected facility maintenance planning can cause real problems. Large capital investments are squandered when buildings and equipment deteriorate to where the repair is structural rather than cosmetic or warranties become invalidated. Failing to maintain public safety facilities adequately also discourages future public investment for new stations in the local fire protection system. Why would a community spend money on a new facility when it has not taken care of the plant it already has?

However, facilities maintenance is concerned about more than just resource management. It is also about providing clean and safe environments for personnel and the public.

The goal is to create a comprehensive process, one that identifies and quantifies areas requiring the necessary actions and costs to renovate, retrofit, restore, modernize, or maintain an "existing building" to a "like new" condition. To accomplish this, first a set of program needs and objectives should be established to provide a framework for the process. Many organizations do not even have a master property schedule, which is the first step in the process. This entails determining program needs and program objectives.

Program Needs

- Organize a facilities audit procedure.
- Set up a schedule to accomplish the audit.
- Identify the buildings to be evaluated.
- Select project manager and/or team leader.
- Determine the type of reports to be generated.

Program Objectives

- Identify deficiencies and provide a written condition status of each building.
- Provide a building evaluation report to assist administrators and managers in their long-range planning and budgeting activities.
- Provide the data that will aid in prioritizing building renewal and deferred maintenance projects.

WHAT IS FACILITIES AUDIT?

A **facilities audit** program is a continuous, systematic approach of identifying, assessing, prioritizing, and maintaining the specific maintenance and repair requirements for all facility assets to provide valid documentation, reporting mechanisms, and budgetary information in a detailed database to address facility concerns. Such an audit contains several maintenance and usage elements, including preventive maintenance, planned maintenance, capital renewal, and facilities adaptation.

Preventive Maintenance

Preventive maintenance refers to the day-to-day efforts to control deterioration of facilities as funded by the annual operating budget.

- Periodic scheduled work (preventive maintenance) that has been planned to provide adjustment, cleaning, minor repair, and routine inspection of equipment to reduce service interruptions
- Call-in requests for service on an as-needed basis
- Scheduled repetitive activities; for example, housekeeping, groundskeeping, site maintenance, and certain types of general or preventive service contracts

Planned Maintenance

Accumulated deferred maintenance (catch-up maintenance/controlled maintenance) refers to expenditures for repairs not accomplished as part of normal maintenance, which have accumulated to the point facility deterioration is evident and could impair the proper functioning of the facility. This backlog is usually caused by either a perceived lower priority status than those funded within available funding, or the lack of funds all together. Costs estimated for the deferred maintenance projects should include code compliance even if such compliance requires expenditures additional to those essential to effect needed repairs.

Capital Renewal

A systematic management process needs to be in place to plan and budget for known future cyclical repair and replacement requirements, extending the life and retaining the usable condition of facilities and systems. These repairs and replacements are not normally contained in the annual operating budget. **Capital renewal** refers to the replacement or rebuilding of major facility components (i.e., replacing roofs, boilers, chillers, and transformers) and major remodels of facilities to modernize and meet current code requirements.

Facilities Adaptation

Facilities adaptation consists of expenditures required to adapt the facilities to the evolving needs of the department and to meet changing standards. These

facilities audit
■ A continuous, systematic approach of identifying, assessing, prioritizing, and maintaining the specific maintenance and repair requirements for all facility assets to provide valid documentation, reporting mechanisms, and budgetary information in a detailed database to address facility concerns.

capital renewal
■ The replacement or rebuilding of major facility components (i.e., replacing roofs, boilers, chillers, and transformers) and major remodels of facilities to modernize and meet current code requirements.

expenditures are in addition to normal maintenance. Examples include facility alterations required by changing codes—that is, handicap accessibility, vehicle exhaust removal, OSHA-compliant washing stations—adding length to apparatus bays to accommodate larger equipment, and the like.

LIFE CYCLE

A building's useful life is limited by the durability of its systems (a building does not typically fail as a whole, but by its components). A building's production life is limited by its continuing ability to meet the needs of the occupants, which may change over time.

Each building consists of a set of subsystems, as shown in Table 7.7. Each of these subsystems has a projected useful life span. Determining what it is can help to ascertain how much money should be allocated for the replacement of those systems annually.

TABLE 7.7	Buildings Consisting of Subsystems			
	RANGE	USEFUL LIFE CHOSEN	CURRENT REPLACEMENT COST	PROJECTED REPLACEMENT YEAR
Foundations and structure	NA			
Exterior roofing material	15–50	20		
Exterior cladding	50 Up	50		
Interior partitions	25 Up	25		
Interior finishes	5 Up	12		
Elevators	20–75	75		
Plumbing	20–80	40		
HVAC–moving	15–25	20		
HVAC—static	30–75	40		
Electrical—moving	20–50	35		
Electrical—static	30–75	50		
Fire protection	40–100	50		
Special equipment	10–50	15		
Site improvements	20–40	25		
Site utilities	30–50	40		
Foundation/substructure	40–60	50		
Superstructure	40–60	50		
Exterior wall system	20–35	25		

Exterior windows	25–40	30		
Exterior doors	18–25	20		
Roof systems	40–60	20		
Interior structural partitions	40–60	50		
Interior doors	20–30	30		
Interior floor finishes	12–18	15		
Interior wall finishes	10–15	25		
Interior ceiling finishes	20–30	25		
Specialties	30–50	40		
Plumbing piping	25–40	30		
Plumbing fixtures	20–35	30		
HVAC distribution	30–50	40		
HVAC controls	15–25	20		
Electrical services/general	30–50	40		
Electrical distribution	30–50	50		
Electrical lighting	20–30	25		
Special electrical	10–20	15		
Equipment and furnishings	20–30	25		

Whether you are operating a few facilities or 50, the determination of useful life and projected replacement cost are key factors in the organization's ability to maintain useful facilities. It is also critical from a budget standpoint to protect the organization's long-term fiscal health.

The characteristics of facilities subsystems are that they collectively make the building usable and each subsystem has a definable life. Establishing the anticipated useful life of major components is important in the development of a long-term capital replacement plan. In addition, routine replacement of building components is necessary to maintain the facility's appearance and to provide a quality working environment. Examples are shown in Table 7.8.

Planned periodic maintenance is essential to preserve the appearance, cleanliness, and proper work environment. Without a comprehensive CIP maintenance, renewal, and replacement schedule used to develop a facility capital plan for the department, a department will soon find its infrastructure falling into disrepair very quickly. Unfortunately, a department that has multiple facilities will find it very difficult to catch up once this occurs.

For example, your department has eight existing facilities with two additional stations scheduled to be built in the next 10 years. The existing facilities range from 5 years to 40 years of age. You have conducted your building assessment for **facility renewal**, adaptation of planning, and preventive maintenance. Table 7.9 outlines the organization's cost for each area.

facility renewal
■ A comprehensive approach toward the development of a long-term strategic facilities plan. Facility renewal incorporates a facilities audit, preventative maintenance, planned maintenance, capital renewal, facilities adaptation, and building life cycles of critical components into the capital improvement plan.

	RANGE	USEFUL LIFE CHOSEN	CURRENT REPLACEMENT COST	PROJECTED REPLACEMENT YEAR
TABLE 7.8	**Life Cycle**			
Carpet	3–5	4		
Paint (interior)	3–5	5		
Wall coverings	5–7	5		
Window coverings	7–10	7		
Linoleum	8–12	8		
IT wiring/cabling support	5–7	7		
Paint (exterior)	5–7	5		

TABLE 7.9	**Preventive Maintenance**
Preventive maintenance ($6,200 annually per facility × 10)	$62,000
Planned maintenance	326,000
Renewal and adaptation	1,500,000
New fire stations (2)	6,500,000
Total 10-year capital cost	$8,388,000

This would require an annual capital investment of $838,800 if the department were to pay as you go and not bond any of the projects. Without adequate assessment of the infrastructure and effective planning, a department can find itself behind the infrastructure curve very quickly.

THE STATE OF U.S. FIRE APPARATUS

Many departments today are facing a fleet crisis as their replacement schedules have been pushed back due to several economic factors. One indicator of the engine/pumper fleet inventory is the number of vehicles over 15 years old. Although vehicle age is not sufficient to confirm a need for replacement, it does indicate a potential need. With the change and improvements seen in the last two decades in fire apparatus design, it can also become a firefighter safety issue as well (see Figure 7.12).

Figure 7.13 indicates that in larger communities, those with at least 50,000 population, one-sixth to one-fourth (17 to 25 percent) of engines are at least 15 years old, except for communities of 500,000 to 999,999 population, where the percentage is only 12 percent. In smaller communities, those with less than 5,000 population, roughly one-half to two-thirds (52 to 65 percent)

FIGURE 7.12 Fire Apparatus *(Courtesy of Fresno Fire Department)*

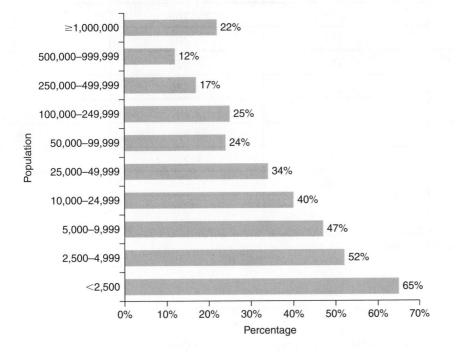

FIGURE 7.13
Percentage of Engines and Pumpers at Least 15 Years Old *(Source: U.S. Fire Administration,* Four Years Later—A Second Needs Assessment of the U.S. Fire Service, *A Cooperative Study Authorized by U.S. Public Law 108-767, Title XXXVI, FA-303/October 2006, p. 63)*

of engines are at least 15 years old. Table 7.10 indicates there are more than 40,000 engines in use that are at least 15 years old, including nearly 11,000 at least 30 years old. Most of these engines aged 15 years old or more are in use in smaller communities, with less than 5,000 population, but hundreds are in use in departments for every community size. Table 7.11(page 245) shows the average number of engines and ambulances in service and the age of the engine by community size.

As with the fire service's fixed facilities, its rolling stock is another significant investment that must be planned for. Every department should have an ongoing replacement schedule for not only the apparatus but also all capital equipment such as turnouts, SCBAs, radios, extrication equipment, and any item meeting the definition of a capital item in the organization. Again, without proper planning, it is very easy to fall behind quickly to the point it impacts firefighter safety and delivery of service.

TABLE 7.10 **Number of Engines in Service, Limited to Engines at Least 15 Years Old by Age of Equipment and Size of Community Protected**

POPULATION PROTECTED	TOTAL NUMBER OF ENGINES IN SERVICE OF THIS AGE IN FIRE DEPARTMENTS PROTECTING COMMUNITIES OF THIS POPULATION SIZE		
	15 TO 19 YEARS OLD	20 TO 29 YEARS OLD	30+ YEARS OLD
1,000,000 or more	135	131	2
500,000–999,999	166	18	0
250,000–499,999	141	47	15
100,000–249,999	523	109	61
50,000–99,999	511	202	22
25,000–49,999	1,073	579	139
10,000–24,999	2,057	1,470	588
5,000–9,999	2,033	2,108	1,129
2,500–4,999	2,142	2,775	1,606
Under 2,500	5,225	8,250	7,288
Total	14,006	15,688	10,851
Percent of U.S. Total	17%	19%	13%

Source: U.S. Fire Administration, *Four Years Later—A Second Needs Assessment of the U.S. Fire Service*, A Cooperative Study Authorized by U.S. Public Law 108-767, Title XXXVI, FA-303/October 2006, p. 64.

A sample of an apparatus replacement schedule for a three-station department is in Figure 7.14(page 246). The spreadsheet shows the balance of the replacement fund based upon when units are projected to be replaced. The schedule is built upon forecasting assumptions, which must be updated annually to reflect current economic and market conditions.

THE STATE OF U.S. PERSONAL PROTECTIVE EQUIPMENT

The cost of maintaining sufficient and reliable personal protective equipment (PPE) has risen quite substantially in the past decade. It is now common for firefighters to be supplied two sets of turnouts, a radio for each person on duty, PASS devices, and personal SCBA masks, all which directly relate to firefighter safety. If not planned for effectively, the costs to maintain this equipment can make a significant impact on the budget and result in deferring need replacement. It is often looked at as the easy fix when budgets get tight to extend replacement cycles, but this practice can come back to haunt a department two to three years later.

The 2006 needs assessment provides an overview of several pertinent issues regarding protective equipment. Many in the fire service and the public take for granted that all firefighters are fully equipped with the latest technology and

TABLE 7.11 Average Number of Engines/Pumpers and Ambulances* in Service and Age of Engine/ Pumper Apparatus by Community Size

POPULATION OF COMMUNITY	AVERAGE NUMBER OF ENGINES	ENGINES 0–14 YEARS OLD	ENGINES 15–19 YEARS OLD	ENGINES 20–29 YEARS OLD	ENGINES 30 OR MORE YEARS OLD	AVERAGE NUMBER OF AMBULANCES*
1,000,000 or more	83.13	65.25	9.00	8.75	0.13	24.20
500,000–999,999	39.67	34.81	4.37	0.48	0.00	10.93
250,000–499,999	22.28	18.51	2.62	0.87	0.28	4.98
100,000–249,999	12.88	9.71	2.41	0.50	0.28	3.23
50,000–99,999	6.73	5.10	1.14	0.45	0.05	1.84
25,000–49,999	4.87	3.20	1.00	0.54	0.13	1.16
10,000–24,999	3.49	2.09	0.70	0.50	0.20	0.80
5,000–9,999	2.98	1.59	0.54	0.56	0.30	0.40
2,500–4,999	2.60	1.27	0.44	0.57	0.33	0.17
Under 2,500	2.33	0.83	0.38	0.60	0.53	0.07
Total	2.99	1.50	0.51	0.58	0.40	0.34

*Ambulances include other patient-transport vehicles.

Source: U.S. Fire Administration, *Four Years Later—A Second Needs Assessment of the U.S. Fire Service, A Cooperative Study Authorized by U.S. Public Law 108-767, Title XXXVI,* FA-303/October 2006, p. 74.

Fiscal Year	2008–2009	2009–2010	2010–2011	2011–2012	2012–2013	2013–2014	2014–2015	2015–2016	2016–2017	2017–2018	2018–2019	2019–2020	2020–2021	2021–2022	2022–2023	2023–2024	2024–2025	2025–2026
Cash balance forward	1,744,774	2,130,841	2,405,421	2,170,690	2,057,778	2,329,801	2,044,306	1,890,910	2,157,092	1,812,973	2,076,428	2,349,102	1,954,249	1,715,240	1,975,273	271,994	481,514	698,367
Interest accrued	61,067	74,579	84,190	75,974	72,022	81,543	71,551	66,182	75,498	63,454	72,675	82,219	68,399	60,033	69,135	9,520	16,853	24,443
Transfer from EMS	225,000	100,000	100,000	100,000	100,000	100,000	100,000	100,000	100,000	100,000	100,000	100,000	100,000	100,000	100,000	100,000	100,000	100,000
Transfer from suppression	100,000	100,000	100,000	100,000	100,000	100,000	100,000	100,000	100,000	100,000	100,000	100,000	100,000	100,000	100,000	100,000	100,000	100,000
CDBG funding																		
Apparatus purchases			518,920	388,886		567,038	424,946		619,617			677,072	507,408		1,972,414			808,460
Ending balance	2,130,841	2,405,421	2,170,690	2,057,778	2,329,801	2,044,306	1,890,910	2,157,092	1,812,973	2,076,428	2,349,102	1,954,249	1,715,240	1,975,273	271,994	481,514	698,367	114,349

Interest rate assumption @ 3.5%.
Inflation rate assumption @ 3%.
Replacement based on a 15-year life for fire engines and ladder truck, 18-year life for water tenders.
First major purchase in year 2011—fire engine.

FIGURE 7.14 Apparatus/Capital Equipment Schedule, Fiscal Years 2008–2026

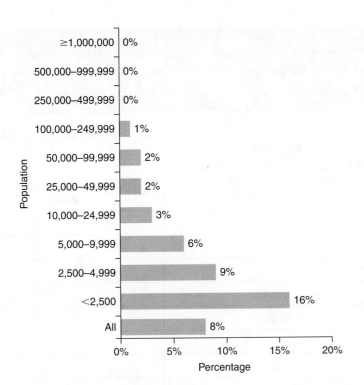

FIGURE 7.15

Estimated Percentage of Departments Where Not All Firefighters Have Personal Protective Clothing *(Source: U.S. Fire Administration, Four Years Later—A Second Needs Assessment of the U.S. Fire Service, A Cooperative Study Authorized by U.S. Public Law 108-767, Title XXXVI, FA-303/October 2006, p. 84)*

equipment. The protective ensemble consists of the basic pieces of equipment for all firefighters. Yet the 2006 survey done by the USFA indicated not only that this is the case but also that the age of this vital piece of equipment is also an issue for many departments.

Nearly all of the firefighters in departments estimated not to have personal protective clothing for all firefighters serve in fire departments that protect communities with less than 10,000 population. Seven out of ten are in communities with less than 2,500 population.

Figure 7.15 indicates how many emergency responders are equipped with their own personal protective clothing. For communities with at least 10,000 population, 3 percent or less of departments are estimated not to have personal protective clothing for all firefighters. For communities of less than 2,500 population, the percentage is 16 percent. The overall percentage is 8 percent.

Two-thirds of departments have at least some personal protective clothing that is at least 10 years old (see Table 7.12). For departments protecting at least 25,000 population, less than half of departments have at least some personal protective clothing that is at least 10 years old.

THE STATE OF U.S. SELF-CONTAINED BREATHING APPARATUS

For communities with at least 50,000 population, at most 5 percent of departments do not have enough self-contained breathing apparatus (SCBA) units to equip all emergency responders on a shift. This percentage rises to three-fourths for departments protecting communities with less than 2,500 population. For larger communities, roughly one-fourth to one-third of departments have at least some SCBA units at least 10 years old. For smaller communities, the percentage

TABLE 7.12	Firefighters in Departments Where Not All Firefighters Are Equipped with Personal Protective Clothing and Percent of Personal Protective Clothing That Is at Least 10 Years Old by Size of Community

POPULATION PROTECTED	ESTIMATED FIREFIGHTERS IN DEPARTMENTS THAT DO NOT HAVE PERSONAL PROTECTIVE CLOTHING FOR ALL FIREFIGHTERS	ESTIMATED PERCENT OF DEPARTMENTS WITH AT LEAST SOME PERSONAL PROTECTIVE CLOTHING THAT IS AT LEAST 10 YEARS OLD
1,000,000 or more	0	20
500,000–999,999	0	16
250,000–499,999	0	41
100,000–249,999	0*	32
50,000–99,999	1,000	32
25,000–49,999	1,000	45
10,000–24,999	4,000	53
5,000–9,999	8,000	63
2,500–4,999	15,000	68
Under 2,500	64,000	72
Total	93,000	66

*Rounds to zero but is not zero

Source: U.S. Fire Administration, *Four Years Later—A Second Needs Assessment of the U.S. Fire Service*, A Cooperative Study Authorized by U.S. Public Law 108-767, Title XXXVI, FA-303/October 2006, pp. 70–71.

rises to two-thirds. Overall, the percentage of departments with at least some SCBA units at least 10 years old is three-fifths (59 percent) (see Figure 7.16).

The breakdown of need by community size is given in Figure 7.17 and Table 7.13, in terms of the percentage of departments where not all personnel on a shift have SCBA and the percentage where some SCBA units are at least 10 years old, both by size of community protected.

THE STATE OF U.S. PERSONAL ALERT SAFETY SYSTEM (PASS)

Table 7.14 indicates what percentage of emergency responders on a single shift are equipped with personal alert safety system (PASS) devices. The breakdown of need is given in Figure 7.18 in terms of percentage of departments in which not all emergency responders on a shift have PASS devices, by size of community protected.

For communities with populations of 50,000 or more, at most 5 percent of departments have insufficient PASS devices to equip all necessary responders on a shift. This rises to one in five for communities with 10,000 to 24,999 population,

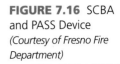

FIGURE 7.16 SCBA and PASS Device *(Courtesy of Fresno Fire Department)*

TABLE 7.13	Departments Where Not All Firefighters on a Shift Have SCBA and Where at Least Some SCBA Units at Least 10 Years Old by Size of Community	
POPULATION PROTECTED	**PERCENT OF DEPARTMENTS WHERE ALL FIREFIGHTERS ON A SHIFT ARE EQUIPPED WITH SCBA**	**PERCENT OF DEPARTMENTS WHERE AT LEAST SOME SCBA UNITS ARE AT LEAST 10 YEARS OLD**
1,000,000 or more	0	27
500,000–999,999	0	18
250,000–499,999	4	26
100,000–249,999	2	31
50,000–99,999	5	32
25,000–49,999	11	40
10,000–24,999	23	45
5,000–9,999	47	53
2,500–4,999	67	60
Under 2,500	77	67
Total	60	59

Source: U.S. Fire Administration, *Four Years Later—A Second Needs Assessment of the U.S. Fire Service*, A Cooperative Study Authorized by U.S. Public Law 108-767, Title XXXVI, FA-303/October 2006, pp. 67–68.

one-third for communities with 5,000 to 9,999 population, over half for communities with 2,500 to 4,999 population, and three-fifths in the departments protecting communities with less than 2,500 population.

THE STATE OF U.S. RADIO EQUIPMENT

Table 7.15 indicates what percentage of emergency responders on a single shift are equipped with portable radios and what fractions of those radios are

FIGURE 7.17

Percentage of
Departments Where
Not All Firefighters on
a Shift Have Self-
Contained Breathing
Apparatus (SCBA)
*(Source: U.S. Fire
Administration,* Four Years
Later—A Second Needs
Assessment of the U.S. Fire
Service, *A Cooperative
Study Authorized by U.S.
Public Law 108-767, Title
XXXVI, FA-303/October
2006, pp. 67–68)*

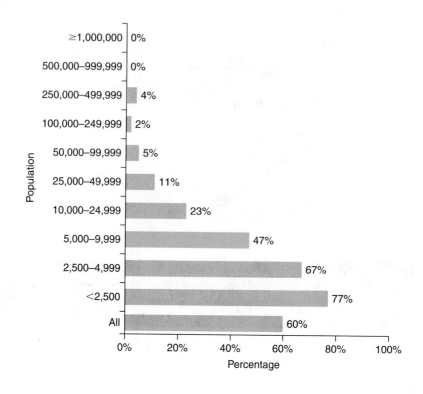

TABLE 7.14 | **Estimated Average Percentage of Emergency Responders per Shift Not Provided with PASS Devices, by Size of Community**

POPULATION PROTECTED	EMERGENCY RESPONDERS PER SHIFT NOT PROVIDED WITH PASS DEVICES
1,000,000 or more	0%
500,000–999,999	0%
250,000–499,999	4%
100,000–249,999	5%
50,000–99,999	4%
25,000–49,999	9%
10,000–24,999	19%
5,000–9,999	35%
2,500–4,999	54%
Under 2,500	62%
Total	48%

Source: U.S. Fire Administration, *Four Years Later—A Second Needs Assessment of the U.S. Fire Service*, A Cooperative Study Authorized by U.S. Public Law 108-767, Title XXXVI, FA-303/October 2006, p. 69.

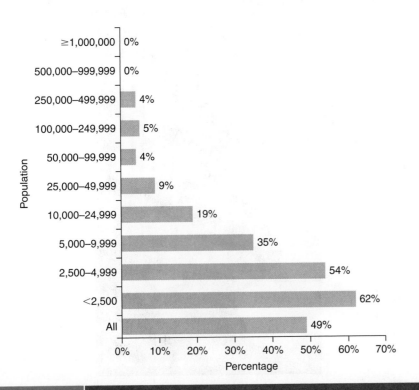

FIGURE 7.18
Percentage of
Departments Where
Not All Emergency
Responders on a Shift
Have Personal Alert
Safety System (PASS)
Devices   (Source: U.S.
Fire Administration, Four
Years Later—A Second
Needs Assessment of the
U.S. Fire Service, A
Cooperative Study
Authorized by U.S. Public
Law 108-767, Title XXXVI,
FA-303/October 2006,
p. 68)

TABLE 7.15	Departments Where Not All Emergency Responders on a Shift Have Radios and Radios Lacking Water Resistance or Intrinsic Safety in an Explosive Atmosphere by Size of Community

| POPULATION PROTECTED | DEPARTMENTS WHERE NOT ALL EMERGENCY RESPONDERS ON A SHIFT HAVE RADIOS | DEPARTMENT WHERE ONLY SOME OR NONE OF RADIOS | |
		HAVE WATER RESISTANCE	HAVE INTRINSIC SAFETY IN EXPLOSIVE ATMOSPHERE
1,000,000 or more	2%	40%	27%
500,000–999,999	12%	38%	42%
250,000–499,999	19%	40%	41%
100,000–249,999	13%	51%	51%
50,000–99,999	12%	42%	45%
25,000–49,999	17%	51%	61%
10,000–24,999	20%	58%	65%
5,000–9,999	30%	67%	71%
2,500–4,999	39%	74%	77%
Under 2,500	43%	77%	83%
Total	36%	71%	75%

Source: U.S. Fire Administration, *Four Years Later—A Second Needs Assessment of the U.S. Fire Service*, A Cooperative Study Authorized by U.S. Public Law 108-767, Title XXXVI, FA-303/October 2006, p. 67.

FIGURE 7.19 Radio
(Courtesy of Fresno Fire
Department)

water-resistant and intrinsically safe in an explosive atmosphere, respectively. In addition, the ability for agencies and jurisdictions to communicate with one another during an emergency can be limited when the equipment is outdated (see Figure 7.19). A substantial element of the ability to communicate is the availability of portable radios to the first responder (see Figure 7.20).

FIGURE 7.20
Percentage of
Departments Where
Not All Emergency
Responders on a Shift
Have Radios   (Source:
U.S. Fire Administration,
Four Years Later—A
Second Needs Assessment
of the U.S. Fire Service, A
Cooperative Study
Authorized by U.S. Public
Law 108-767, Title XXXVI,
FA-303/October 2006,
p. 66)

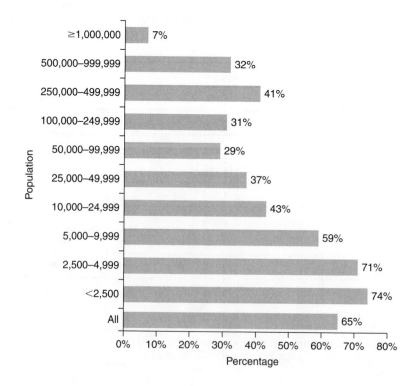

Economic Forecasting

Developing an economic organizational forecast provides the chief executive with strategic and operational insight and forms the basis of budgeting. Without an economic forecast, local government is simply responding to the day-to-day operating environment and has limited capacity to maximize future opportunities or minimize potential risk.

An economic forecast is a prediction of future financial performance. It includes projections for revenue and expenses. It is not a detailed breakdown of revenue and expense items (as in a budget), but a higher level view of the business considering the main drivers of the anticipated revenues and expenses.

Building an economic forecast provides insight into both the current and future financial health of the organization. Structuring this forecast should involve an assessment of the dynamic environment in which the organization operates. Simply considering these issues often allows senior leadership to view the organization in a more strategic manner, which has the capacity to improve not only decision making but also the overall strategic development of the department.

An economic forecast can (and should) be the precursor to budget development. Developing an overall forecast and using this as a framework for budget creation improves budget accuracy because a valid forecast applies the quantifiable approach. It is not simply changing the revenue or expense by some arbitrary percentage; instead, a forecast considers changes in the factors that influence the revenues and expenses and uses these to calculate the future value.

A forecast can also analyze the impact of economic decisions on organizational performance. For example, the long-term forecast of health care and pension costs indicates that if revenue does not keep pace with the forecasted increase in cost, services will need to be reduced, which impacts organizational performance.

Sensitivity analysis applied to a business forecast permits the review of a range of possible outcomes. Providing worst-case to best-case scenarios allows the leadership team to assess, monitor, and implement actions to deal best with these possibilities.

In essence an economic forecast provides a high-level strategic budget overview, assists in the identification of threats and opportunities, and provides a quantifiable framework for the development of the organization's strategies and actions.

Economic forecasting methods range from arbitrary year-on-year variations to complex data-driven algorithms. The best choice depends upon the department being analyzed; the quality and quantity of data available; the purpose of the analysis; the analytical expertise of the user; and the time, usability, and cost constraints.

Although local government typically operates in a very stable environment and has demonstrated consistent incremental variations over a number of years, it is possible to successfully apply year-on-year variations directly to current performance data. This can easily be undertaken using a spreadsheet.

At the other extreme is the volatility experienced by both the public and private sectors during the past decade. Therefore, this forecasting has become an important element in leading and managing at all levels of government.[21]

Conclusion

Local governments around the world find their budgets in the throes of a major recession, one that began as a crisis in the subprime mortgage market but has morphed into a full-blown economic crisis of frightening proportions. Tough economic times require making difficult decisions in order to preserve a balanced budget. What can local managers do to minimize the spreading impact of this crisis on their operating and capital budgets?

Leaders in local government know a balanced budget is a moving target. Sometimes budget plans go awry, even when a local government has vigilantly pursued all the right financial management practices. Economies go through cycles brought on by recessions, by inflation, or in recent years by overexpansion in a sector of the economy.

They also experience sectoral shifts brought on by changes in technology, economic development, and competition from domestic and international sources. Then there are the imponderables: natural disasters, epidemics, weather events, and terrorism. The current recession has pressed managers to the limit in mustering the skills needed to navigate the hardest-hit communities through these turbulent economic waters.

In 2008, the U.S. gross domestic product (GDP) was approximately $14.8 trillion, of which 10.9 percent was produced by local governments. In other words, $1 of every $10 produced by the U.S. economy that year was generated by the nation's local governments (see Figure 7.21).[22]

As the economy experiences this budget crisis, it is apparent the fire service will be challenged for some time with a myriad of difficult fiscal decisions unlike those many senior officials have faced in the past (see Figure 7.22).

FIGURE 7.21 Local Government Expenditures as a Percentage of GDP *(Source: Adapted from usgovernmentspending. com)*

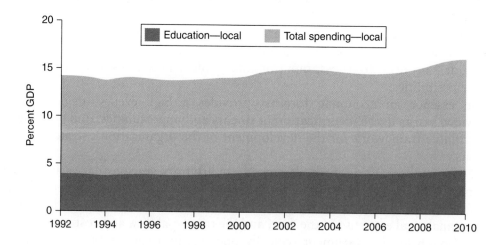

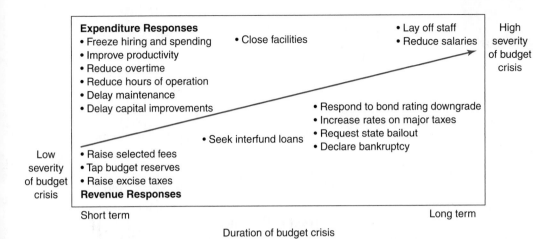

FIGURE 7.22
Strategies for Responding to Budget Crisis *(Source: Robert L. Bland, A Budgeting Guide for Local Governments, 2nd ed. [Washington, DC: ICMA, 2007], 97.)*

As leaders direct the way in the 21st century, the capability of their leadership will dictate what the fire service will become in the future. It is becoming increasingly evident that past practices may not be sustainable in the future. How leaders lead and manage their organizations over the course of the next decade will lay a foundation for the future fire service. The short-term stability and long-term industry economic health will be forged by those in leadership positions today.

PEARSON
myfirekit

Visit MyFireKit Chapter 7 for Perspectives of Industry Leaders.

Review Questions

1. Describe the various budget methods used in government and explain each one.
2. Describe what GASB is and explain its role and importance.
3. Describe what GAAP is and explain its role and importance.
4. Conduct an analysis of your own department and determine the costs of health care and pension to the budget.
5. Conduct a trend line analysis of all personnel cost for the past 10 years to determine percentage of increase during this time.
6. Evaluate the adequacy of your agency's capital replacement plan for facilities and equipment.
7. Conduct a building life cycle audit on one of your agency's facilities to determine annualized replacement costs.
8. Conduct an audit of PPE, SCBAs, PASS, and radios for your department, and develop a replacement schedule for each.

PEARSON

myfirekit™

For additional review and practice tests, visit www.bradybooks.com and click on MyBradyKit to access book-specific resources for this text!

Register for MyFireKit by following directions on the MyFireKit student access card provided with this text. If there is no card, go to www.bradybooks.com and follow the MyBradyKit link to buy access from there.

References

1. A. E. Buck, *Public Budgeting: A Discussion of Budgetary Practice in the National, State and Local Governments of the United States* (New York: Harper and Brothers, 1929).
2. Charlie Tyer and Jennifer Williard, "Public Budgeting in America: A Twentieth Century Perspective," *Journal of Public Budgeting, Accounting, and Financial Management*, 9, no. 2 (Summer 1997).
3. David Osborne and Ted Gaebler, *Reinventing Government: How the Entrepreneurial Spirit Is Transforming the Public Sector* (Reading, MA: Addison-Wesley, 1992).
4. Charlie Tyer and Jennifer Willand, "Public Budgeting in America: A Twentieth Century Retrospective," *Journal of Public Budgeting, Accounting, and Financial Management*, 9, no. 2 (Summer 1997).
5. Governmental Accounting Standards Board, *Facts about GASB*. Retrieved October 6, 2010, from http://www.gasb.org; www.ipspr.sc.edu/publication/Budgeting_in_America.htm; http://www.gasb.org/cs/BlobServer?blobcol=urldata&blobtable=MungoBlobs&blobkey=id&blobwhere=1175821352715&blobheader=application%2Fpdf
6. http://www.gasb.org/st/summary/gstsm34.html; http://www.iaff.org/gasb45/gasbBrief.pdf
7. GASB Governmental Accounting Standards Board States Summary of Statement 45. Retrieved October 31, 2010, from http://www.gasb.org

8. Tiffany Bradford, *GAAP and Accounting Standards, An Explanation of Generally Accepted Accounting Principles (GAAP)*, August 6, 2007. Retrieved October 6, 2010, from http://www.suite101.com/content/what-is-gaap-a28142

9. Federal Emergency Management Agency, United States Fire Administration, *Funding Alternatives for Fire and Emergency Services*, chap. 1, pp. 1–3. Retrieved October 31, 2010, from http://www.usfa.dhs.gov/downloads/pdf/publications/fa-141.pdf

10. Ibid., chap. 7, p. 1.

11. U.S. Fire Administration, *Four Years Later—A Second Needs Assessment of the U.S. Fire Service*, A Cooperative Study Authorized by U.S. Public Law 108-767, Title XXXVI, FA-303/October 2006.

12. S. Keehan et al., "Health Spending Projections Through 2017," *Health Affairs Web Exclusive W146*, February 21, 2008.

13. Ibid.

14. For an establishment-level analysis of health costs using the NCS data, see Christine Eibner, Kanika Kapur, and M. Susan Marquis, *Snapshot: Employer Health Insurance Costs in the United States*, California HealthCare Foundation, July 2007. Retrieved October 6, 2010, from http://www.chcf.org/publications/2007/07/snapshot-employer-health-insurance-costs-in-the-united-states

15. Robert Pear, "U.S. Health Care Spending Reaches All-Time High: 15% of GDP," *New York Times*, January 9, 2004, p. 3.

16. John Ritter, "San Diego Now 'Enron by the Sea,'" *USAToday.com*, October 24, 2004. Retrieved October 6, 2010, from http://www.usatoday.com/news/nation/2004-10-24-sandiego-_x.htm

17. Alicia H. Munnell and Dan Muldoon, *Are Retirement Savings Too Exposed to Market Risk?* (Brief Number 8-16) (Chestnut Hill, MA: Center for Retirement Research at Boston College, October 2008). Retrieved October 31, 2010, from http://crr.bc.edu/briefs/are_retirement_savings_too_exposed_to_market_risk.html

18. Robert Barkin, "Planning to Retire? Maybe Not," *American City & County*, July 1, 2009.

19. *American City & County*, 124, no. 3 (March 2009).

20. James E. Spiotto, Chapman and Cutler LLP, Presentation to the Civic Federation, The Federal Reserve Bank of Chicago, and The National Tax Association, A Forum on Public Pension Funding, *If the Pension Bomb Stops Ticking, What Happens Next?* 2008.

21. BusinessPerformanceAnalysis.com, *Business Forecasting*. Retrieved October 31, 2010, from www.businessperformanceanalysis.com/Business-Forecasting.html

22. Robert Bland, "Managing Your Budget: Making Tough Decisions in Tough Times," *ICMA Public Management Magazine*, 91, no. 3 (April 2009).

OBJECTIVES

After completing this chapter, you should be able to:

- Describe the societal impacts on the fire service.
- Explain how demographic shifts in the United States have impacted the fire profession.
- Explain how generational values impact how one leads and manages the workforce.
- Describe the effect of legislative mandates on the fire service.
- Describe the labor–management initiative.
- Explain the various methods used in contract negotiation.
- Describe the proper methods for performance appraisals and the discipline process.

PEARSON
myfirekit™

For additional review and practice tests, visit www.bradybooks.com and click on MyBradyKit to access book-specific resources for this text!

Introduction

As historic events and technological inventions have helped to shape the fire profession today, so have the societal impacts of an ever-changing nation. With those changes, the socioeconomic environment of the fire service has also changed. Today, the social factors affecting the fire service seem to be shifting more rapidly than ever before; therefore, leaders in fire service need to examine and understand such factors, which have helped to shape its identity today and will continue to do so in the future. The fire profession must also be aware of prominent legislation that affects the responsibilities of fire officers and has changed the way business is conducted on a daily basis.

This chapter provides an overview of the trends impacting the workforce today and into the foreseeable future. Diversity demographics, values, education, and legislation all affect the fire service. These influences have impacted the way labor and management work together in respect to negotiations, evaluations, and discipline. For leaders to be effective, they must keep abreast of these trends and influences, not only to be successful in managing their departments but also to minimize the litigation risk to the organization.

The Changing Composition of the United States

The composition of the U.S. population is changing. Society is vastly different today from what it was when the country was founded and, in fact, has experienced dramatic shifts and changes in just the last 30 years. Here is a quick history review.

1770 TO 1870

In the late 1700s, American society and government were predominantly made up of Anglo-Saxon settlers struggling to establish livelihoods, businesses, and families.

1870 TO 1970

Between 1870 and 1970, the power structure of society began to shift. Former slaves gained full citizen status. The early civil rights movement began. The Industrial Revolution fueled an enthusiasm among Americans and those who wished to become Americans. Immigrants primarily from northern Europe poured into the country by the thousands and settled in the major cities, forming neighborhood enclaves and establishing a network of support.

Shortly after the turn of the century, women secured the right to vote. Women also played a prominent role in the massive work effort undertaken during World War II when they moved out of the home and into the factory, taking on responsibilities that had previously belonged to men.

The hundred years following the Civil War saw significant changes wrought with judicial and legislative initiatives. Ten amendments to the Constitution were passed during this period, many of which dealt with the provision of rights to women and minorities. During the later part of this period, the modern civil rights movement was born. The Civil Rights Act of 1957 (the first federal civil rights legislation to be passed since 1875) authorized the federal government to take legal measures to prevent a citizen from being denied voting rights. The passage of the Civil Rights Act of 1964 introduced the age of equal opportunity. In 1964 the Twenty-Fourth Amendment to the Constitution banned the poll tax, and in 1965 the Voting Rights Act eliminated all discriminatory qualifying tests for voter registrants. During the 1960s and early 1970s, guidelines for nondiscriminatory hiring were developed, and court orders and consent decrees combined to assist in affirmative action.

1970 AND BEYOND

Since the early 1970s, the nation has seen new power struggles in American society. African Americans, Hispanics, Asians, Native Americans, and women all have emerged as powerful voting blocs and economic forces, which have influenced not only the political dynamic at all levels of government but also the economy. Civil disturbances erupted in the larger cities, and the strain of growth coupled with new diversity presented challenges for many communities. Judicial cases have seen the courts issue major rulings in regard to racial, sexual, age, and religious discrimination.

The result of this historic social evolution is a country that has become a kaleidoscope of diversity in the workplace and society. Such diversity, once found only in metropolitan areas, is now seen throughout the nation, even in the most rural areas and smallest communities. Diversity is those characteristics that make people distinct—age, gender, race, ethnicity, ability, and religion.

The workplace must be reflective of the society in which it operates, and this is no different for the fire service. In fact, the fire service may have a larger obligation to mirror society because it functions as an element of government. Today Caucasian males are no longer the primary group from which department leadership is drawn. The pool of fire department entrants now contains fewer Caucasian males, in part because of past legal mandates that have provided equity and fairness in the entry-level process, which has opened competition to women and minorities.

In 1990, Caucasian males composed 61 percent of the workforce, and by the year 2000, they made up approximately 50 percent of the total workforce. With these economic and demographic trends, what will these percentages be in the future? Today a common expectation of the fire service is to be representative of the community served; such representation in the fire department workforce is a critical component of future organizational design. To represent the diversity of the community it serves, the fire service must adopt an open, supportive environment for diversity. As such, its organizational policies and practices should demonstrate this commitment.

Trends Impacting the Workplace

Just as the private sector has been drawn into a competitive world market during the past three decades, the fire service is now confronted with the need to compete. Being publicly funded, the fire service has struggled to gain a diminishing share of citizen dollars—through either taxation or donation. The volunteer fire service faces the same critical competitive demands as does the career fire service. The old adage, "Those who wear the gold make the rules," may still be true in some organizations; but to be successful today, chief executives know they must provide for participation and inclusion of the workforce.

EMPLOYEE EMPOWERMENT

The fire service must also empower its employees, an alien concept to management structures in many fire departments prior to the 1990s. Empowering employees means allowing employees at their respective levels of responsibility to make decisions regarding service delivery initiatives without hierarchical permission. In the 1970s and 1980s, as the empowerment and collaboration approach began to gain popularity in the public sector, many fire department officers were simply not ready to relinquish their power. Instead of decisions being made solely at the top levels, empowerment requires managers to establish boundaries within which their subordinates can make decisions. As the fire service has evolved, it has come to realize that, to remain competitive, it must move forward with empowered employees. Empowerment is also an important element in the design and implementation of any organizational change of significance. The enabling of employees to think, behave, take action, control work, and make decisions autonomously is an important element in all of this.

THE CHANGE IN THE TRADITIONAL POWER STRUCTURE

As empowerment of employees has been encouraged, the traditional power structure of the fire service culture has had to change Although the fire service is still a paramilitary organization, the rigid command and control techniques that were in use when many of today's senior fire service leadership entered the profession are no longer in use. The command structure may still be intact, but how leaders manage on a day-to-day basis is much different. The one-way directives in nonemergent situations are normally not an effective way to deal with the various generations of leaders working in the fire service. Yet, at times, this approach is still needed. The change in the traditional power structure has created a lack of respect in some organizations for the rank achieved by its officers and command staff. When the pendulum swings too far, this lack of respect will play out in many negative and unproductive ways in an organization, as discussed in Chapter 6.

Demographic Shifts

A social factor affecting today's fire service is the demographic shifts of people in the community. According to U.S. Census figures, most individuals move to a different location every four to five years. For a long span of time, urban areas saw the abandonment of the inner city by middle-class families, who had sought the quiet refuge of the suburbs and chosen to commute to work (in some cases, a long commute). Because of this movement to the suburbs, city neighborhoods have seen dramatic transformations that, in some cases, changed a neighborhood from an area of prosperity to one of decline. Although these alterations have been stimulated for a variety of reasons, today many cities have been or are in the process of revitalizing their inner city, as people are moving back to the metropolitan centers.

CHANGING FACE OF THE FAMILY

Another societal factor having an impact on modern society is the changing concept of family with the number of single-parent and divorced-parent families increasing. According to Morris Massey, the family is the single most important factor in forming values and behaviors. Many children today live either with a single parent or in a nontraditional home arrangement in which one or both of their biological parents or guardians are not present. In managing a diverse workforce, many fire departments now have assumed the task of accommodating nontraditional parents in the department's daily work demands. The fire service is also impacted by the instability that some home lives often create: department personnel act as first responders to domestic incidents or as a source of primary health care provision for those without other means to access such services.

THE 21ST-CENTURY WORKPLACE

The changing labor force is yet another significant factor affecting the fire service. The Department of Labor sponsored a study of the labor issues confronting the United States as it moved into the 21st century. In 1987, Hudson Institute published *Workforce 2000: Work and Workers for the 21st Century,* a study of the changing American workforce. *Workforce 2000* issued four simple predictions; although none of them were obvious in 1987, all of them have proved largely correct.

1. The U.S. economy would grow at a healthy pace, fueled by a rebound in U.S. exports, productivity growth, and a strong world economy.
2. Because of productivity gains, manufacturing would shrink as a share of employment in the United States. But it would not "wither away."
3. The workforce would grow slowly, becoming older and more female and including more minorities.
4. New jobs in service industries would demand much higher skill levels. *Workforce 2000* stated very few new jobs would be available for those who could not read, follow directions, and use mathematics—another prediction that was clearly on the money.

Workforce 2000 also missed a few trends, which now seem obvious.

1. *The digital revolution.*   In 1987, IBM's first personal computer (PC) was barely five years old. The PC was still only a new and improved typewriter for everyone but a few hackers. But PC prices soon tumbled as computing capability soared. Although *Workforce 2000* did not completely miss the digital revolution, it did not fully anticipate its breadth and speed.
2. *Geographic disparities.*   *Workforce 2000* looked more at overall trends than at developments in specific geographical regions.
3. *The diversity industry.*   *Workforce 2000* was "credited" with creating a diversity craze. To prepare for the increasingly diverse workforce, which it foresaw, entrepreneurs responded by offering sensitivity training to accommodate cultural differences in the workplace.[1]

WORKFORCE 2020

America's voyage from the workforce of 2000 to the workforce of 2020 is unique; this is not merely because of the heights to which some will climb or the difficulties others will endure. Two particular qualities give a truly unprecedented character to the roads ahead. First, what little conscious discrimination remains will be swept away soon—not by government regulation but by the enlightened self-interest of employers. Second, more and more individuals now undertake their own journeys through the labor force, rather than "hitching rides" on the traditional mass transportation provided by unions, large corporations, and government bureaucracies. For most workers, this "free agency" will be immensely liberating.[2] The *Workforce 2020* report notes four areas of significant change: technology, globalization, an older workforce, and ethnic diversity.

First, the pace of technological change in today's economy has never been greater. It is unknown what innovations will transform the global economy by 2020, any more than analysis in the mid-1970s could have foreseen the rise of the personal computer or the proliferation of satellite, fiber optic, and wireless communications. However, the computer and telecommunications revolutions enable us to speculate in an informed manner on the implications of the current innovation age for the American workforce.

- Automation will continue to displace low-skilled or unskilled workers in America's manufacturing firms and offices.
- Experience suggests the development, marketing, and servicing of ever more sophisticated products—and the use of those products in an ever richer ensemble of personal and professional services—almost certainly will create more jobs than the underlying technology will destroy.
- The best jobs created in the innovation age will be filled by Americans (and workers in other advanced countries) to the extent workers possess the skills required to compete for them and carry them out.
- The best new jobs will demand brains rather than brawn; and because physical presence in a specific location at a particular time will become increasingly irrelevant, structural barriers to the employment of women and older Americans will continue to fall away.

Second, from the U.S. perspective, the rest of the world matters to a much greater degree than it did in the past. Economically, what made sense in the past for American workers will not suffice in today's marketplace if we consider only the U.S. economy or the characteristics of the U.S. labor force.[3]

The implications of this globalization for U.S. workers are no less complex than the implications of new technology.

- Manufacturing will continue to dominate U.S. exports. Almost 20 percent of U.S. manufacturing workers now have jobs that depend on exports, and that figure will only continue to escalate.
- Globalization will have a significant effect on low-skilled or unskilled American workers. They will compete for jobs and wages not only with their counterparts across town or in other parts of the United States but also with low-skilled workers around the globe.
- Manufacturing's share of total U.S. employment will continue to decline, due to the combined effects of automation and globalization.

- Globalization and technological change will make most segments of the U.S. economy extremely volatile, as comparative advantages in particular market segments rise and then fall away. We have witnessed this impact in the financial markets. When one financial market has a day of large losses or gains, it has a tendency to impact other such markets around the world.

Third, America is getting older. Many employers have yet to come to grips with the full implications of America's aging. The oldest among America's so-called baby boomers—the massive cohort born between 1945 and 1965—will begin to reach age 65 in 2010. America's aging baby boomers will decisively affect the U.S. workforce.

- America's taxpayer-funded entitlements for its aging population—Medicare and Social Security—are likely to undergo profound changes in the next two decades.
- Depending on how the funding of such programs is resolved and how well individual baby boomers have prepared for retirement, some who reach 65 will continue to require outside income and will be unable to retire.
- Whether they continue working or simply enjoy the fruits of past labors, America's aging baby boomers will constitute a large and powerful segment of the consumer market.

Fourth, the U.S. labor force continues its ethnic diversification, though at a fairly slow pace. Most white non-Hispanics entering America's early 21st-century workforce simply will replace existing white workers; minorities will constitute slightly more than half of net new entrants to the U.S. workforce. Minorities will account for only about a third of total new entrants over the next decade. Whites constitute 76 percent of the total labor force today and will account for 68 percent in 2020. The share of African Americans in the labor force probably will remain constant, at 11 percent, over the next 20 years. The Asian and Hispanic shares will grow to 6 and 14 percent, respectively. Most of this change will be due to the growth of Asian and Hispanic workforce representation concentrated in the South and the West, not on a national scale. The aging of the U.S. workforce will be far more dramatic than its ethnic shifts.[4]

Summary of *Workforce 2020*

- The aging of the large cohort of baby boomers means Social Security and Medicare benefits are almost certain to be reduced substantially by the time the boomers begin to turn 65.
- Slow population growth and the retirements of baby boomers ensure the workforce will grow only slowly in years to come. Two factors will determine its growth rate: the extent of immigration and the labor force participation of men and women (particularly older ones).
- As they age, some boomers will want to keep working and some—particularly those involved in tedious and physically demanding work—will want to retire. But many if not most white-collar baby boomers will discover their private savings and Social Security benefits fall short of replacing their former earnings. Thus, they will want or perhaps need to keep working. Furthermore, many employers will need them to do so, because the labor force will have

grown very slowly in the preceding decades. Public policies should change to encourage older people to stay in the workforce longer if they so choose, and corporations may need to offer new inducements to retain productive older workers.

- The nation's population and workforce will continue to become more ethnically diverse, but only gradually. White non-Hispanics will still account for 68 percent of the workforce in 2020. In Western states, though, and particularly in California, diversification will be more significant, as the Hispanic and Asian shares of the population and workforce rise rapidly.

- The gender diversification of the workplace will also proceed. Women will comprise half of the 2020 workforce.

- In the 1990s, immigration accounted for fully half of the increase in the labor force; if immigration policy remains unchanged, immigrants will constitute an increasing share of workers in the early 21st century. Thus, the job qualifications of immigrants will have an increasingly important impact on the skill and education levels of the workforce. Unless they acquire more schooling in the United States than they did in their native countries, recent immigrants will account for a rapidly rising share of the otherwise dwindling number of Americans who lack a high school education.

- Because economic growth will depend on increased worker productivity, the educational attainments of today's students raise an important concern for tomorrow's workforce. Educational levels need to be raised for all, and the continuing disparities between white and minority students are particularly worrisome. Overall, minority students are making greater educational gains than whites; but because their gains are only slightly greater, and the existing gap in educational levels between white and minority students is substantial, the gap is expected to remain.[5]

FIREFIGHTER DIVERSITY

A January 2006 report on firefighter diversity commissioned by the IAFF provided a snapshot of the current gap between the general population and the fire service. There is a growing awareness of the effect on diversity of the huge wave of retirements that is happening now and will continue for the next few years. In the literature review of this study, various researchers explored the business, financial, and legal reasons for supporting diversity. In the fire service, in addition to all the practical issues related to diversity, strong social and emotional reasons also exist. The fire service serves each and every community member, regardless of ethnicity, gender, race, background, economy, or any other factor, and also depends on the entire community and country for its funding and support. As public safety and public service organizations, most fire departments want to better understand, communicate with, and enlist cooperation in their multicultural communities.[6]

Not surprisingly, most prior research about diversity recruiting has been conducted with a focus on the private sector, especially on how to attract limited top talent to the leading corporations in what is perceived to be a competitive environment for employers. In fact, even in the private sector, much of the research is about reaching management-level candidates and competing for a limited number of qualified candidates. By contrast, the fire service generally operates under entirely different parameters. Candidates for the fire service are entry level, usually high school graduates; and when a test is given, thousands show up to compete! Traditionally, there are many more applicants for any one job than are needed. In fact, the whole idea of recruiting is somewhat new for many fire departments. Fire agencies have traditionally considered how to narrow or decrease their applicant pools rather than to increase them. The idea of attempting to recruit even more people, putting them through a selection process that hires only a few, can at first seem nonsensical.

Also, as the economy continues to pressure public safety departments to reduce their budgets, many fire and human resources executives, as well as city managers and civil service board members, ask why they should spend money to do any recruiting. However, diversity in the workforce does not occur simply because a department is located in a diverse metropolitan area or because a department follows the law regarding nondiscrimination. Instead, to achieve diversity, a department must have a strong commitment, which includes an active recruiting and outreach strategy.[7]

The IAFF study used national statistics to make the same comparisons. Table 8.1 shows national (U.S.) statistics comparing U.S. population to reported representation in the fire service. (Note: Two Canadian fire departments were also included in the study but either did not complete demographic information or did not meet or exceed average Canadian representation.)

Similar to an adverse impact ratio, the closer the ratio is to 1.00, the closer the representation of the target group is to the actual population. Using these ratios, a successful department would be one whose ratio exceeds the national ratio. This would mean for Blacks, a ratio between .68 and 1.00 exceeds the national percentage and comes closer to actual population representation. Likewise, for women, ratios above .10 would demonstrate better than average progress toward diversity[8] (see Table 8.1).

TABLE 8.1	Demographic Representation (U.S. Census Data Compared to Bureau of Labor Statistics Representation in the Fire Service)			
	BLACK	HISPANIC	ASIAN	WOMEN
U.S. Census 2000	12.3%	12.5%	3.6%	50.9%
Firefighting profession U.S. Bureau of Labor Statistics	8.4%	8.6%	1.3%	5.1%
Group ratio comparing percent represented in profession to percent of each group in population	.68	.69	.36	.10
Average percent of each group in best practice departments	11.8%	8.5%	1.6%	7.8%

Source: Kathryn A. Fox, Chris W. Hornick, and Erin Hardin, *Achieving and Retaining a Diverse Fire Service Workforce*, International Association of Fire Fighters Diversity Initiative, January 2006, p. 18.

The study analyzed the ratio of the top 10 departments deemed best practice departments for diversity recruitment. For the purpose of this analysis, researchers measured success in best practice departments by the percentage of minorities and women represented in the local population based on census data compared to the percent represented in the department. The average ratios in best practice departments are below the average ratios nationally for minorities. This is because other departments in the United States demonstrate equal or better representation of minorities compared to their communities. The goal of this IAFF study was not to identify *all* departments with good diversity but to identify those departments that actively and successfully recruit for diversity, whose numbers are expected to improve over time as they continue pursuing their diversity goals.[9]

Another study, entitled *A National Report Card on Women in Firefighting,* released in April 2008, stated the following:

NEW DATA CHALLENGE LOW FEMALE NUMBERS

Among the 350,000 paid firefighters in the nation today, the 2000 Census reports women number slightly more than 11,000, or 3.7 percent. This figure places firefighting in the lowest 11 percent of all occupations in terms of women employees. Even more striking is the large number of departments in which the number of women is zero or nearly so. Not one paid woman firefighter has ever worked in more than half the nation's departments. Among the 291 metropolitan areas in the 2000 Census, 51.2 percent had no paid women firefighters in the entire metropolitan area, typically including multiple departments. In 2005, departments in jurisdictions as large as Garden Grove, California, population 165,000, remained entirely male. New York City counts less than .25 percent women among its uniformed force, and Los Angeles employs 2.5 percent.

When fire department leaders are challenged about these numbers, they traditionally respond that women do not want and cannot handle the job, so the low numbers are to be expected. Are they right?

To answer this question, a benchmark was developed for expected female representation using the 2000 Census. It computed the percentage of women in the nation's labor force of typical firefighter age (20 to 49) and educational background (high school graduate but no college degree), working full time in one of 184 occupations resembling firefighting in requiring strength, stamina, and dexterity, or involving outdoor dirty or dangerous work. These comparison occupations include bus mechanics, drywall installers, enlisted military personnel, highway maintenance workers, loggers, professional athletes, refuse collectors, roofers, septic tank servicers, tire builders, and welders. The proportion of women among the employees in these 184 occupations is 17 percent.[10]

According to the *National Report Card on Women* report, "The study validates decades of anecdotal wisdom about the inclusion, acceptance, training, testing, and promotion of women in fire and emergency services. It also points to a future where, barring continued cultural and traditional resistance, women should comprise 17 percent (up from the current 3.7 percent national average) of the first-responders workforce."[11]

Both studies show that the fire service still has much work to do in this area. Over the course of the next 10 to 15 years, because the majority of the retiring workforce will be white male, the opportunity to diversify the fire profession has never been better.

Differing Generational Values

An important factor posing a challenge for fire officers is the issue of leading and managing a multigenerational workforce with conflicting values and expectations. Depending on circumstances, age differences of 10 to 40 years present significantly different perspectives on how to view problems and solve them. Leaders today understand and must appreciate the elements combined to cause these conflicts.

An individual's basic value structure is often imprinted by the age of seven. These **values** are influenced by family, religion, school, friends, and experiences. Once imprinted, these values are reinforced through modeling the behaviors of those around us and are solidified by the age of 20. The values do not change except under the most extreme circumstances, which are often referred to as "significant emotional events." Even then, values are rarely altered totally. In a recent survey conducted by the Society for Human Resource Management, 40 percent of human resource professionals have observed conflict among employees as a result of generational differences![12] In organizations with 500 or more employees, 58 percent of human resource professionals reported conflict between younger and older workers, largely due to differing perspectives on work ethics and work–life balance. These data suggest a huge potential for miscommunication, low morale, and poor productivity unless the generations learn to handle conflict successfully.[13] These generational issues are basic business matters that every industry will face in the next decade. Age has taken its place beside gender, race, and culture as a way that defines experiences, whether someone matured during the Great Depression, witnessed the turbulent times of the 1960s, or grew up with the Internet and the latest technology.

When it comes to the workplace, generational differences offer a complex variety of perspectives and talents. Each generation may have a slightly different view on work, family, and priorities. Different generations also bring differing skill sets to the workforce; for example, the ability to work with tools versus technology. Managing these differences poses critical issues for every organization. Yet the particular generational issues faced today in the fire service are important reflections of what the fire service is to become; as new generations enter the service, they will begin to influence the culture, the methods of work, and the internal dynamics of the organization. This is not merely an issue of young versus old or what generation one belongs to. The fire service is truly living through a historic shift in the framework of its workforce. Throughout time, each generation has matured in a very different world, and these distinct societal conditions have helped to define how people respond to life in general, in the workplace, and with their coworkers. Leaders and managers must have a level of understanding what frames the issues for each generation. Their challenge is to bring together this diverse generational workforce and help shape it into a collaborative team. This will occur only when leaders can utilize the strengths each brings to the table, with an understanding that, when leaders and managers bridge the generations, they are building the future of their organizations. Today, there are four generations in the workforce; and as people live and work longer, it may be five in the near future. They include the following:

■ *The traditionalists,* born 1922–1945, are much more likely to attend symphonies than rock concerts and eat steak instead of tofu. They like consistency and uniformity, and prefer detail and logic when approaching projects. Computers

values
■ Enduring beliefs and assumptions about specific modes of conduct or states of existence that are preferable to opposite or converse modes of conduct or states of existence.

were considered foreign objects and once feared. The traditional values include hard work, dedication, sacrifice, respect for rules, duty before pleasure, and honor. As leaders, they typically use a directive style.

■ *The boomers,* born 1945–1960, grew up with Elvis and the Beatles and watched the Vietnam War on television. They are good team players and typically are reluctant to go against peers. This group's values include optimism, teamwork, involvement, and personal growth. As leaders, they are typically collegial and consensual.

■ *The Gen Xers,* born 1960–1980, are the most recent group integrated into the workplace and are the first to be techno-literate. They loathe paying dues and want companies to be loyal to them. They view job portability as a necessity to acquiring skills and experience, and will stay with a company only if conditions are right. They bring great adaptability and creativity to their work and as leaders are fair, competent, and straightforward. Their values include diversity, fun and informality, self-reliance, and pragmatism.

■ *The Millennials,* born 1980–2000, are a group that has dealt with a number of significant events in their young lives. Columbine, Jonestown, and 9/11 are some of the major events that have impacted this generation's psyche. This generation also grew up at the same time as revolutionary developments were occurring in all technological applications. So for this group, change is a part of daily life and they are multitasking masters. They want and prefer to handle multiple jobs at once. Forget about paying dues; this group wants learning opportunities and will combine teamwork with can-do attitudes. Values for the millennials include optimism, civic duty, confidence, achievement, and respect for differences. Their leadership style will be collaborative.

These generational differences bring a range of thought, values, importance placed on work, and expectations of those in authority, and of what work owes to the employees. Each can be a future strength to organizations and the fire service if managed effectively or else can create serious conflict and disharmony if not managed well.

PEARSON

Visit MyFireKit Chapter 8 for information regarding Generational Issues.

Changing Education/Experiential Levels

The final factor to consider is the changing educational and experiential background of those who will compose the fire service in the future. The fire service, because of generational diversity, will require officers with a high level of interpersonal skills. Whereas the role of today's officer is that of leader, manager, coach, mentor, and facilitator, the future fire officer will need to possess a higher level of education and breadth of experience to meet the demands of the job. The educational challenge for the fire service is twofold. First, many entrants now have higher education levels. In numerous fire departments, a majority of the new employees or volunteers have some college education or already possess a degree. Second, many entrants have fewer technical or hands-on skills. Fewer entrants possess military experience or trade backgrounds; however, this may change in the coming years due to the war effort in the Middle East. Thus, the fire service must meet the challenge of dealing with a more intellectually talented group of employees who, at the same time, may require more extensive training to gain the necessary hands-on firefighting technical skills.

Legislative Mandates

legislative mandates
■ Laws that make mandatory such things as equal opportunities to the groups that have been deprived of these opportunities in the past: voting rights, rights for those with disabilities, etc.

Legislative mandates have been one of the driving forces that have compelled not only the fire service but also the private sector toward acceptance and inclusion of diversity. These laws have provided equal opportunities to the groups deprived of these opportunities in the past. The following overview discusses the laws shaping today's workplace. It is imperative for all those in supervisory positions to have a basic understanding of pertinent laws and legislative acts that influence personnel management.

BASIS FOR EMPLOYMENT LAW

Employment law has its basis in federal and state laws as well as in how the courts have interpreted these laws (case law). The foundation of a body of law called common law is in general principles of fairness. The fire service manager must have a working knowledge of the laws, rules, and regulations and familiarity with decisions rendered by federal and state courts in order to avoid problems of unfair (or illegal) treatment of employees.

EQUAL EMPLOYMENT OPPORTUNITY

The days are long past when employers and supervisors can manage their workforces any way they wish. Federal and state laws have been passed prohibiting discrimination against employees in the areas of race, national origin, color, creed, gender, age, disability, military experience, religion, and status with regard to public assistance. The term *discrimination* is used in many ways. *Webster's Collegiate Dictionary* defines it as "the act, practice, or an instance of discriminating categorically rather than individually" and "prejudiced or prejudicial outlook, action, or treatment." Employers must choose (discriminate) among applicants for a job (or promotion) on the basis of each candidate's qualifications. However, discrimination is also used to mean unequal treatment of members of a particular group when compared to members of another group. Individuals covered under equal employment laws are referred to as "members of a protected class." To implement laws barring discrimination, several regulatory agencies have developed guidelines and regulations.

PEARSON
myfirekit
Visit MyFireKit Chapter 8 for information on Employee Laws and Regulations.

Employment Selection and Testing

An employer may not ask obviously illegal questions about a prospective employee on job applications, such as questions about an applicant's color, race, creed, religion, sex, national origin, and disability. Some employers may require a potential new hire to submit to a polygraph, voice stress test, or any other test that purports to measure "truth." These instruments are used in conjunction with a thorough background investigation to help in evaluating the integrity of what the candidate has reported. The Employee Polygraph Protection Act of 1988 (29 USC 2002) is a federal statute that generally prohibits private employers from requiring, requesting, suggesting, or causing any employee or applicant to take a lie detector test. The statute, enforced by the U.S. Department of Labor, does not apply "with respect to the United States government, or any state or local government, or any political subdivision. . . ." The statute and regulations make it clear that it does not apply to most government employers: 29 USC 2006(a); 29 CFR 801.10. In the fire service, use of lie detector tests may be restricted by state statutes or by collective bargaining agreements.[14]

UNIFORM GUIDELINES ON EMPLOYMENT SELECTION PROCEDURES

To implement the provisions of the Civil Rights Act of 1964 and the interpretations of it based upon court decisions, the Equal Employment Opportunity Commission (EEOC) and other federal agencies developed their own compliance guidelines and regulations, each having a slightly different set of rules and expectations. The Equal Employment Opportunity Commission, the Civil Service Commission, the Department of Labor, and the Department of Justice have jointly adopted these uniform guidelines to meet that need, and to apply the same principles to the federal government as are applied to other employers. These guidelines affect almost all phases of employment practices, including the following: hiring, promotions, recruiting, demotion, performance appraisal, training, labor union membership requirements, and license and certification requirements.

The purpose of fair employment laws is to achieve equality of employment opportunities through eradication of barriers that discriminate on the basis of race, gender, religion, and other protected classifications.

IMPLEMENTING EQUAL OPPORTUNITY

Human resources play a very important role in the success of any organization. It is the responsibility of the organization to treat its workers efficiently by creating equal opportunities for them all and recognizing people as individuals in the workplace to obtain the best performance. In order to meet any business objectives, organizations need to make the best use of their human resources by treating them equally. Employers are bound by the law and by the requirement to operate in a moral way in terms of their employment policies, which means not discriminating on the basis sex, race, or disability when recruiting and employing.

Sex Discrimination/Discrimination in Job Assignments and Conditions

Title VII of the Civil Rights Act prohibits discrimination in employment on the basis of sex. As with racial discrimination, it has taken a series of court decisions and EEOC rulings to determine exactly how broad this prohibition is.

Height/Weight Restrictions

Many legal cases involving the use of height and weight restrictions were actually discrimination cases designed to bar women and certain races from jobs. Citing height and weight requirements as a substitute for strength may be determined to be discriminatory. The courts have generally found that, if an employer has a strength or agility requirement for a job, the employer must test for it and not use a height/weight restriction.

Conviction and Arrest Records

Should the issue of conviction and arrest records arise, legal advice should be sought. Most jurisdictions have established policies on methods of dealing with candidates who have a criminal record. Felony convictions are generally not acceptable for firefighter appointment.

FIREFIGHTER SELECTION

The selection of new firefighters is one of the most important functions of departments today. The personnel hired today will be the ones to lead the organization in the future. Departments use a variety of processes to ensure they are attracting and hiring the best candidates.

Job-Related Validation Strategy

Validity simply means the applicant's test score actually is a predictor of the person's ability to perform the job for which he or she is being hired. (Courts will look closely at tests administered to prospective candidates to ensure validity is present.) In March 1975, the Supreme Court case of *Albermarle Paper Co. v. Moody*, 422 U.S. 405, reaffirmed the idea that any real "test" used for selecting or promoting employees must be a valid predictor or performance measure for a particular job. The burden of proof for showing the test score is valid falls on the employer. It is obvious that selecting the appropriate test for firefighter selection is of the utmost importance. Many tests meeting the required criteria (and that may already have been "court tested") are available commercially. Departments should not consider writing their own tests unless a person trained in test construction is available on staff, and there is a means to validate his or her efforts.

Personality and Psychological Tests

Psychological tests attempt to measure personality characteristics. The Rorschach ("inkblot") test, the Thematic Apperception Test (TAT), and the Minnesota Multiphasic Personality Inventory (MMPI) are some examples. These types of tests are difficult to relate to the jobs firefighters are required to perform and may be determined to be invalid if challenged in the courts. The main difficulty is in connecting particular personality traits to specific job requirements.

Selection Interviews

Selection can be subjective, and courts view subjective decision making with some suspicion because, without some method of control, an employer may indulge in a preference for one class of applicants or against another class. Attorneys often recommend the following controls to reduce equal employment opportunity (EEO) concerns with the interview process: identify specific job criteria to be looked for in the interview, put those criteria in writing, and provide a review process for those difficult or controversial decisions.

Types of References

Background information can be obtained from academic records, prior work references, and personal references given by the applicant. Personal references are often of minimal value, because the applicant usually will not give a negative source as a personal reference. Investigating academic credentials, prior work references, and places of employment is worthwhile, because information on the application form may be incorrect. Academic records will usually not be released without the applicant's written permission. Many employers are hesitant to give negative information about former employees, because doing so may increase their liability, so the potential employer has to weigh the worth of prior work references.

Probation Periods

Most fire departments require a new employee to complete some type of probationary period. Usually, the department uses this period to evaluate the suitability of the employee and requires that he or she complete some education requirements (First Responder, Firefighter I, etc.). Again, objective criteria must be utilized in determining what the employee is expected to accomplish during this period. Some examples include attendance at a particular number of training sessions, responses to fire calls, and successful completion of specific education and skills requirements. The most often used probationary period for career and volunteer firefighters is one year. At the successful completion of the probationary period, the firefighters become vested in the organization and are subject to all personnel procedures and employee rights. Probationary periods may be extended if personnel procedures or employee contract language establishes a process for extending probation. Usually the probationary extension period is no longer than six months.

Employee and Labor Relations

Fire chiefs are often caught between the bargaining interests of personnel and the political and economic realities of the world in which they work. Because these interests often conflict, it is the fire chief and the leadership team of the organization who must bridge the interests of both the employees and the political and economic reality of the community served.

Whereas the earliest fire departments in the United States were all volunteer, eventually the paid personnel in some places began to organize. In 1918, the International Association of Fire Fighters (IAFF) was formed, with members in both Canada and the United States. Chartered by the American Federation of Labor on February 28, 1918, the IAFF represents over 70 percent of the career firefighters in the country, with more than 250,000 members, predominantly in the United States. An overview of the union movement is helpful when attempting to understand contemporary labor relations.[15]

PEARSON
myfirekit
Visit MyFireKit
Chapter 8 for the
History of Labor
Relations.

LABOR RELATIONS

Several factors in our national climate influence employee and labor relations. Undoubtedly, national economics and the availability of jobs, especially in those

FIGURE 8.1 Labor-Management Initiative Logo *(Source: IAFC)*

occupations traditionally unionized, affect labor relations and union membership. Other factors include worker health and safety, as well as a need to safeguard the environment. These concerns have brought changes in legislation and additional federal and state legislation.

In the last several years, efforts have increased to improve the labor–management relations in the fire profession at the national and state levels. In August 2008, the IAFC and IAFF signed a **labor–management initiative** guiding principle document, which states:

> The following principles were developed and agreed to by the International Association of Fire Chiefs (IAFC) and International Association of Fire Fighters (IAFF) in the true spirit of cooperation for the enhancement of the fire/EMS service and the communities it serves. It should be recognized mutual adherence to these values requires the constant effort of both labor and management representatives.

The importance of good labor relations is recognized by both the fire chiefs and the labor leaders that have jointly created an effort that focuses on the development of positive labor–management relations (see Figure 8.1).

labor–management initiative
■ An initiative between the IAFC and IAFF to promote and facilitate positive labor and management relationships.

LMI LABOR-MANAGEMENT INITIATIVE

GUIDING PRINCIPLES

- To recognize labor and management have a mutual goal of ensuring the well-being and safety of fire/EMS personnel and providing high quality service to the public.
- To work together to improve communications, enhance training, increase participative decision-making, and promote a labor–management relationship based upon mutual trust, respect, and understanding.
- To create labor–management partnerships by forming labor–management committees at appropriate levels, or adapting as necessary, existing councils or committees if such groups exist.
- To provide systemic training to labor and management leaders on collaborative methods of dispute resolution, recognizing this process allows management and union leaders to identify problems and craft solutions to better serve their members and the public.
- To promote these principles to our respective members at all levels of both organizations.[16]

LABOR–MANAGEMENT PRACTICES

Some states have enacted legislation that supports union missions, primarily through the establishment of laws mandating union membership for workers in unionized establishments or favoring union membership. These "shops" provide union security and are termed the following:

- *Closed shop.*   Job applicants must join the union.
- *Open shop.*   Employees have a choice regarding the need to join the union.
- *Union shop.*   All workers must join following a probationary period.
- *Preferential shop.*   Union members have job preference.
- *Agency shop.*   If worker is not a union member, a "service fee" must be paid to the union for its representation.
- *Checkoff.*   Union dues are deducted from the paycheck and transmitted directly to the union.

As permitted by federal legislation (Taft–Hartley Act), 22 states have passed laws to prohibit compulsory union membership. These "right-to-work" states allow only for the checkoff and not for the other forms of union security listed here.

ORGANIZED LABOR ACTIONS

Employees of the federal government are prohibited from striking by statute, and most states also bar public safety employees from striking. Public sentiment about and political reaction to public safety strikes usually do not promise good for future contract improvements for the firefighter, and they create a public relations nightmare for the department. Other types of job actions that labor might make include sick-outs, work slowdowns, picketing, work-to-rule, and modifications of these. As an alternative to the strike option, and as a type of trade-off for statutory prohibition against striking, many states have passed binding **arbitration** laws, forcing both labor and management to accept a final ruling of a third party's arbitration finding.

arbitration
■ The act or process of arbitrating; a process through which two or more parties use a disinterested third party, an arbitrator or an arbitration board, in order to resolve a dispute; the third party, an arbitrator, hears the evidence brought by both sides and makes a decision. Sometimes the decision is binding on both parties.

LABOR RELATIONS IN VOLUNTEER AND NONUNIONIZED DEPARTMENTS

It is important to keep in mind that labor relations exist in all types of fire departments: those with all-volunteer members, combination departments, and nonunionized departments. Full-time paid members of combination departments may be unionized and follow typical contract agreements. At this time, union affiliation is not widely offered to volunteer or paid-call firefighters. However, certain departments do have union representation for these groups. Nonunionized firefighters may work under the same type of agreement or formal understanding, but they do not have the impact of affiliation with a national labor organization that unionized firefighters have. In many volunteer and part-paid departments, the **grievance procedures**, negotiation process, and other personnel policies and procedures may be less formalized than in union-represented agencies. Combination departments present special managerial challenges regarding labor relations. Certain combination departments, where the paid presence is strong, create a climate that is negative to the volunteers. The opposite may exist if the department is predominantly volunteer. Departments that believe both career and volunteer members are highly valuable resources, demonstrate this belief to all members, and encourage a mutual respect for each other's contributions will create a labor relations climate that is mutually supportive of career and volunteer members.[17]

grievance procedures
■ Step-by-step process an employee must follow to get his or her complaint addressed satisfactorily. In this process, the formal (written) complaint moves from one level of authority (of the firm and the union) to the next higher level. Grievance procedures are typically included in union collective bargaining agreements.

Collective Bargaining

The term **collective bargaining** was developed from legal terminology focused on labor law of the 1930s and 1940s. Earlier, at the turn of the 20th century, with the beginning of the Industrial Revolution and the development of factories and assembly lines becoming prevalent, more employees were victimized by unfair labor practices. Union representation had been held in abeyance by the Sherman Act (1890). Federal courts ruled that the antitrust protection offered by the Sherman Act prohibited union representation of multiple labor groups. Then, in 1932, the Norris–La Guardia Act revisited the Sherman Act and determined it was in the public's best interest to allow employees to bargain collectively. Unions began to flourish, but business and industry started developing "sweetheart" contracts with company unions. These mutually beneficial labor agreements left the rank-and-file worker without work. Strikes, boycotts, and other costly labor conflicts caused the federal government to develop the National Labor Relations Act (Wagner Act) in 1935. This act created the National Labor Relations Board, which established procedures for collective bargaining. Using this method, workers organize together (usually in unions) to meet, converse, and negotiate about the work conditions with their employers; it normally results in a written contract setting forth wages, hours, and other conditions. The right to representation and the working conditions that were subject to negotiation became governed by specific written procedures and guidelines. In 1947 the Taft–Hartley Act further defined union and management rights. Right-to-work laws were tempered with restrictions on strikes.

The Landrum-Griffin Act, the Civil Service Reform Act, the Federal Labor Relations Authority, the Fair Labor Standards Act, and, most recently, the Americans with Disabilities Act have further defined the responsibilities of labor and management. The collective bargaining process affords all employees, regardless of age, sex, race, or disability, a better-defined set of legal standards to operate within. When labor and management discover and use the best methods of negotiation to solve disputes productively, federal and state labor laws will have accomplished their purpose. Today, many examples of effective labor–management relationships prove the value of effective, positive negotiation skills. Over the last 10 to 20 years, arbitration awards and court decisions have further defined management and employee rights. In certain cases management and employee rights have been specifically written into the employee–employer contract. Management rights usually include the right to do the following:

- Direct the workforce.
- Hire, promote, transfer, and assign without interference.
- Suspend, demote, discharge for cause, or take other disciplinary action.
- Take action necessary to maintain a department's efficiency.
- Make reasonable rules and regulations.

Settling Disputes

When settling disputes, management should always function within the parameters of management rights, which may be articulated in state law, a city or county charter, or a memorandum of understanding with the labor group. When questioned by labor regarding a contract issue or a problem with a policy as it relates to the current scope of coverage, managers do best to communicate openly with labor representatives and solve the dispute. Otherwise, employee–employer contracts become overburdened with detailed processes, procedures, and operational limitations when management and labor lose trust and are forced to protect their "turf" with written limitations. When labor relations fail, the attorneys and arbitrators are kept busy solving the problems that normal negotiation circumstances would usually resolve (with good labor–management relations). Breaking down labor–management mistrust requires patient and persistent leadership skills. The team feeling is developed when labor experiences leadership that openly communicates the entire game plan and provides a motivation for progress. Leadership integrity and credibility are enhanced as the organization accomplishes the plan of action and the community responds positively to the organization's efforts.

When agreement simply cannot be reached by the negotiating parties, several methods are available to reach a conclusion. These fall into three general categories: (1) use of a third party, (2) power plays (legal and illegal) by the union, and (3) power plays (legal and illegal) by management. The use of a neutral third party, called arbitration or mediation, may come in one of several forms. Compulsory binding arbitration, often considered the alternative to a strike, has the arbitrator (or arbitration board) considering the issue and reaching a decision that is binding on both sides. This form of arbitration typically is used with public employees who are forbidden to strike; obviously, it constrains the disputing parties. "Final" or "best offer" binding arbitration has the arbitrator selecting one or the other of the two parties' final, best offers. Knowledge of impending binding arbitration may prompt the disputing sides to renegotiate on their own, thus maintaining some control over the outcome. The form of arbitration lacking the final power of binding arbitration is called fact-finding, or nonbinding arbitration. This version uses a fact-finding individual or panel to examine the issue and suggest a solution. Neither side is bound by the suggestion, but often the weight of outside opinion forces the parties to settle.

The mildest form of third-party involvement is called **mediation**. In this process, a neutral mediator works with each side, either separately or together, in an attempt to get the parties to resolve the issue. The mediator acts as a neutral third party whose sole purpose is to open communication between the negotiating parties. Each party states its bottom-line requests, and the mediator attempts to find a way to develop an agreement. If third-party attempts to reconcile differences fail, either side may try to convince the other through a show of power. These actions by a union may take the form of picket, boycott, work slowdown, public petitions, work-to-rule, sick-out, or strike, where permissible. They are designed, of course, to call the public's attention to the situation. Slowdowns and work-to-rule actions may resemble some aspects of a strike, with firefighters responding to emergency calls but doing little else. Fire unions have used sick-outs, but management may counter them by enforcing medical checks.

Actions by management designed either to counter the power plays of the union or to demonstrate its own power may take several forms. These include no retroactive pay, impasse, last and best offer, compulsory contract agreement, salary freeze, and so on. Disciplinary action, injunctions, fines, and jail sentences may be imposed on labor union representatives and employees who participate in illegal labor activities. For example, contracts that have compulsory arbitration

mediation
■ The mildest form of third-party involvement in settling disputes. In this process, a neutral mediator works with each side, either separately or together, in an attempt to get the parties to resolve the issue. The mediator acts as a neutral third party whose sole purpose is to open communication between the negotiating parties. Each party states its bottom-line requests, and the mediator attempts to find a way to develop an agreement.

also will make striking or picketing illegal labor activities. Although extreme actions may bring a contractual impasse to a quicker end, the long-term repercussions will not prove worthwhile. Avoiding bitter confrontations will always better serve the department and the community.

POWERS AND MUTUAL GAINS BARGAINING

During the negotiation process, the parties can exercise power with several motives. Controlling power results in one party using organizational power to force the weaker party to obey. This tactic, referred to as win-lose or zero-sum theory, often results in solving a small, immediate concern only to fuel the weaker party's motivation to prepare for a much larger "war." As an alternative, the more advisable use of power involves a cooperative spirit that strives for win-win or non-zero-sum agreements. The logic behind the win-win theory is the belief that collective agreement results in a better solution. The collective interests of both labor and management can be melded together for a better, more productive net result.

The term *mutual gains bargaining* refers to negotiations that are satisfying to all parties. Contract bargaining between the firefighter bargaining group and management using mutual gains bargaining often results in both parties gaining; organizational goals and objectives become a guiding principle. The good of the community becomes more of a real issue than an excuse for agreement with one side or the other (employee versus employer). Thus, service to the community should be enhanced. Because the need to become more productive and accountable to the public is a reality in the 21st century, future labor agreements will focus on mutual gains bargaining issues.

THE GENERAL PROCESS OF NEGOTIATING

The process of negotiation between two parties requires communication with the intent to settle a given matter or issue. In order to initiate negotiations constructively, each party first should communicate its goals (desired outcomes). Then each party needs to review the other's goals, and the first discussion should be about those issues that both sides agree upon. It also is appropriate for either side to indicate any issues that are totally unacceptable. Negotiators must learn how to say "no" reasonably, without intimidation or emotional outbursts. Those unresolved issues (the topics that fall between yes and absolutely no) should be the first ones to be negotiated. Compromise, middle ground, or alternate options may be found during the negotiation discussions.

When dealing with controversial issues and one party says, "absolutely not," the negotiator should typically expect four stages in the negotiation process: conflict, containment, accommodation, and cooperation. The conflict stage can be damaging to long-term relations. When feelings have been vented adequately and the clashing parties realize that continued conflict will bring further harm, the containment stage is reached. The accommodation stage begins when the parties settle some of the side issues, even though there is still disagreement regarding the entire solution. During the cooperation stage, labor and management begin to believe each side has truthfully expressed all feelings, motives, and bottom-line needs; and both sides start to focus on reaching an agreement on the major issues. A cooperative spirit allows both sides to save face and develop a compromise solution.

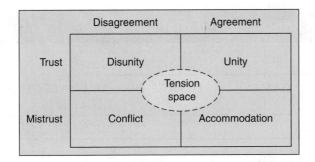

FIGURE 8.2 Group Development Climates

A similar concept of negotiation logic is depicted graphically in Figure 8.2. The four quadrants of group development climates portrayed in the illustration are defined as follows:

1. *Unity.* Both parties agree and trust each other.
2. *Accommodation.* The parties agree over certain issues but do not trust each other.
3. *Conflict.* The parties neither agree nor trust each other.
4. *Disunity.* The parties trust each other but are in disagreement over certain issues.

In organizational life, "professional" disagreements often arise. By discussing these disagreements rationally, better decisions can be reached. If either or both parties begin to show or perceive mistrust, the climate moves to the conflict quadrant. Groups in the conflict quadrant can move to the accommodation quadrant, because this movement does not require trust. Trust is easy to lose but difficult to develop. Typically there is no movement directly from conflict to disunity or unity, because each of those quadrants requires the reestablishment of trust. The center area, labeled tension space, changes in volume and can be increased by either side through precipitous acts, thoughtless words, disregard of goodwill efforts, rumors, ego feeding, personalization of arguments, and the like. Increases in tension space push the climate into the conflict quadrant, so efforts to reduce the volume of tension space are well advised. At times the parties trust each other but are in disagreement over certain issues. This is called disunity. Remaining in the disunity quadrant for long periods is difficult because mistrust can develop.

Principled Negotiation

One need not be a professional negotiator to experience bargaining. Everyday life is full of successful and unsuccessful attempts to reach agreement with others in matters of business, pleasure, and ordinary living. Negotiations range from deciding where to have dinner to buying a car. Most of us recognize two types of negotiations: hard or soft. Hard negotiators are tough minded and do not give in until they win their way. Soft negotiators do give in, perhaps too quickly and easily, and thus lose in the bargaining process. Many negotiations take place in a win-lose, zero-sum environment, in which the struggle is big and the objective is to emerge a winner. However, a third form of negotiating, **principled negotiation**, is hard on merits and soft on people. It advocates deciding issues on merit, rather than on what the parties indicate they will or will not do. It emphasizes mutual

principled negotiation
■ The name given to the interest-based approach to negotiation set out in the best-known conflict resolution book, *Getting to Yes*, first published in 1981 by Roger Fisher and William Ury. The book advocates four fundamental principles of negotiation: (1) separate the people from the problem; (2) focus on interests, not positions; (3) invent positions for mutual gain; and (4) insist on using objective criteria.

TABLE 8.2	The Basics of Principled Negotiating	
PROBLEM		**SOLUTION**
POSITIONAL BARGAINING: WHICH GAME SHOULD YOU PLAY?		**CHANGE THE GAME—NEGOTIATIONS ON THE MERITS**
SOFT	**HARD**	**PRINCIPLED**
Participants are friends.	Participants are adversaries.	Participants are problem solvers.
The goal is agreement.	The goal is victory.	The goal is wise outcome reached efficiently and amicably.
Make concessions to cultivate the relationship.	Demand concessions as a condition of the relationship.	Separate the people from the problem.
Be soft on the people and the problem.	Be hard on the people and the problem.	Be soft on the people and hard on the problem.
Trust others.	Mistrust others.	Proceed independently of trust.
Change your position easily.	Dig in to your position.	Focus on interests, not positions.
Make offers.	Make threats.	Explore interests.
Disclose your bottom line.	Mislead as to your bottom line.	Avoid having a bottom line.
Accept one-sided losses to reach agreement.	Demand one-sided gains as the price of agreement.	Invent options for mutual gain.
Search for the answer: the one the other will accept.	Search for the answer: the one you will accept.	Develop multiple options to choose from; decide later.
Insist on agreement.	Insist on your position.	Insist on using objective criteria.
Try to avoid a contest of will.	Try to win a contest of will.	Try to reach a result based on standards independent of will.
Yield to pressure.	Apply pressure.	Reason and be open to reason; yield to principle, not pressure.

gains and looks for results based on fair standards. It works best when both sides understand its concepts. The basics of principled negotiating are expressed in Table 8.2.

Principled negotiating, or bargaining, uses six guidelines:

1. Do not bargain over positions.
2. Separate the people from the problem.
3. Focus on interests, not positions.
4. Invent options for mutual gain.
5. Insist on using objective criteria.
6. Develop your best alternative to the negotiated agreement you want to reach.

Maintaining the bargaining process so that cooperative relationships result demands negotiators who can do the following:

1. Deal well with differences.
2. Disentangle relationship issues from substantive ones.
3. Be constructive.
4. Balance emotions with reason.
5. Learn how the other side sees things.
6. Listen well and consult before deciding.
7. Be wholly trustworthy but not wholly trusting.
8. Persuade, not coerce.
9. Deal seriously with those with whom we differ.[18]

Performance Appraisals

In the absence of a carefully structured system of appraisal, people tend to judge the work performance of others, including subordinates, naturally, informally, and arbitrarily. The human inclination to judge can create serious motivational, ethical, and legal problems in the workplace. Without a structured appraisal system, there is little chance of ensuring the judgments made will be lawful, fair, defensible, and accurate.

Performance appraisal systems began as a method of income justification. Appraisal was often used to decide whether the salary or wage of an individual employee was reasonable. The appraisal process was linked to material outcomes. If an employee's performance were found to be less than ideal, a cut in a pay would follow. On the other hand, if his or her performance were better than the supervisor expected, a pay raise would be in order. Initially, little, if any, consideration was given to the professional developmental possibilities the appraisal could be used for. It was felt that only a cut in pay or a raise would motivate an employee either to improve or to continue to perform well. Sometimes this basic system succeeded in getting the intended results; but, more often, it did not.

Early motivational researchers were aware that different people with roughly equal work abilities could be paid the same amount of money and yet demonstrate quite different levels of motivation and performance. These observations were confirmed in several empirical studies. Although pay rates were important, they were not the only element that had an impact on employee performance. Researchers found other issues, such as morale and self-esteem, also exercised major influence on employee performance. As a result, the traditional emphasis on rewarding employees based solely upon the outcomes achieved evolved in the 1950s as a tool for motivation and development of the employee.

Performance appraisal may be defined as a structured formal interaction between a subordinate and supervisor, usually in the form of a periodic (annual or semiannual) interview, in which the work performance of the subordinate is examined and discussed, with a view toward identifying weaknesses and strengths as well as opportunities for improvement and skills development.

Many but not all organizations use appraisal results, either directly or indirectly, to help determine monetary rewards and to identify the better-performing employees who should get the majority of available merit pay increases, bonuses,

performance appraisal
■ The process of evaluating an individual's performance by comparing it to existing standards or objectives. Process by which a supervisor (1) examines and evaluates an employee's work behavior by comparing it with preset standards, (2) documents the results of the comparison, and (3) uses the results to provide feedback to the employee to show where improvements are needed and why.

and promotions. By the same token, organizations utilize appraisal results to identify the poorer performers who may require some form of counseling or, in more extreme cases, demotion, dismissal, or decreases in pay. One of the best ways to appreciate the purposes of performance appraisal is to look at it from the different viewpoints of the main stakeholders: the employee and the organization.

EMPLOYEE VIEWPOINT

From the employee viewpoint, the purpose of performance appraisal is fourfold:

1. Tell me what you want me to do.
2. Tell me how well I have done it.
3. Help me improve my performance.
4. Reward me for doing well.

ORGANIZATIONAL VIEWPOINT

From the organization's viewpoint, one of the most important reasons for having a system of performance appraisal is to establish and uphold the *principle of accountability*. For decades researchers have known that one of the chief causes of organizational failure is "nonalignment of responsibility and accountability." Nonalignment occurs when employees are given responsibilities and duties, but are not held accountable for the ways in which they perform them. Typically, several individuals or work units appear to have overlapping roles. The overlap allows each individual or business unit to "pass the buck" to the others. Ultimately, in the severely nonaligned system, no one is accountable for anything, and so the principle of accountability breaks down completely. Organizational failure is the only possible outcome. In cases in which the nonalignment is not as severe, the organization may run like a poorly made or badly tuned engine: it will be sluggish, costly, and unreliable.

One of the principal aims of performance appraisal is to make people accountable by aligning responsibility and accountability at every level of the organization. Therefore, performance appraisal must be both developmental and administrative (see Table 8.3).

CONDUCTING THE PERFORMANCE APPRAISAL

Organizations use several methods and formats to conduct a performance appraisal. Appraisals are carried out either to improve employee performance and productivity or to determine whether some reward in salary, status, and benefits should be given. Performance appraisals also ascertain whether a remedial or corrective program is warranted.

The impact of a good performance appraisal is that it will conclude with the employee's feeling that he or she received fair and honest treatment. When this is achieved, the appraisal, describing both strengths and weaknesses observed by the appraiser over a set time period, should help to lead to improved performance. A critical appraisal should be explained clearly to the employee, along with a self-improvement plan that outlines how the employee can improve to acceptable levels. Department members must know and believe not only that an appraisal is based on the duties of the position but also that it provides an accurate measurement of the member's performance in the major dimensions of the job.

TABLE 8.3	Developmental and Administrative Appraisal

DEVELOPMENTAL	ADMINISTRATIVE
Provide performance feedback.	Document personnel decisions.
Identify individual strengths/weaknesses.	Determine promotion candidates.
Recognize individual performance.	Determine transfers and assignments.
Assist in goal identification.	Identify poor performance.
Evaluate goal achievement.	Decide retention or termination.
Identify individual training needs.	Validate selection criteria.
Determine organizational training needs.	Meet legal requirements.
Reinforce authority structure.	Evaluate training programs/progress.
Allow employees to discuss concerns.	Personnel planning.
Improve communication.	Make reward/compensation decisions.
Provide a forum for leaders to help.	

The appraisal period must be a predetermined period of time and reflect the entire time frame evaluated.

Whatever the instrument or method that evaluators use, the employee needs evidence that the method and the person conducting the appraisal are fair, honest, unbiased, and objective. Employees often make their own judgments about who works hard, is loyal, and so forth; and if appraisals do not reflect this, the appraisals and the evaluators lose credibility.

In certain cases, evaluators may be incapable of making an accurate appraisal. They may be biased or prejudiced, or may act on friendship. Evaluators may not want to appraise fellow employees negatively and thus "score" everyone high. They may not know what the performance standards require, or they may not have been trained to use the evaluation system. Evaluators who have not appraised performance over the stipulated period may inappropriately make judgments based on only one or two incidents.

Evaluating Employee Performance

The appraisal process is the third of three critical elements of the organizational performance cycle. The first element is a *job analysis* that describes each personnel classification (job responsibility and minimum qualifications). The second element is *performance standards* that explain the job expectations and the desired quality of the accomplished job. The third element is the *appraisal system* that identifies how successful the individual and the team have been in accomplishing the desired results (job standards, mission, goals, and objectives). In other words, the job analysis describes the employees' position, the performance standards explain the quantity and quality of job expectations, and the appraisal gives a description of measured accomplishment for the employee who is attempting to achieve desired results (measured in part by conformance to job standards). The entire process can be termed the *organizational performance cycle*. The organization must properly plan its management strategy for implementing an appraisal process. This process should focus on measuring progress toward accomplishing desired results (as targeted by the goals and objectives) and personal ability to meet job standards. Key points in the development of a good job appraisal process include the following:

- The appraisal process should focus on measurable job functions (as established in the job description of the job analysis) and preestablished job standards to which the employee appraised is expected to perform.
- The appraisal form should be developed so that the measurement criteria are designed clearly, in order for the employees involved with the appraisal process to understand how to use the forms. Each performance factor should be described clearly in a reference guide. It is especially important to emphasize the appraisal measurement expectations.
- The appraisal system should then be packaged as a program, and the entire program should be presented to the management team (all supervisors). The appraisal program should be included in the ongoing management training program; training is essential in order to prevent misguided use of the process. All new supervisors must also be well prepared to perform their first job appraisal.
- The appraisal system should encourage written follow-up for measured appraisal factors, so that evaluators can record feedback on observations of job performance and judge employee performance effectively.

Foremost in the decision to include certain employee appraisal measurement factors in the departmental evaluation process is the concept of job relatedness. The appraisal factors should relate back to job requirements (identified in the job analysis and documented job standards). The federal Equal Employment Opportunity Commission (EEOC) guidelines stress the importance of valid appraisal processes that reference specific job situations when measuring job-related employee performance. This is especially important when the performance appraisal becomes the subject of a personnel action such as promotion, demotion, disciplinary action, or termination.[19] The bottom line regarding an appraisal process is that the measurement criteria be reliable and valid. The appraisal system should provide an ongoing, reliable, and predictable method (as used by all evaluators).

PERFORMANCE APPRAISAL METHODS

Performance appraisals take many forms, including competency and behavioral appraisal, comparative rankings to other employees, and accomplished objectives based upon the employee job description.

- The essay method employs written narratives assessing an employee's strengths, weaknesses, past performance, and potential, and provides recommendations for improvement.
- The absolute method requires the rater to compose a statement describing the employee's behavior. Several other types of performance appraisal methods are common in both the public and private sectors.

- Comparative standards or multi-person comparison is a relative, as opposed to an absolute, method that compares one employee's performance with that of one or more others.
- In group rank ordering, also referred to as the forced-choice method, the supervisor places employees into a particular classification such as top one-fifth and second one-fifth. If a supervisor has 50 employees, only five could be in the top fifth, and five must be assigned to the bottom fifth (see Table 8.4).

The mixed-standard scale method is an approach to performance appraisal similar to other scale methods but is based on comparison with (better than, equal to, or worse than) a standard (see Table 8.5).

TABLE 8.4	Forced-Choice Distribution Scale			
LOWEST PERFORMERS	**NEXT LOWEST**	**MIDDLE**	**NEXT HIGHEST**	**HIGHEST PERFORMERS**
10%	20%	40%	20%	10%
(5 employees)	(10 employees)	(20 employees)	(10 employees)	(5 employees)

TABLE 8.5	Mixed-Standard Scale

Directions: Please indicate whether the individual's performance is above (+), equal to (0), or lower (-) than each of the following standards:

1. ___ Employee uses good judgment when addressing problems and provides workable alternatives; however, at times does not take actions to prevent problems. (medium PROBLEM SOLVING)

2. ___ Employee lacks supervisory skills; frequently handles employees poorly and is at times argumentative. (low LEADERSHIP)

3. ___ Employee is extremely cooperative; can be expected to take the lead in developing cooperation among employees; completes job tasks with a positive attitude. (high COOPERATION)

4. ___ Employee has effective supervision skills; encourages productivity, quality, and employee development. (medium LEADERSHIP)

5. ___ Employee normally displays an argumentative or defensive attitude toward fellow employees and job assignments. (low COOPERATION)

6. ___ Employee is generally agreeable but becomes argumentative at times when given job assignments; cooperates with other employees as expected. (medium COOPERATION)

7. ___ Employee is not good at solving problems; uses poor judgment and does not anticipate potential difficulties. (low PROBLEM SOLVING)

8. ___ Employee anticipates potential problems and provides creative, proactive alternative solutions; has good attention to follow-up. (high PROBLEM SOLVING)

9. ___ Employee displays skilled direction; effectively coordinates unit activities; is generally a dynamic leader and motivates employees to high performance. (high LEADERSHIP)

TABLE 8.6	Competency Appraisal Form

Communication: Clearly conveys and receives information and ideas through a variety of media to individuals or groups in a manner that engages the listener, helps the listener understand and retain the message, and invites response and feedback. Keeps others informed as appropriate. Demonstrates good written, oral, and listening skills.

KEY ELEMENT	GREATLY EXCEEDS EXPECTATIONS	EXCEEDS EXPECTATIONS	MEETS EXPECTATIONS	OCCASIONALLY MEETS EXPECTATIONS	UNSATISFACTORY
Organization and clarity	❑	❑	❑	❑	❑
Listening skills	❑	❑	❑	❑	❑
Keeping others informed	❑	❑	❑	❑	❑
Written communication	❑	❑	❑	❑	❑
Sensitivity to others	❑	❑	❑	❑	❑
Comments:					

Competency Appraisal

Appraising competencies, those work-related skills and behaviors needed to perform a job effectively, involves rating proficiency on each of the key elements of the competency. The appraisal form is designed to require the supervisor to assess each key element on the five-level rating scale as shown above (see Table 8.6).

A critical element of any competency appraisal is the identification and description of the range of behaviors observed within each of the key elements of the competency. These examples are aligned with the rating scales and rewritten into concise statements (see Table 8.7).

Graphic Rating Scale

The graphic rating scale is a trait approach to performance appraisal whereby each employee is rated according to a scale of individual characteristics (see Figure 8.3). This scale lists a set of performance factors, such as job knowledge, work quality, and cooperation, which the supervisor uses to rate employee performance using an incremental scale.

Graphic rating scales are the most easily developed, administered, and scored format. They consist of a listing of desirable or undesirable personality traits in one column and beside each trait is a scale (or box), which the rater marks to indicate the extent to which the rated employee demonstrates the trait. An example of a graphic rating appraisal appears in Table 8.8.

Behavioral Methods of Appraisal

The behaviorally anchored rating scale (BARS) combines elements from an emergency incident and uses a graphic rating scale approach to provide an

TABLE 8.7 Sample Competency Appraisal Examples

Communication: Clearly conveys and receives information and ideas through a variety of media to individuals or groups in a manner that engages the listener, helps the listener understand and retain the message, and invites response and feedback. Keeps others informed as appropriate. Demonstrates good written, oral, and listening skills.

KEY ELEMENT	GREATLY EXCEEDS EXPECTATIONS	EXCEEDS EXPECTATIONS	MEETS EXPECTATIONS	NEEDS DEVELOPMENT	UNSATISFACTORY
Organization and clarity	Conveys thoughts and ideas so as to avoid misunderstandings. Communicates with needs and expectations of audience in mind.	Provides clarity of thought and needs.	Conveys thoughts clearly and concisely.	Occasionally rumbles; not concise.	Has difficulty expressing thoughts.
Listening skills	Listens with demonstrated understanding and empathy. Thoughtfully explores topic as appropriate.	Excellent listener; does not interrupt.	Listens actively and attentively and asks appropriate questions.	Sometimes is distracted; misses conversation.	Fails to listen and share feedback.
Keeping others informed	Continuously fulfills all knowledge requirements of supervisors, coworkers, and others.	Provides substantial communication; keeps others well informed.	Keeps supervisors, coworkers, and others well informed.	Periodically will forget to share information.	Fails to share important information or passes on trivia.
Written communication	Communications are error free, have positive tone, and seem professionally written.	Sets a standard, which others emulate.	Communicates well in writing.	Report format sometimes confusing and does not flow well.	Written communications are unclear, disorganized, lack substance; contain grammatical and/or spelling errors.
Sensitivity to others	Continuously tailors communications to match the listener/s; uses appropriate style, level of detail, grammar, and organization of thoughts to actively engage the listener/s.	Always looking to help others and is consistently sensitive to others when communicating.	Consistently sensitive to cultural, gender, educational, and other individual characteristics when communicating with others.	Typically sensitive; occasionally makes inappropriate remarks.	Insensitive to cultural, gender, educational, and other individual characteristics when communicating with others.

TABLE 8.8 Graphic Rating Appraisal of Personnel

PERSON EVALUATED: _____ **POSITION:** _____

DIVISION: _____

	OUTSTANDING	VERY GOOD	GOOD	SATISFACTORY	UNSATISFACTORY	UNKNOWN
Performance Factors						
1. Effectiveness						
2. Use of time and materials						
3. Prompt completion of work						
4. Thoroughness						
5. Initiative						
6. Perseverance						
Abilities, Skills, and Faculties						
7. Technical skills						
8. Communication skills						
9. Judgment						
10. Analytical ability						
11. Ability to organize						
12. Ability to inspire and influence staff						
13. Ability to inspire and influence others than staff						
14. Flexibility and adaptability						
15. Imaginativeness and creativity						
16. Ability to develop subordinates						
17. Breadth of concepts						

Ethical Considerations

18. Loyalty to department

19. Loyalty to peers

20. Loyalty to subordinates

21. Sense of ethics

22. Cooperativeness

23. Responsibility

24. Commitment of service

25. Open-mindedness

Date Evaluated: _____   Evaluator: _____

The above appraisal was reviewed with me on _____

_____
(Signature of Person Evaluated)

Comments: _____
_____

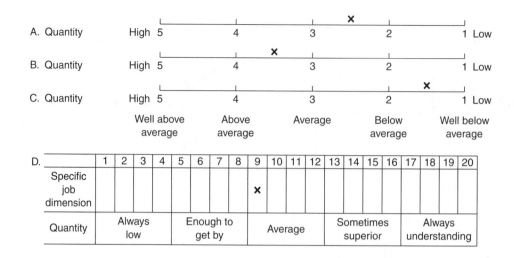

FIGURE 8.3 Graphic Rating Scale

assessment. The supervisor rates employees according to items on a numerical scale (see Table 8.9).

The behavior observation scale (BOS) is a performance appraisal method that measures the frequency of observed behavior. This scale of 1 (almost never) to 5 (almost always) is used to appraise an employee's job performance based upon key job tasks (see Table 8.10).

Management by Objectives

Management by objectives (MBO) evaluates how well an employee has accomplished objectives determined to be critical in job performance. This method aligns objectives with quantitative performance measures such as sales, profits, and zero-defect units produced. The use of management objectives was first widely advocated in the 1950s by the noted management theorist Peter Drucker.

MBO is a philosophy of management that rates performance on the basis of employee achievement of goals set by mutual agreement of employee and manager. Key elements of MBO follow:

- Employee involvement creates higher levels of commitment and performance.
- MBO encourages employees to work effectively toward achieving desired results.
- Performance measures should be measurable and should define results.

Figure 8.4 illustrates this process.

Unique Features and Advantages of Management by Objectives The principle behind MBO is to create empowered employees who are clear about their roles and responsibilities, understand the objectives they are to achieve, and thus help in the achievement of organizational as well as personal goals. Some of the important features and advantages of MBO follow:

- *Clarity of goals.* With MBO came the concept of SMART goals; that is, goals that are *s*pecific, *m*easurable, *a*chievable, *r*ealistic, and *t*ime bound. The goals thus set are clear, motivating, and they link organizational goals with performance targets for the employees.

TABLE 8.9 | Behaviorally Anchored Rating Scale (BARS)

FIREFIGHTING STRATEGY: Knowledge of Fire Characteristics

HIGH	7	–Uses tactics to promote the most effective fire extinguishment and the highest level of firefighter safety
	6	–Correctly assesses best point of entry for fighting fire
	5	–Uses type of smoke as indicator of type of fire
AVERAGE	4	–Understands basic hydraulics
	3	–Cannot tell the type of fire by observing the color of flame
	2	–Cannot identify location of the fire
LOW	1	–Will not change firefighting strategy once a course of action is established

TABLE 8.10 | Behavior Observation Scale (BOS)

5 represents *almost always*   95–100% of the time
4 represents *frequently*   85–94% of the time
3 represents *sometimes*   75–84% of the time
2 represents *seldom*   65–74% of the time
1 represents *almost never*   0–64% of the time

SALES PRODUCTIVITY	ALMOST NEVER				ALMOST ALWAYS
1. Reviews individual productivity results with manager	1	2	3	4	5
2. Suggests to peers ways of building sales	1	2	3	4	5
3. Formulates specific objectives for each contact	1	2	3	4	5
4. Focuses on product rather than customer problem	1	2	3	4	5
5. Keeps account plans updated	1	2	3	4	5
6. Keeps customer waiting for service	1	2	3	4	5
7. Anticipates and prepares for customer concerns	1	2	3	4	5
8. Follows up on customer leads	1	2	3	4	5

FIGURE 8.4 The
MBO Process

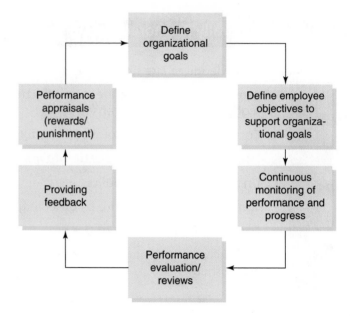

- *Focus on future rather than on past.*   Goals and standards are set for the performance for the future with periodic reviews and feedback.
- *Motivation.*   Involving employees in the whole process of goal setting and increasing employee empowerment encourages employee job satisfaction and commitment.
- *Better communication and coordination.*   Frequent reviews and interactions between superiors and subordinates help to maintain harmonious relationships within the enterprise and also solve many problems faced during the period.

The MBO approach overcomes some of the problems that arise as a result of assuming that the employee traits needed for job success can be reliably identified and measured. The MBO method recognizes that it is difficult to neatly dissect all the complex and varied elements that make up employee performance. Therefore, its focus is on the observation of direct results instead of on the traits and attributes of employees (which may or may not contribute to performance) that can at times be subjective (see Table 8.11). If the employee meets or exceeds the set objectives, then he or she has demonstrated an acceptable level of job performance. Therefore, employees are judged according to real outcomes, and not on their potential for success or on someone's subjective opinion of their abilities.

Disadvantages of Management by Objectives   Although the MBO method of performance appraisal can give employees a satisfying sense of autonomy and achievement, the downside is it can lead to unrealistic expectations about what can be reasonably accomplished.

Supervisors and subordinates must have an open line of communication to use MBO appraisal methods. They will need the ability to communicate expectations clearly not only during the initial stage of setting objectives but also for self-auditing and self-monitoring purposes.

Although one of the strengths of the MBO method is the clarity of purpose that flows from a set of well-articulated objectives, this can also be a source of weakness.

TABLE 8.11	MBO Performance and Development Plan

Employee Name:		Date:	
Position:		Department:	
Appraised By:		Date Started:	

Performance Plan

Major Responsibilities/Goals	Appraisal of Goal Achievement
Major Responsibilities: Goals:	
Major Responsibilities: Goals:	
Major Responsibilities: Goals:	

Development Plan

Development Goals	Development Activities	Due Date	Status	Comments
1.				
2.				
3.				

Overall Comments

Supervisor Comments:

Employee Comments:

Sign Off

Planning		Review	
Supervisor	Date	Employee	Date
Manager	Date	Employee	Date

Source: What Makes a Good Leader. *MBO Employee Performance Evaluation Form.* Retrieved October 12, 2010, from http://www.whatmakesagoodleader.com/Employee-Performance-Evaluation-Form.html

FIGURE 8.5 The 360°
Appraisal

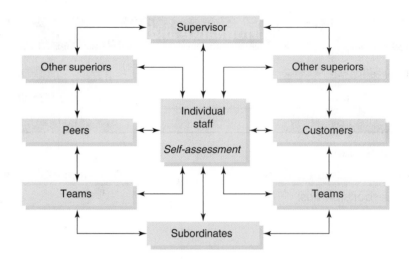

It has become very apparent that the modern organization must be flexible to survive. Conversely, objectives, by their very nature, tend to impose a certain rigidity.

Of course, the obvious answer is to make the objectives more fluid and yielding, but the penalty for fluidity is loss of clarity. Variable objectives may cause employee confusion. It is also possible that fluid objectives may be distorted to disguise or justify failures in performance.[20]

The 360-Degree Feedback

The 360-degree feedback method provides a comprehensive perspective of employee performance by utilizing feedback from the full circle of people with whom the employee interacts: supervisors, subordinates, and coworkers (see Figure 8.5).

This multilevel method of feedback is effective for both career coaching and identifying strengths and weaknesses. It is also a great tool for senior leadership to use to gain a perspective on how well their subordinates think they are doing (see Table 8.12).

TABLE 8.12	360-Degree Professional Development Assessment

Please complete this Professional Development Assessment on me. Your confidential, but candid, evaluation of my performance and behaviors will help me understand how others perceive my performance and will aid in my professional development.

Please indicate your level of agreement with each of the following statements on a scale of 1 (complete disagreement) to 5 (complete agreement). Leave blank any categories you are unable to assess.

EVALUATION CATEGORIES

Based on your interactions, observations, and working with or for me, I:

1. Am clearly receptive and responsive to requests from others and encourage a cooperative and responsive attitude among my departmental staff (e.g., convey and foster a "can-do" attitude).

1   2   3   4   5

Comments: _____

TABLE 8.12 360-Degree Professional Development Assessment (*Continued*)

2. Demonstrate through my personal actions and through the conduct of my employees my understanding and support of the department's mission, vision, and organizational principles (e.g., promote organizational goals or meet organizational needs).

   1   2   3   4   5

   Comments: _____

3. Understand and appreciate different and opposing perspectives on issues, solicit input, and easily adapt to changes in the organization and/or the department.

   1   2   3   4   5

   Comments: _____

4. Communicate openly and clearly with peers and subordinates.

   1   2   3   4   5

   Comments: _____

5. Am knowledgeable and embrace contemporary management concepts regarding people and process; facilitate implementation of innovative methods that increase morale, efficiencies, and customer service.

   1   2   3   4   5

   Comments: _____

6. Am a supportive, active listener, and contributor in discussions and meetings.

   1   2   3   4   5

   Comments: _____

7. Follow through on commitments.

   1   2   3   4   5

   Comments: _____

8. Demonstrative initiative in responding to emerging needs (for the organization) and effectively communicate to those around me how we will respond to new needs, collectively and individually.

   1   2   3   4   5

   Comments: _____

9. Based on your observations, list what you think I should:

   a. do more of: _____

   b. do less of: _____

   c. do the same: _____

PROBATIONARY APPRAISAL

The evaluation of new firefighters often continues past the graduation from the recruit academy through the probationary year. The **probationary appraisal** process provides for monthly appraisal of the employee to assess progress from

probationary appraisal
■ A monthly appraisal of the employee to assess progress from month to month for the first year of employment.

month to month for the first year of employment. Many departments often transfer these employees to several officers through the probationary period to gain a clear perspective on the employee's ability. A sample probationary appraisal form is shown in Table 8.13.

TABLE 8.13	Probationary Firefighter Appraisal

Member:_____ Date _____ Assignment:_____

Captain:_____ Period Covered: ___ To:____ Evaluation No:_____

QUALITY OF INFORMATION

Number of shifts captain worked with probationary firefighter during this evaluation period: _____
Number of shifts probationary firefighter missed (i.e., illness, vacation, holiday, etc.): _____
Number of times this probationary firefighter was late during this evaluation period: _____
Number of shift trades the probationary firefighter had during this evaluation period: _____

RATING CATEGORY	ABOVE STANDARD	STANDARD	BELOW STANDARD	UNSATIS-FACTORY

Place an **X** in the appropriate box. Any rating other than Standard must be accompanied by comments.

1. PERFORMANCE AT MEDICAL-AID AND RESCUE EMERGENCIES

a) Patient assessment	❑	❑	❑	❑
b) Cooperation with EMS units	❑	❑	❑	❑
c) Knowledge and operation of medical equipment	❑	❑	❑	❑
d) Remains calm and shows professionalism as a firefighter	❑	❑	❑	❑

2. PERFORMANCE DURING FIRE GROUND OPERATIONS

a) Calm and cooperative	❑	❑	❑	❑
b) Compliance with instructions	❑	❑	❑	❑
c) Stamina during fire ground operations	❑	❑	❑	❑
d) Uses appropriate skills in various fire ground functions (ventilation, overhaul, salvage, rescue, etc.)	❑	❑	❑	❑
e) Performs firefighting functions in a safe manner	❑	❑	❑	❑

3. COMMUNICATION/PUBLIC RELATIONS

a) Demonstrates good communications skills with supervisors and public	❑	❑	❑	❑
b) Completes all written assignments in a timely and efficient manner	❑	❑	❑	❑
c) Handles public relations matters in a courteous manner	❑	❑	❑	❑

4. PARTICIPATION IN DRILL AND TRAINING FUNCTIONS

a) Cooperation, motivation, and concern for safety	❑	❑	❑	❑
b) Correctly places SCBA in service (60 sec)	❑	❑	❑	❑
c) Safely and proficiently operates truck equipment	❑	❑	❑	❑
d) Performance during simulated emergency incidents	❑	❑	❑	❑
e) Prepares and delivers accurate station schools	❑	❑	❑	❑

| TABLE 8.13 | Probationary Firefighter Appraisal (*Continued*) |

5. STUDY HABITS
a) Ability and willingness to learn ☐ ☐ ☐ ☐
b) Initiative and time management ☐ ☐ ☐ ☐

6. PHYSICAL FITNESS ACTIVITIES
a) Actively participates ☐ ☐ ☐ ☐
b) Physical condition (endurance) ☐ ☐ ☐ ☐
c) Physical strength and ability ☐ ☐ ☐ ☐
d) Wears proper uniform ☐ ☐ ☐ ☐

RATING CATEGORY	ABOVE STANDARD	STANDARD	BELOW STANDARD	UNSATIS-FACTORY
7. PERFORMANCE OF STATION MAINTENANCE DUTIES				
a) Efficient and willing in performing station maintenance duties	☐	☐	☐	☐
b) Plans, organizes, and completes work effectively	☐	☐	☐	☐
c) Initiative in performing duties	☐	☐	☐	☐
8. PERFORMANCE OF EQUIPMENT MAINTENANCE DUTIES				
a) Assists in routine equipment maintenance	☐	☐	☐	☐
b) Properly cares for equipment during fire ground operations and sees that all equipment is left in a ready condition	☐	☐	☐	☐
c) Mechanical dexterity	☐	☐	☐	☐
9. OBSERVANCE OF CITY/DEPARTMENT RULES, REGULATIONS, AND POLICIES				
a) Demonstrates knowledge of city and dept. rules, regulations, and policies	☐	☐	☐	☐
b) Keeps informed on all directives, bulletins, and department communications	☐	☐	☐	☐
c) Personal appearance conforms to uniform guidelines	☐	☐	☐	☐

Evaluator's Comments:

Firefighter Comments:

Performance plan for next rating period:

Signature of Member:_____ Date:_____ Assignment:_____

☐ I agree with evaluation   ☐ I disagree with evaluation   ☐ I request a review with next higher level

Signature of Evaluator:_____ Date:_____ Rank:_____

Signature of Battalion Chief:_____ Date:_____

Signature of Administrator:_____ Date:_____ Position:_____

Source: Fresno Fire Department, Fresno, California.

The Promotional Process

Today, many fire departments use assessment processes for promotion. These assessment processes are designed to evaluate specific performances. Although assessment centers utilize many different approaches, most share some common testing elements. Unlike the promotional tests previously used in the fire service, assessment processes focus on behaviors rather than simply knowledge. The characteristics of assessment processes follow:

- Assessment processes evaluate performance rather than knowledge. Successful performance requires technical knowledge, but it is based on what the participant *does* rather than what the participant *knows*.
- Assessment processes are based on the principle that "behavior predicts behavior." In other words, a participant who successfully performs the behaviors in a process will probably be able to perform them in the position for which he or she is being tested.
- Job analysis, as cited earlier, guides the development of the assessment process. Each valid assessment process is based on a thorough analysis of the identified minimum job qualifications. The dimensions, behaviors, and tasks included in the assessment reflect those required in the position.
- An assessment process is a systematic process for measuring the identified behaviors and dimensions.
- Preparation for an assessment process requires practice and professional development. An assessment process is more than regurgitating knowledge; it is evaluation of a person's ability to apply knowledge. Unlike a written test, if a person has not taken the time to develop those skills and behaviors evaluated in the assessment process, reading and studying will not be of any value.

See "Common Components of the Assessment Process" on page 300.

A dimension is an attribute or quality necessary to perform a task or a job. For any given job in the fire service, specific dimensions are the foundation for being able to perform the job successfully.

During the development of an assessment process, the specific dimensions required for the position are also developed. Based on those dimensions, the components are identified that will be used in the assessment. It is essential that every promotional candidate have a clear understanding of the dimensions for the job being sought because it makes preparation for the assessment more straightforward. Dimensions are different than personality traits. Personality traits such as honesty, initiative, and so on are embedded in each dimension, and those traits become part of how each person applies the dimension. For example, an important personality trait is respect for others. This trait is a key ingredient in such dimensions as interpersonal relations and decision making. Many different dimensions are used in evaluation processes. However, 11 dimensions are commonly found in fire service assessment processes.

The Disciplinary Process

Effective discipline can help to eliminate ineffective employee behavior. An employee should be disciplined when he or she chooses to break the rules or is not willing to perform the job to standards. Discipline is corrective actions taken by a

1. **Oral communication.** The ability to communicate orders and information to others clearly and concisely, and the ability to listen to messages others are sending.
2. **Written communication.** The ability to create, complete, and disseminate written materials such as reports, memos, and letters.
3. **Problem analysis.** The ability to break a problem into individual parts and identify the relationships between the parts and the problem.
4. **Interpersonal relations.** The ability to create and maintain positive working relationships with other people. These relations should foster honesty, trust, and open communication.
5. **Delegation.** The ability to identify tasks that can be given to others and the ability to communicate expectations and specifics about the tasks.
6. **Decision making.** The ability to identify solutions to problems, evaluate the pros and cons of each solution, select the best solution, and then communicate that solution to others.
7. **Decisiveness.** The readiness to make decisions and stay committed to a course of action.
8. **Organization.** The ability to efficiently establish a plan of action for self and subordinates to achieve a specific goal. This includes effective use of personnel and resources.
9. **Technical expertise.** The skills necessary to perform technical operations safely, effectively, and efficiently. For example, as a company officer and battalion chief, technical expertise is required in fireground and emergency operations.
10. **Evaluation.** The ability to assess personnel performance or progress toward a goal objectively.
11. **Time management.** The ability to effectively and efficiently use time as a resource in accomplishing assigned tasks and duties.[21]

supervisor when an employee does not abide by organizational rules and standards. Common categories of disciplinary problems are attendance, poor performance, and misconduct. Attendance problems include unexcused absence, chronic absenteeism, unexcused or excessive tardiness, and leaving without permission. Poor performance consists of failure to complete work assignments, producing substandard products or services, and failure to meet established production requirements. Misconduct includes theft, falsifying an employment application, willfully damaging organizational property, and falsifying work records.

Discipline is often thought of as the application of punishment for the infraction of a rule. Children are disciplined at home or in school for misbehaving. Some managers carry that same logic forward when supervising their employees. The results of such a simplistic view of discipline can be devastating for individual employees and the team as a whole. Discipline should be viewed instead as a corrective experience with positive expectations. Leaders develop discipline through quality training; clear and understandable standards of operation; team-supported policies and procedures; and constant, trusted communication between supervisors

- *Oral presentation.* Candidates prepare and then deliver short talks on assigned topics. Examples may be instructing the staff of a business on the safe use of fire extinguishers, talking to kids about fire safety in the home, and presenting to senior staff on a new deployment concept.
- *Written presentation.* Candidates prepare a memo or report, per instructions. These instructions may be to prepare a report to the chief regarding a complaint from another city department.
- *Role-playing.* Candidates interact with a subordinate with some hidden problem requiring interpersonal skills, supervisory skill, knowledge of department policy, and so on. The subordinate might be starting to miss work or be under some suspicion of improper activity.
- *Tactical scenario.* Candidates respond to real-time, unfolding, and difficult tactical scenarios that test their ability to think quickly and accurately. Such scenarios include emergency events, often with one or more usual circumstances present.
- *Group problem solving.* Candidates work together on an assigned problem or prepare an agenda or plan a public relations event. Often, the written exercise mentioned above will form the foundation of this common exercise.
- *Structured interview questions.* Candidates respond to a uniform set of questions such as why they want this promotion, what they view as contemporary issues in the fire service, and the like.

and employees. They use disciplinary action to correct an employee's acknowledged violation of the standards of operation. A proper level of discipline should be in place before taking punitive disciplinary action. Employees cannot be expected to comply with a standard of operation or behavior that has not been properly explained. Discipline as an organizational function associated with controlling and teaching involves goal setting, team building, peer pressure, conflict resolution, employee assistance, due process, grievance processes, and intervention strategies.

progressive discipline
■ Attempts to correct the employee's behavior by imposing increasingly severe penalties for each infraction.

PEARSON
myfirekit

Visit MyFireKit Chapter 8 for more information on the disciplinary process.

The purpose of discipline is to provide a corrective action with personnel who have made a mistake. This can be accomplished through an honest and thoughtful communication between the supervisor and the employee. If the mistake was deliberate, repeated, or a serious violation of known standards, the supervisor may decide that a stronger disciplinary action is needed. The first level of disciplinary action begins with an oral reprimand. Disciplinary action may progress to a written reprimand, time off, demotion, and, possibly, termination. Most disciplinary problems can be managed through a step-by-step process called **progressive discipline**.

Employee Assistance Programs

Modern personnel management focuses on the wise use of human resources and demonstrates concern for individuals. Organizations often use expert programs to assist employees and volunteers who are experiencing problems affecting job performance and their lives. The **employee assistance program** (EAP) is available for use by employees without the knowledge of the organization's officers or staff. Professional counselors help employees with personnel problems, marital problems, drug or alcohol abuse, and other mental and physical problems. In addition to expert support offered through the employee assistance program, employers may organize a peer support group. Peer groups, with support from a professional counselor, are helpful for employees who request group support for organizational problem counseling. An employee may desire peer feedback on how to deal with a personnel problem or a distasteful relationship with a supervisor. The peer group should be well trained and prepared to offer constructive advice. Peer groups also are useful for critical incident stress debriefings. Many times a group replay of a horrendous employee experience is helpful for the employee's releasing stress and emotional concerns. Participation in these programs is entirely confidential, and the employer should not place a notation in the personnel folder unless attendance at the EAP was mandated by disciplinary action. Employee assistance programs provide a useful resource for managers who realize that professional help can improve a worker's nonconforming behavior.

employee assistance program
■ Programs offered by employers that provide an opportunity for troubled employees to seek professional support to solve personal and/or work-related problems.

Conclusion

To lead and manage in the today's complex environment is often difficult, and it will not get any easier. The management of personnel provides the greatest potential for organizational success and presents the greatest risk for the organization and for the leader personally if done incorrectly. As covered in this chapter, the fire service is balancing a multitude of changes from societal impacts, diversity, and generational issues; ensuring it is meeting all laws and regulations; and effectively managing personnel on a day-to-day basis. As leaders manage in this ever-changing environment, it is essential to keep abreast of the changes occurring in each of these areas. Anything less will place both the leaders and the organization at risk.

PEARSON
myfirekit

Visit MyFireKit Chapter 8 for Perspectives of Industry Leaders.

CHAPTER REVIEW

Review Questions

1. Explain how demographic shifts in the United States have impacted the fire service.
2. How do the different generational values affect how we manage our personnel?
3. List four important federal laws and decisions that influence labor relations, and any state and local regulations in your area significantly affecting labor relations.
4. Select a prominent trend in society, and describe how it has influenced the labor–management climate and the goals of managers and members.
5. Identify and describe a law from your state that affects fire department labor relations positively or negatively.
6. Explain the IAFC/IAFF labor–management initiative.
7. Explain the meanings and processes of collective bargaining, mediation, and binding arbitration.
8. List three examples of soft and hard positional bargaining and principled bargaining.
9. Describe one employee appraisal form, technique, or program that fosters the development of desired knowledge, skills, and attitudes.
10. Explain the concept of discipline as it applies both to the operations of a fire department and to the need for human resource development within the same organization.
11. Compare any differences in the application of discipline to volunteer members and career members.
12. Describe the importance of due process as it relates to official charges and discipline.
13. Describe how discipline can be a positive function of human resource development as well as a necessity in personnel administration.

PEARSON
myfirekit™

For additional review and practice tests, visit www.bradybooks.com and click on MyBradyKit to access book-specific resources for this text!

Register for MyFireKit by following directions on the MyFireKit student access card provided with this text. If there is no card, go to www.bradybooks.com and follow the MyBradyKit link to buy access from there.

References

1. Richard W. Judy and Carol D'Amico, *Workforce 2020: Work and Workers in the 21st Century* (Indianapolis, IN: Hudson Institute, 1997).
2. Ibid.
3. Ibid.
4. Ibid.
5. Ibid.
6. Kathryn A. Fox, Chris W. Hornick, and Erin Hardin, *Achieving and Retaining a Diverse*

Fire Service Workforce, International Association of Fire Fighters Diversity Initiative, January 2006, p. 1.

7. Ibid., p. 4.

8. Ibid., p. 18.

9. Ibid., p. 36.

10. Denise M. Hulett, Marc Bendick, Jr., Sheila Y. Thomas, and Francine Moccio, *A National Report Card on Women in Firefighting* (Madison, WI: International Association of Women in Fire & Emergency Services, April 2008), p. 1.

11. Letter to Members from President of International Association of Women in Fire & Emergency Services, April 2008.

12. Linda Gravett, *Managing Conflict Across Generations.* Retrieved October 9, 2010, from http://www.training-modules.com/contributions/managing_conflict.asp

13. David Lee, *HumanNature@Work.* www.humannatureatwork.com

14. Ibid., p. 206.

15. *Personnel Management for the Fire Service Course Guide,* National Fire Academy Degrees at a Distance Program.

16. Memorandum of Understanding, IAFC/IAFF Labor-Management Initiative Guiding Principles, August 2008.

17. *Unit 11: Employee and Labor Relations,* United States Fire Administration, National Fire Academy.

18. *Unit 12: Collective Bargaining,* United States Fire Administration, National Fire Academy.

19. U.S. Equal Employment Opportunity Commission, *Employment Tests and Selection Procedures.* Retrieved October 10, 2010, from http://www.eeoc.gov/policy/docs/factemployment_procedures.html

20. Archer North & Associates, *Results Method: Management by Objectives (MBO).* Retrieved October 10, 2010, from http://www.performance-appraisal.com/results.htm

21. Ed Kirtley, *Preparing for Fire Department Assessment Centers.* Retrieved January 1, 2010, from www.osufst.org/leaders-corner

Firefighter Health and Safety: Making a Difference

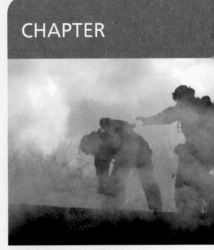

(Courtesy of Tony Escobedo)

OBJECTIVES

After completing this chapter, you should be able to:

- Describe the trend of firefighter fatalities for the past 20 years.
- Describe the leading cause of death of on-duty firefighters.
- Describe the geographical distribution of firefighter fatalities.
- Explain the prominent causes of firefighter injuries.
- Describe the initiatives focused on firefighter safety.
- Describe the initiatives focused on firefighter health and wellness.

PEARSON
myfirekit™

For additional review and practice tests, visit www.bradybooks.com and click on MyBradyKit to access book-specific resources for this text!

Introduction

Many in the fire service have had the opportunity to attend the **National Fire Academy** in Emmitsburg, Maryland. As they walk through the grounds of the academy, they often pass the National Fallen Firefighters Memorial. Constructed in 1981, Congress officially designated the memorial as the National Memorial for career and volunteer firefighters in 1990. It is a symbol of honor for those who carry on the traditions of providing service to their communities. The full impact of this site is not felt until one has visited during the **National Fallen Firefighters Foundation** Memorial weekend and participated in the ceremony honoring those lost in the line of duty. During this weekend, the families of those who have been lost in the line of the duty the preceding year attend to recognize the loss of their loved ones and to remember the sacrifices that have been made in service to others. If everyone in the fire service, from the newest recruit to the most senior chief officer, could have the opportunity not only to attend the ceremony but also to present a commemorative flag to one of these families, it would change the culture of the fire profession overnight. The opportunity to look into the eyes and observe the grief of those family members who have been left behind, to experience what the loss of a firefighter means to the fire service, and to witness the impact of that loss on the members of the department, is incredibly moving. If everyone in the fire service were exposed to this, one would do everything in his or her power to eliminate line-of-duty deaths in the U.S. fire service.

This chapter covers a range of statistical information and studies as an overview of the fatality and injury trends the fire service has experienced over the last 10 to 20 years. In reading through this chapter, realize that every statistic involves an individual, a family left behind, a department, and a community that were impacted greatly by the loss of the individual. Although the numbers can tell a story, they cannot capture the true impact of what occurs when a firefighter fatality or significant injury is experienced.

National Fire Academy ■ In the United States, the national institute of education for the fire service, located in Emmitsburg, Maryland.

National Fallen Firefighters Foundation ■ Abbreviated as NFFF, it was created to lead a nationwide effort to honor America's fallen firefighters and provide resources to assist their survivors in rebuilding their lives.

National Fallen Firefighters Foundation

Congress created the National Fallen Firefighters Foundation (NFFF) (see Figure 9.1) to lead a nationwide effort to honor America's fallen firefighters and provide resources to assist their survivors in rebuilding their lives. When Congress established the NFFF (see Figure 9.2), it provided neither funding nor financial assistance to carry out its mission. However, since 1992, the nonprofit Foundation has developed and expanded programs that fulfill that mandate. The Foundation's recent emphasis on preventing line-of-duty deaths is a natural extension of those efforts, which are directed equally toward all firefighters and involve no other mission or constituency.

THE ANNUAL NATIONAL FALLEN FIREFIGHTERS MEMORIAL WEEKEND

Each October, the Foundation sponsors the official national tribute to all firefighters who died in the line of duty during the preceding year. Thousands attend

FIGURE 9.1 National Fallen Firefighters Memorial *(Courtesy of Ron J. Siarnicki, National Fallen Firefighters Foundation)*

FIGURE 9.2 National Fallen Firefighters Foundation Logo *(Courtesy of Ron J. Siarnicki, National Fallen Firefighters Foundation)*

the weekend activities held at the National Fire Academy in Emmitsburg, Maryland. This weekend features special programs for survivors and coworkers along with moving public ceremonies. The Foundation provides travel, lodging, and meals for immediate survivors of the fallen firefighters being honored. This allows survivors to participate in Family Day sessions conducted by trained grief counselors and in the private and public tributes.

SUPPORT PROGRAMS FOR SURVIVORS

When a firefighter dies in the line of duty, the Foundation provides survivors with a place to turn. Families receive emotional assistance through a Fire Service Survivors Network, which matches survivors with similar experiences and circumstances. Families receive a quarterly newsletter and specialized grief publications. The network's Web site provides extensive information on survivor benefits, Foundation programs, and other resources.

Awarding Scholarships to Fire Service Survivors

Spouses, children, and stepchildren of fallen firefighters are eligible for scholarship assistance for education and job training costs.

Helping Departments Deal with Line-of-Duty Deaths

Under a Department of Justice grant, the Foundation offers training to aid fire departments in handling a line-of-duty death. Departments receive extensive pre-incident planning support. Immediately after a death, a Chief-to-Chief Network provides technical assistance and personal support to help the department and the family.

Working to Prevent Line-of-Duty Deaths

With the support of fire and life safety organizations, the Foundation has launched a major initiative to reduce firefighter deaths. The goal of the Firefighter Life Safety Initiatives Program is to reduce line-of-duty firefighter deaths by 25 percent within 5 years and by 50 percent within 10 years. Part of the effort is the promotion of research, which can positively impact this goal.

RON SIARNICKI, EXECUTIVE DIRECTOR, NATIONAL FALLEN FIREFIGHTERS FOUNDATION

(RB) You have been the executive director of the National Fallen Firefighters Foundation for how many years now?

(RS) It will be eight years July 2009. It's hard to believe.

(RB) Share a little bit about how important the foundation is.

(RS) The organization is committed to families of fallen firefighters through a congressional mandate in 1992. The foundation was formed to honor all the fallen firefighters and help the families rebuild their lives. It truly tries to reach out to all aspects of the fire service community and bring them together to carry out the mission. We try to work for programs that reach out to the families to help them put things back together. The foundation is a small group and its uniqueness in the fire service community also carries out our mission. It has been a phenomenal journey. People reach out and do what they need to do to be of service to the families.

(RB) I know one of the objectives of the foundation is the elimination of firefighter fatalities.

(RS) When the foundation was created, that was our mission. It is a funny story how it happened. In 2002, the memorial in Emmitsburg was for the 9/11 deaths and the other firefighters who died in 2001. The memorial was originally constructed in the 1980s. There was space to put individual markers in a semicircle around the memorial. In 2002, more names had to be added, and we realized that we could only add names for at least one more year before we had to create a third row or redesign the memorial. What struck us was we were running out of room for names of fallen firefighters, and our first response was we need to build a bigger memorial. As we thought through the process, we said, no, what we need to be doing is putting fewer names on the wall. This is where the idea started for creating prevention as a part of our mission. We did renovate the memorial, as the American fire service community knows. We added enough space for the names for at least the next 50 years, understanding that hopefully line-of-duty deaths will decrease. At this same time, the birth of **Everyone Goes Home** created the 16 life safety initiatives, which are the footprint for whatever goes on in the prevention and research part of our program. I feel it is a true representation of the issues identified by the American fire service and how we can, in fact, have a 50 percent reduction in line-of-duty deaths over the next 10 years.[2]

Everyone Goes Home ■ A national program instituted by the National Fallen Firefighters Foundation to prevent line-of-duty deaths and injuries. In March 2004, a Firefighter Life Safety Summit was held to address the need for change within the fire service. Through this meeting, the 16 firefighter life safety initiatives were produced and a program was born to ensure *Everyone Goes Home*.

Creating a National Firefighter Memorial Park

The Foundation is expanding the national memorial site in Emmitsburg, Maryland, to create the first permanent national park honoring all firefighters. The park includes a brick Walk of Honor that connects the Fallen Firefighters Memorial Chapel and the official national monument.[1]

Firefighting Fatalities

Each year in the United States, approximately 100 firefighters are killed while on duty and tens of thousands are injured. Although the number of firefighter fatalities has steadily decreased over the past 20 years, the incidence of firefighter fatalities per 100,000 incidents has actually risen over the last several years (see Figures 9.3 and 9.4).

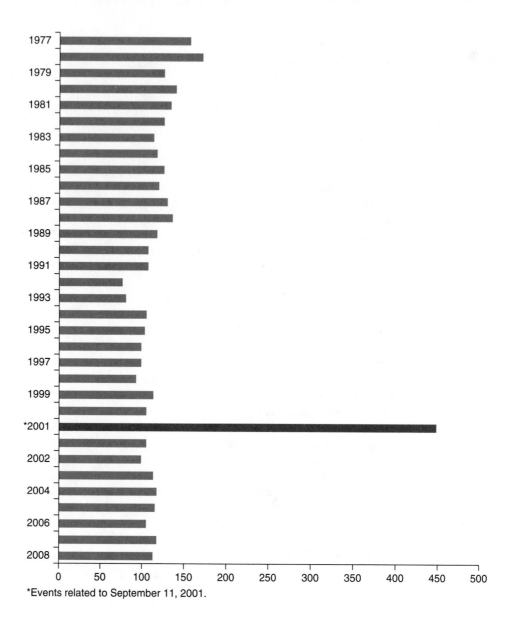

FIGURE 9.3 On-Duty Firefighter Fatalities, 1977–2008 *(Source: USFA, Historical Overview, http://www.usfa.dhs.gov/ fireservice/fatalities/ statistics/history.shtm, October 2010)*

*Events related to September 11, 2001.

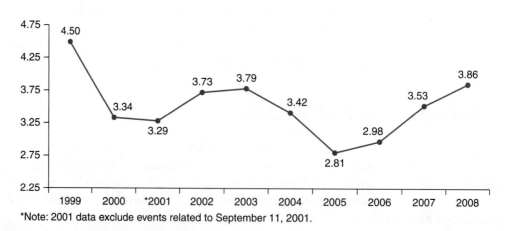

FIGURE 9.4 Firefighter Fatalities per 100,000 Fire Incidents, 1999–2008 *(Source: USFA, Historical Overview, http://www.usfa.dhs.gov/ fireservice/fatalities/ statistics/history.shtm, October 2010)*

*Note: 2001 data exclude events related to September 11, 2001.

Trends in Firefighter Mortality

In the last two decades, several high-profile incidents involving firefighter fatalities have brought national attention to the issue of firefighter mortality in the United States. Although the attention from the national media may be short-lived, the awareness of the continued high level of fatalities has begun to change the fabric of the fire service and prompted many organizations and fire departments to initiate programs to protect firefighters. In 2002, FEMA, the USFA, and the National Fire Data Center released a study prepared for them by TriData Corporation entitled *Firefighter Fatality Retrospective Study*. Its basis was data from a number of sources, including USFA files from the National Fire Data Center (NFDC), the National Fallen Firefighters Foundation (NFFF), and the Public Safety Officers' Benefits (PSOB) Program. The study's objectives were to analyze and identify trends in mortality and examine relationships among data elements. To this end, data were collected on firefighter fatalities between 1990 and 2000 with the objective of better targeting prevention strategies in keeping with the USFA's goal to reduce firefighter deaths 25 percent by 2005. In contrast to the annual USFA firefighter fatality reports, the 2002 analysis allowed for comparisons over time to determine changes in firefighter mortality, with a depth of scrutiny not present in earlier analyses.

Unfortunately, certain forces and circumstances leading to firefighter fatalities are simply beyond human control. However, through research, study, training, improved operations, development of new technologies, the appropriate use of staffing, and a change in the fire profession attitude, it should be possible to reduce the number of firefighters killed each year significantly.

PREVENTION AND RESOURCES

Sometimes unforeseen circumstances exist or catastrophic fires occur that can lead to firefighter deaths. However, a number of firefighter fatalities are the result of a chain of events, for which, if detected early, the potential exists to prevent many of them from occurring. Prevention strategies include increased emphasis on physical fitness, dietary changes, behavior modification, changes in operational strategies and tactics, more stringent adherence to standard operating procedures (SOPs), emphasis on **Crew Resource Management**, and a change in the profession's attitude in respect to the risk taken versus the gain to be achieved. An example of this is the unacceptable loss of firefighters in abandoned structures.

THE IMPACT

As has been mentioned, the deaths of firefighters profoundly affect not only the families they leave behind but also the communities in which they lived, the firefighters with whom they served, and the fire service as a whole. Each year deaths and fireground injuries have a dramatic effect on the profession. Table 9.1 provides an overview of the number of on-duty deaths and injuries that occurred from 1999 to 2008.

The number of firefighter fatalities annually differs slightly depending on the criteria used to define an on-duty fatality. A firefighter fatality could possibly be declared eligible as an on-duty death some years after the firefighter's injury. As a result, it is not uncommon to find fluctuations of from one to five fatalities annually, depending on the methodology used to compile the USFA annual report. In

Crew Resource Management
■ Training originating from a NASA workshop in 1979 that focused on improving air safety. The NASA research presented at the meeting found the primary cause of the majority of aviation accidents was human error; and the main problems were failures of interpersonal communication, leadership, and decision making in the cockpit. A variety of CRM models have been successfully adapted to different types of industries and organizations, all based on the same basic concepts and principles. This concept has been adopted by the fire service to help improve situational awareness on the fireground.

TABLE 9.1	Firefighter Casualties, 1999–2008		
YEAR	DEATHS[1]	FIREGROUND INJURIES[2]	TOTAL INJURIES[2]
1999	114	45,550	88,500
2000	105	43,065	84,550
2001	105[3]	41,395	82,250
2002	101	37,860	80,800
2003	113	38,045	78,750
2004	119[4]	36,880	75,840
2005	115	41,950	80,100
2006	107	44,210	83,400
2007	188	38,340	80,100
2008	118	36,595	79,700

[1] This figure reflects the number of deaths as published in USFA's annual report on firefighter fatalities. All totals are provisional and subject to change as further information about individual fatality incidents is presented to USFA.

[2] This figure reflects the number of injuries as published in NFPA's annual report on firefighter injuries.

[3] In 2001, an additional 341 FDNY firefighters, three fire safety directors, two FDNY paramedics, and one volunteer from Jericho (NY) Fire Department died in the line of duty at the World Trade Center on September 11.

[4] The Hometown Heroes Survivors Benefits Act of 2003 resulted in an approximately 10 percent increase to the total number of firefighter fatalities counted for the annual USFA report on firefighter fatalities in the United States beginning with CY2004.

Source: U.S. Fire Administration, *Firefighter Reports and Statistics* (Emmitsburg, MD: Author, February 2010).

the last two decades, several high-profile incidents involving firefighter fatalities have brought national attention to the issue of firefighter mortality in the United States (e.g., six firefighters killed in Worcester, Massachusetts, in 1999; 14 killed at Storm King Mountain, Colorado, in 1994; and 9 killed in Charleston, South Carolina, in 2007). The events of 9/11 dramatically highlighted the heroic activities of fire service community of the United States. As a result, a growing awareness of the continued level of fatalities has begun to change the thought process of the fire service and prompted many departments and fire service organizations to initiate new programs to protect firefighters. Fire departments throughout the country are adjusting their tactics to promote firefighter safety and reduce firefighter deaths and injuries. One such adjustment is the designation of rapid intervention teams/crews (RITs/RICs) for working fires. These teams or crews stand by outside of a structure fire and are deployed immediately to initiate a rescue attempt after a firefighter calls for help or is declared missing.

To reduce firefighter mortality, organizations such as the International Association of Fire Fighters (IAFF), International Association of Fire Chiefs (IAFC), National Volunteer Fire Council (NVFC), and the United States Fire Administration (USFA) have begun to develop a variety of programs and new initiatives intended to promote firefighter health, safety, and survival. Similar efforts are underway to develop training programs to teach firefighters how to rescue themselves and their fellow firefighters in the event they become trapped or disoriented in a fire. Ultimately, through research, study, training, improved operations, development of new

The *Firefighter Fatality Retrospective* Study, April 2002, which was prepared by TriData Corporation for the Federal Emergency Management Agency, United States Fire Administration, and National Fire Data Center, reported the following:

NATURE OF FATAL INJURY

- The leading nature of fatal injuries to firefighters is heart attack (44 percent).
- Trauma, including internal and head injuries, is the second leading type of fatal injury at 27 percent.
- Asphyxia and burns combined account for 20 percent of fatalities. More firefighters die from trauma than from asphyxiation and burns combined.

Firefighters under the age of 35 are more likely to be killed by traumatic injuries than they are to die of medical causes (e.g., heart attack, stroke). After age 35, the proportion of deaths due to traumatic injuries decreases, and the proportion of deaths due to medical causes rises steadily.

AGE

- Approximately 60 percent of firefighter fatalities were over the age of 40 when they were killed; one-third were over 50.
- Firefighters over the age of 40 comprise 46 percent of the fire service.
- Those over the age of 50 account for only 16 percent of firefighters.
- Although older firefighters possess a wealth of invaluable knowledge and experience, they are killed while on duty at a rate disproportionate to their representation in the fire service.
- These older firefighters tend to be affiliated with volunteer agencies.
- Forty percent of volunteer firefighters are over the age of 50, compared to only 25 percent of career firefighters.

DEPARTMENT AFFILIATION

- The majority of firefighter fatalities (57 percent) were members of local or municipal volunteer fire agencies (including combination departments comprised of both career and volunteer personnel).
- Full-time career personnel account for 33 percent of firefighter fatalities; they comprise only approximately 26 percent of the American fire service.
- Numerically, more volunteer firefighters are killed than career personnel, yet career personnel are killed at a rate disproportionate to their representation in the fire service.

EMERGENCY MEDICAL SERVICES (EMS) FATALITIES

- In many fire departments, EMS calls account for between 50 and 80 percent of emergency call volume.
- These incidents resulted in only 3 percent of firefighter fatalities.
- Trauma (internal/head) accounts for the deaths of 50 percent of firefighters who were involved in EMS operations at the time of their fatal injury.
- Of those involved in EMS operations, 38 percent died from heart attacks.

(continued)

TYPE OF DUTY

- Of those firefighters killed while en route to an incident, 85 percent were volunteers.
- For firefighters killed performing in-station duties, 69 percent were career personnel; the majority of those deaths were the result of heart attacks. These variations can be attributed to differences between career and volunteer agencies. Generally, unless they are on a call or other fire department business, career personnel are required to be in the fire station for the duration of their shift, which is generally between 10 and 24 hours long. As a result, volunteers are more likely than career firefighters to die while responding.

MOTOR VEHICLE COLLISIONS (MVCs)

- Since 1984, MVCs have accounted for between 20 and 25 percent of firefighter fatalities annually.
- One-quarter of firefighters who died in MVCs were killed in private/personally owned vehicles (POVs). Following POVs, the apparatus most often involved in fatal collisions were tankers, engines/pumpers, and airplanes.
- More firefighters are killed in tanker collisions than in engines and ladders combined.
- About 27 percent of fatalities killed in MVCs were ejected from the vehicle at the time of the collision; only 21 percent of firefighters were reportedly wearing their seat belts prior to the collision.

Most volunteer departments do not require personnel to stand by in the fire station; members are allowed to respond directly to incidents from their homes or workplaces, often in their POVs. As a result, volunteers are more likely than career firefighters not only to be killed in POV collisions but also to be involved in collisions involving tankers, which are predominantly used in rural areas without hydrants or other readily available sources of water. Such areas are almost exclusively protected by volunteer fire departments.

TRAINING

The study indicated approximately 6 percent of firefighter fatalities occurred during training activities, a larger proportion than in the previous decade. Over time, the leading type of training activity resulting in fatalities has remained physical fitness, followed by equipment/apparatus drills and live fire exercises.

MULTIPLE FIREFIGHTER FATALITY INCIDENTS

Between 1990 and 2000:

- Eight percent of fatal incidents involved the death of more than one firefighter; these incidents accounted for 18 percent of firefighter fatalities.
- About 14 percent of firefighters were killed in incidents that resulted in the deaths of two or three firefighters.
- Incidents involving the death of more than four firefighters are rare and accounted for only 3 percent of fatalities. However, these findings represent an increase from an earlier USFA study that found that, between 1982 and 1991, only 4 percent of incidents involved the death of more than one firefighter; those incidents accounted for 13 percent of firefighter fatalities. Approximately 90 percent of firefighters killed in multiple-fatality incidents die from traumatic injuries. In contrast, only 37 percent of those killed in single-fatality incidents die from traumatic injuries.[3]

technologies, the appropriate level and use of staffing, and other critical factors, it should be possible to reduce the number of firefighters killed each year substantially.

THE GOALS AND OBJECTIVES OF THE 2002 STUDY

The 2002 *Firefighter Fatality Retrospective Study*, prepared by TriData Corporation for the Federal Emergency Management Agency, United States Fire Administration, and National Fire Data Center, sought to analyze and identify trends in mortality, examine relationships among data elements, and aid in targeting prevention strategies for the USFA's goal to reduce firefighter deaths 25 percent by the year 2005. In contrast to the annual USFA firefighter fatality reports, this analysis allowed for comparisons over time to determine substantive changes in firefighter mortality, with a depth of scrutiny not present in earlier analyses. Examples of questions explored included the following:

- Given the increase in emergency medical services (EMS) call volume over the past 20 years, has the number of firefighter fatalities associated with EMS calls also increased?
- Has the introduction of technologies such as PASS devices and integrated PASS/SCBA affected the trends in firefighter deaths?

Examples of relationships in the data explored in this analysis include changes to death rates (or the magnitude of deaths) due to enhancements such as self-contained breathing apparatus (SCBA) or personal alert safety system (PASS). The investigation also considered patterns in the deaths of career versus volunteer firefighters and the relationships among age and gender and the causes of firefighter deaths.

FUTURE ANALYSES

Although the 2002 analysis provided a broad perspective on historical trends in firefighter fatalities, the data were not currently available to perform quality analyses in still other areas. For example, in 1999 the Occupational Safety and Health Administration (OSHA) revised its standard on respiratory protection, known as two-in/two-out. It is not yet clear how this law has affected firefighter deaths. Similarly, training program improvements, the development of health and wellness initiatives, and the use of RITs may also affect future trends in firefighter injuries and deaths. The last decade's initiatives (Everyone Goes Home, Near-Miss Reporting System, and Crew Resource Management) all may have a future impact on the number of firefighter injuries and line-of-duty deaths. As the Near-Miss Reporting System database is expanded and updated, it should become possible to determine more clearly the effects of these and other changes and trends in the fire service. After reviewing the following statistical information, these questions need to asked: What can be done from a leadership and administrative perspective to enhance firefighter safety? Are there policies, procedures, new programs, or behavioral adjustments that could be made in a department to impact positively the trends noted in this study?

TREND IN FIREFIGHTER FATALITIES

A **trend in firefighter fatalities** is a pattern that is evident from past events. Determining a trend requires sufficient data to see whether there is a relationship of one (usually) or more (infrequent) variables over time. The number of fatalities annually from 1977 to 2008 has fluctuated from a high of 171 in 1978 to a low of 77 in 1992.

trend in firefighter fatalities
■ A pattern that is evident from past events. To determine a trend, sufficient data are required to see whether there is a relationship of one (usually) or more (infrequent) variables over time. The change in a series of data over a period of years that remains after the data have been adjusted to remove known significant fluctuations, such as those experienced on September 11, 2001.

PEARSON
myfirekit

Visit MyFireKit
Chapter 9 for a
more detailed analy-
sis on Firefighter
Fatalities.

Each year in the United States and its protectorates, approximately 100 firefighters are killed while on duty and tens of thousands are injured. Although the number of firefighter fatalities has steadily decreased over the past 20 years, the incidence of firefighter fatalities per 100,000 incidents has actually risen. Despite a downward dip in the early 1990s, the level of firefighter fatalities is back up to the same levels experienced in the 1980s. In the last decade, several high-profile incidents involving firefighter fatalities have brought national attention to the issue of firefighter mortality in the United States. While the attention from the national media has been fleeting, the awareness of the continued high level of fatalities has changed the fabric of the fire service and prompted many organizations and fire departments to initiate programs to protect firefighters. Through research, study, training, improved operations, development of new technologies, the appropriate use of staffing, and other factors, it should be possible to significantly reduce the number of firefighters killed each year.[4]

FIREFIGHTER FATALITIES AND CARDIOVASCULAR DISEASE

The mortality rate for firefighters, or line-of-duty deaths (LODD), is approximately 50 percent due to cardiovascular disease.

Figure 9.5 shows the trend in percent of deaths due to heart attack from 1984 to 2000. Despite fluctuations, the trend in the proportion of firefighter fatalities from heart attacks has remained constant over the past 16 years.

Where reported by the family or discovered at autopsy, the most common pre-existing condition found for heart attack fatalities was arteriosclerosis, followed by prior heart attack(s) and hypertension.

Firefighters are more likely to suffer a heart attack in the course of performing suppression support duties on the fireground, while in the fire station, or during training exercises. In contrast, deaths due to traumatic injuries are more likely to occur while mitigating or responding to an incident (see Figure 9.6).

Firefighters have one of the highest occupational mortality rates in the country with cardiovascular disease representing the number-one cause of death. Line-of-duty deaths (LODDs) for firefighters are approximately 50 percent, which is higher than the cardiac mortality rate for similar professions such as police officers (22 percent) and EMS professionals (11 percent). In addition, LODDs in firefighters due to heart attacks are much higher than the occupational cardiac mortality rate for the entire nation (15 percent) (see Table 9.2). Clearly stated, firefighters are at a higher risk of dying from a heart attack on the job than the rest of the public. These alarming statistics led researchers to investigate cardiac deaths more closely.

Another study analyzed on-duty cardiac deaths to determine whether certain activities place firefighters at higher risk of suffering a heart attack. Although most

FIGURE 9.5 Percent of Heart Attack Deaths by Year, 1984–2000 (*Source:* Firefighter Fatality Retrospective Study, *prepared by TriData Corporation for Federal Emergency Management Agency, United States Fire Administration, National Fire Data Center, April 2002*)

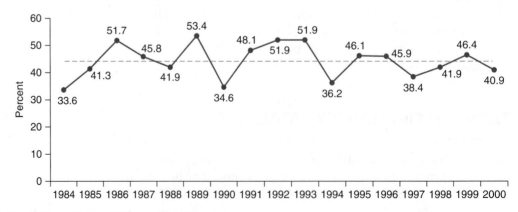

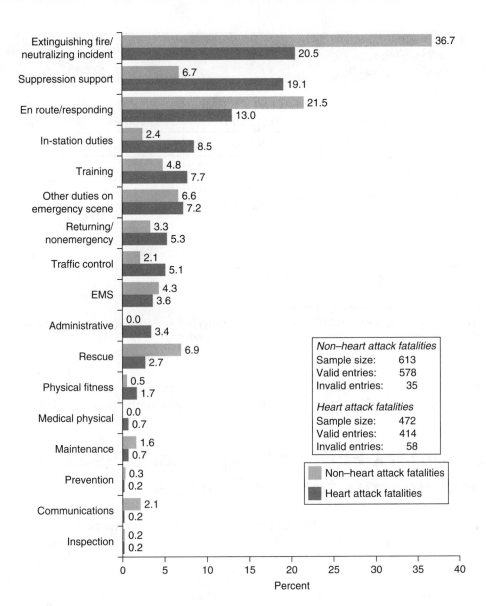

FIGURE 9.6
Comparison of Type of Duty for Heart Attack with Non–Heart Attack Fatalities, 1990–2000
(Source: Firefighter Fatality Retrospective Study, prepared by TriData Corporation for Federal Emergency Management Agency, United States Fire Administration, National Fire Data Center, April 2002)

Non–heart attack fatalities
Sample size: 613
Valid entries: 578
Invalid entries: 35

Heart attack fatalities
Sample size: 472
Valid entries: 414
Invalid entries: 58

Non–heart attack fatalities
Heart attack fatalities

TABLE 9.2	Occupational Mortality Rates
PROFESSION	**CARDIAC MORTALITY RATE**
Firefighting	50%
Police officer	22%
EMS professionals	11%
National cardiac mortality rate	15%

Source: S. N. Kales, E. S. Soteriades, C. A. Christophi, and D. C. Christiani, "Emergency Duties and Deaths from Heart Disease Among Firefighters in the United States," *New England Journal of Medicine*, 356, 2007, 1207–1215.

TABLE 9.3 Risk of Cardiac Deaths in Firefighters for Typical Firefighter Activities

ACTIVITY	OBSERVED DEATHS (N=499)	PERCENT OF TIME PER SHIFT	RISK OF DEATH OVER NONEMERGENCY DUTIES
Fire suppression	144 (32.1%)	2.6%	67%
Alarm response	60 (13.4%)	6.3%	8.1%
Alarm return	78 (17.4%)	11%	6.2%
Physical training	56 (12.5%)	8%	4.9%
Non-fire and EMS emergencies	42 (9.4%)	24%	1.4%*
Nonemergency duties	69 (15.4%)	48%	

*Not a significant increased risk of cardiac death over nonemergency duties.

Source: S. N. Kales, E. S. Soteriades, C. A. Christophi, and D. C. Christiani, "Emergency Duties and Deaths from Heart Disease Among Firefighters in the United States," *New England Journal of Medicine,* 356, 2007, 1207–1215.

cardiac deaths occurred during fire suppression (32 percent), of the activities firefighters undertake, they spent the least amount of time firefighting (see Table 9.3).

Landmark FEMA Study: Heart Disease Is an Epidemic for Firefighters

H. Robert Superko, MD—principal investigator in the landmark Federal Emergency Management Agency (FEMA)–sponsored study of firefighters aged 40 and over conducted at Saint Joseph's Hospital in Atlanta—released preliminary findings in the world's first study of first responders at risk of suffering sudden death or other significant cardiac events. Firefighters are known to have a 300 percent increased risk for cardiac disease as compared to other segments of the population.

"Preliminary findings show that one third of firefighters had heart disease that is unrelated to traditional risk factors, such as high cholesterol," says Dr. Superko. "Those results are astounding and point at job duties and environment as the primary determinants for early death in our country's first responders."

In 2006, Dr. Superko, recognized as a leading expert on lipids, cholesterol, and advanced metabolic markers and their contribution to heart disease, and his team performed a comprehensive scientific battery of sophisticated blood and imaging tests on three hundred firefighters in Gwinnett County, Georgia. First responders in Gwinnett County were identified for the study following an emotional report by Fire Chief Steve Rolader, following the sudden death of one of his firefighters from cardiac arrest while fighting a house fire.

Study volunteers underwent comprehensive genetic screening of more than a million genes including newly identified KIF6 (statin responsiveness gene) and 9p21 (myocardial infarction gene), advanced phenotype (blood) and imaging analyses, diet, and exercise review over the yearlong study. Results and explanations were presented to the groups followed by individual consultations. Complete statistical and comprehensive genetics results are expected this year.

According to Dr. Superko, stress and psychological pressures related to the job, as well as diet, exercise issues, and inherent personality, interacting with a genetic predisposition to heart disease, probably have tremendous impact on the risk of heart attack in these first responders. He said:

> Imagine being awakened from a dead sleep by a loud, shrieking siren several times during the night, responding through the rush of adrenaline, carrying a hundred pounds of equipment on your back, and meeting people at the very worst possible moments in their lives every day and you can begin to understand the toll it takes on the first responders. . . . And, consider the emotional and psychological stress they encounter each day as they respond to society's most brutal moments from murders to car wrecks and death. Finally, those who serve as first responders have a mind-set and a desire to help people. They certainly bring a competitive nature to the job but also a profound desire to help and to do the best for others. All these elements create an environment that puts them at an increased risk for cardiac disease.

In response to the growing awareness of issues of diet and exercise, Gwinnett County Department of Fire and Emergency Services has instituted exercise programs within local firehouses, and the county now reimburses firefighters for fitness club memberships. The department also educates firefighters on proper diet and nutrition with one-on-one opportunities as well as "lunch and learn" programs in the station houses. And, over the years, the traditional firehouse alarms in Gwinnett stations have been replaced with softer alarms and even-voiced prompts to awaken sleeping first responders.

As a result of the research, Saint Joseph's Hospital and Dr. Superko's team implemented a two-month screening program for all Atlanta first responders (firefighters, police, and EMS), regardless of age, to provide them with some basic and advanced diagnostic tests at affordable prices. Several physicians are granting their services free of charge.

Chief Rolader says,

> There are tremendous costs associated with early deaths of our first responders in every community as we lose men and women in their 30s, 40s and 50s who are our first line of defense but who don't live to perform their jobs for very long. . . . With the results of this study, we can implement programs across the country that will save lives.[5]

Firefighter Injuries

The NFPA annually surveys a sample of departments in the United States to make national projections of the fire problem. The sample is stratified by the size of the community protected by the fire department. All U.S. fire departments that protect communities of 50,000 or more are included in the sample because they constitute a small number of departments with a large share of the total population protected. For departments that protect populations of less than 100,000, stratifying the sample by community size permits greater precision in the estimates. Survey returns in recent years have ranged from 2,560 to 3,500 departments annually. The national projects are made by weighting sample results according to the proportion of total U.S. population accounted for by communities of each size.[6]

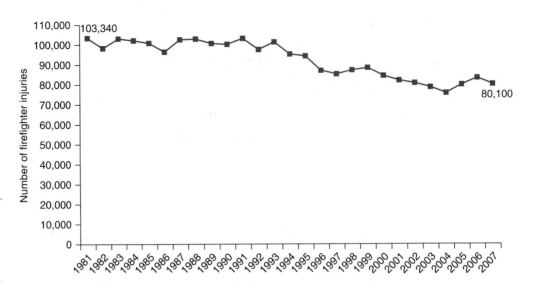

FIGURE 9.7 Total Firefighter Injuries by Year, 1981–2007 (*Source:* Firefighter Fatality Retrospective Study, *Prepared by TriData Corporation for Federal Emergency Management Agency, United States Fire Administration, National Fire Data Center, April 2002)*

In recent years, the number of firefighter injuries has been considerably lower than it was in the 1980s and 1990s (see Figure 9.7), but this is due in part to additional questions on exposures, which allows injuries to be placed in their own categories. Previously some of these exposures may have been included in total injuries under other categories.

The NFPA estimates there were 13,450 exposures to infectious diseases (e.g., hepatitis, meningitis, HIV, other) in 2007. This amounts to 0.9 exposure per 1,000 emergency medical runs by fire departments in 2007.

The NFPA also approximates there were 28,300 exposures to hazardous conditions (e.g., asbestos, radioactive materials, chemicals, fumes, other) in 2007. This amounts to 26.2 exposures per 1,000 hazardous condition runs in 2007. An estimated 16,360 injuries, or 20.4 percent of all firefighter injuries, resulted in lost time in 2007.[7]

Results by type of duty indicate, not surprisingly, the largest share of injuries occurs during fireground operations: 38,340 firefighter injuries, or 47.9 percent of all firefighter injuries.

PEARSON
myfirekit™

Visit MyFireKit Chapter 9 for a detailed analysis of firefighter injuries.

OVERVIEW OF 2007 FIREFIGHTER INJURIES

- A total of 80,100 firefighter injuries occurred in the line of duty in 2007, a decrease of 4.0 percent from the previous year.
- A total of 38,340 firefighter injuries, or 47.9 percent of all firefighter injuries, occurred during fireground operations. An estimated 15,435 occurred at non-fire emergency incidents, whereas 13,665 occurred during other on-duty activities.
- Regionally, the Northeast had the highest fireground injury rate with 4.9 injuries occurring per 100 fires; this was more than twice the rate for the rest of the country.
- The major types of injuries received during fireground operations follow: strain, sprain, muscular pain (45.1 percent); wound, cut, bleeding, bruise (18.2 percent); burns (6.9 percent); and smoke or gas inhalation (5.6 percent). Strains, sprains, and muscular pain accounted for 57.8 percent of all non-fireground injuries[8] (see Figures 9.8 and 9.9).

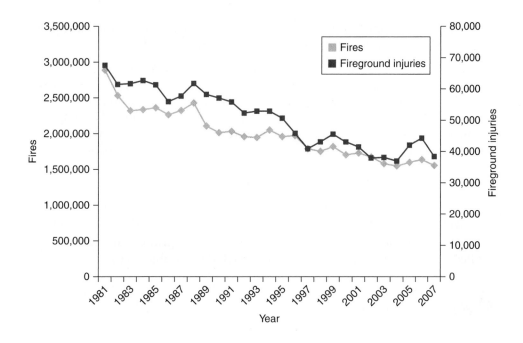

FIGURE 9.8 Decrease in Fireground Injuries Is Similar to the Decrease in Fires *(Source: Firefighter Fatality Retrospective Study, prepared by TriData Corporation for Federal Emergency Management Agency, United States Fire Administration, National Fire Data Center, April 2002)*

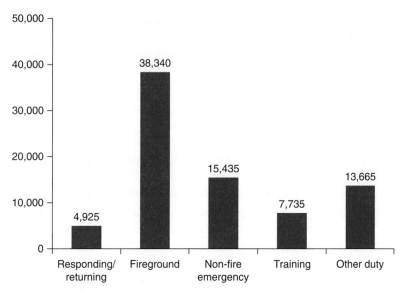

FIGURE 9.9 Number of Firefighter Injuries by Type of Duty, 2007 *(Source: Firefighter Fatality Retrospective Study, prepared by TriData Corporation for Federal Emergency Management Agency, United States Fire Administration, National Fire Data Center, April 2002)*

Everyone Goes Home

On March 10 and 11, 2004, more than 200 individuals assembled in Tampa, Florida, to focus on the troubling question of how to prevent line-of-duty firefighter deaths. Every year approximately 100 firefighters lose their lives in the line of duty in the United States, which equates to about 1 every 80 hours. Hence, the inaugural National Firefighter Life Safety Summit convened to bring fire service leadership together to focus attention on this issue, one of the most critical facing the fire service. The last historic fire service gathering to have had an impact on the line-of-duty death reduction was the original *America Burning* panel in the 1970s. However, since 1990, on average 99 firefighters have died each year in the line of duty, which does not include the 343 firefighters who were lost on September 11, 2001.

FIGURE 9.10
Everyone Goes Home
Logo *(Courtesy of Ron
J. Siarnicki, National Fallen
Firefighters Foundation)*

Of paramount importance in the future will be the ability to institutionalize firefighter safety in the fire profession. As Bill Manning wrote,

> Fifteen years of well-intentioned attempts to institutionalize firefighter safety through broad-based prescriptive measures have failed to push us past the 100-a-year threshold. It is not inappropriate to parallel our situation with the aviation industry, whose leadership, several years back, realized the world's best regulations, standards, research, and technology weren't preventing and, by themselves, couldn't prevent aviation disasters. They came to the conclusion that performance- and attitude-based safety problems demand performance- and attitude-based safety solutions—at the core, it is a people thing.[9]

This focus on attitudes and personal accountability can be seen in the fire service's "Everyone Goes Home" (see Figure 9.10) life safety initiative. This effort is by far the most promising nationally based safety program that the fire service has rolled out in many years. The initiative includes 16 objectives.

health and safety
■ The associated efforts, including fitness, medical screening, awareness, research, staffing, training, and a myriad of other factors, that impact firefighter health and safety.

1. Define and advocate the need for a cultural change within the fire service relating to safety, incorporating leadership, management, supervision, accountability, and personal responsibility.
2. Enhance the personal and organizational accountability for **health and safety** throughout the fire service.
3. Focus greater attention on the integration of risk management with incident management at all levels, including strategic, tactical, and planning responsibilities.
4. Empower all firefighters to stop unsafe practices.
5. Develop and implement national standards for training, qualifications, and certification (including regular recertification) that are equally applicable to all firefighters based on the duties they are expected to perform.
6. Develop and implement national medical and physical fitness standards that are equally applicable to all firefighters, based on the duties they are expected to perform.
7. Create a national research agenda and data collection system that relates to the initiatives.
8. Utilize available technology wherever it can produce higher levels of health and safety.
9. Thoroughly investigate all firefighter fatalities, injuries, and near misses.
10. Grant programs should support the implementation of safe practices and/or mandate safe practices as an eligibility requirement.
11. National standards for emergency response policies and procedures should be developed and championed.
12. National protocols for response to violent incidents should be developed and championed.
13. Firefighters and their families must have access to counseling and psychological support.
14. Public education must receive more resources and be championed as a critical fire and life safety program.
15. Advocacy must be strengthened for the enforcement of codes and the installation of home fire sprinklers.
16. Safety must be a primary consideration in the design of apparatus and equipment.

These initiatives are not necessarily new; rather, they are based on historic information and a set of fundamental truths. By adopting these practices, the fire

FIGURE 9.11
Firefighter Near-Miss
Logo   *(Courtesy of
IAFC—Near Miss Program)*

service of the future will no longer accept that dying on the job is a normal way of doing business. Yes, the work is inherently dangerous, and no, the death toll for firefighters may never be zero. But firefighters are dying unnecessarily now and that must stop. The number of firefighters' injuries is also a significant concern.

Every individual in the fire service has to accept personal responsibility for his or her health and safety and the health and safety of his or her colleagues. Leaders and members of fire departments and fire service organizations must be accountable for themselves and for others by managing the risks to function safely within an unsafe environment. What will play a key role in reducing deaths is risk management; this means identifying situations where predictable risks are likely to be encountered and then making decisions that will reduce, eliminate, or avoid them. Although firefighters are willing to sacrifice their lives, this is not an excuse to take unnecessary risks while trying to save property that is already lost.

Firefighter Near Miss

A **firefighter near miss** is defined as an unintentional, unsafe occurrence that could have resulted in an injury, fatality, or property damage. Only a fortunate break in the chain of events prevented this from happening. Near-miss events include everything from a firefighter almost falling from five feet off a ladder, to an aerial portion of a ladder truck becoming entangled in electrical wires.

The National Fire Fighter Near-Miss Reporting System (see Figure 9.11) is a free, voluntary, confidential, nonpunitive, and secure tool used to learn from other firefighters' experiences and to share one's own. The program's mission is to provide data and case studies for efforts related to the prevention and reduction of firefighter injuries and fatalities.

firefighter near miss
■ An unintentional, unsafe occurrence that could have resulted in an injury, fatality, or property damage.

Search.   Visitors to www.firefighternearmiss.com can search reports by keyword or by specific parameters, such as event type, department type, and contributing factors. This is an invaluable tool for creating drills and raising awareness on topics prevalent in one's station or department.
Learn.   The resources page offers free training tools, such as presentations, grouped reports, videos, and more. The 2010 training calendar and corresponding training modules are also located on the resources page.
Share.   Anyone can submit a near-miss report by visiting www. firefighternearmiss. com. There are 16 questions and two open text boxes for the event description and the lessons learned. The entire process takes approximately 10 to 15 minutes.

PEARSON
myfirekit

Visit MyFireKit Chapter 9 for more information on Submitting a Report.

PERSPECTIVES OF PROMINENT FIRE CHIEFS

In assessing what has been the impact of the initiatives Everyone Goes Home and Firefighter Near Miss on firefighter safety, four prominent fire service leaders were interviewed for their personal views on this subject. Bill Goldfeder, Dave Daniels, Ron Siarnicki, and Alan Brunacini have been involved in the firefighter health and safety movement. Here are their responses to questions posed on the impact of the initiatives Everyone Goes Home and Firefighter Near Miss on firefighter safety.

BILL GOLDFEDER, DEPUTY CHIEF, LOVELAND-SYMMES FIRE DEPARTMENT, OHIO

I'll respond a couple of ways. Number one, we are definitely getting the message across but part 2 to that, not everybody is getting the message. I don't think we do a whole lot more than what we have already done. We have provided every possible lesson, every possible video, every possible tape, and most of this through fire act grants. We have spent millions and millions of dollars providing the resources that fire chiefs, firefighters, and fire officers need to help firefighters get it on the fireground and before the fireground send them the help on safety-related issues. You know, one of the cooler politicians in my day was a guy named Tip O'Neill from Massachusetts. Tip wrote a book that had a saying, All politics is local. My response to you is, right now the number-one person who can solve the firefighter injury and death rate is you and your company officer. You can have a crappy fire chief, but a decent company officer can still make sure things get done in a safe way. You can have a great fire chief with all sorts of great ideas; but, if you have a crummy company officer, your plan is going to fail. If the captain at Engine 9 doesn't agree, you're screwed. So they can make or break you, so I think it starts at the local level. Nationally, we have done so much and there is so much out there. We can send you a program on how to drive safer, how to wear your air pack, how to read the building, how to read smoke; it is out there, but until the fire chief and the company officers decide they are going to change the organization, it's not going to happen. If you as a fire chief, for example, want to solve the seat belt and the apparatus problem, you can solve it 95 percent of the time. Now there's going to be that occasion where a vehicle drives through the red light, slams into you, and you were totally in the right. You can't always avoid that. As far as our people being ejected, as far your apparatus crashing, things like that, when you as a fire chief are ready to solve the problem you can solve problem. You have to develop policy, you have to train your personnel on the policy, you have to educate them on the importance of the policy, and you have to enforce the policy. Where we fail the most is the enforcement. When we have close calls and near misses in most cases it is because many officers would rather avoid confrontation. There is a genetic chip in each of us that made us want to be firefighters, and the reason we want to be firefighters is we like to help people. We generally like to be liked. By doing what we do most of the outcomes are positive and there is a great and easy reward with that. So take this same type of personality profile, that kind of firefighter profile, and now say, OK, let's see, you love being in the firehouse, you have a great time with your brother and sister firefighters, you like to be liked, and now you are in charge of these people. Well, these are my friends, these are my buddies. Yeah, well, you need to be in charge of them. I would rather avoid confrontations if I can; after all we fish together. There is a problem that fire chiefs have to understand. As part of your assessment center and promotional process you have to find people who can get past this hurdle and have the ability to enforce the policies. In a lot of cases in a career department you're enforcing policies with your buddies and pals, and in volunteer fire departments you may be enforcing your policies to your relatives. It is a fascinating issue, but a very real issue. Until that officer at your fire department at Engine 9 decides that the driver is driving like an idiot, there is nothing you can do about it. So it is a local issue, but it is a very solvable issue.

A lot of people are pushing zero line-of-duty deaths. I heard somebody write or somebody wrote me, any line of death is unacceptable; that's ridiculous. Some line-of-duty deaths are acceptable and some of them are very necessary. They were obviously talking about the very small minority on occasion. A firefighter doing all the right things does lose his life when going in to make an attempt to save the life of somebody. So to put it from my standpoint, from a realistic standpoint, not everyone goes home. Once in a while there will be that occasion when we must take extreme risks. We are all who is coming to that fire, that emergency, so we do have to take that risk. Now what leads up to that is the training of the organization, preparedness, the preplanning, the command level supervision—all these can minimize the opportunity of that firefighter having to lose his or her life or get hurt. Yet, sometimes it happens. Now the other side of this is the majority of where the problem can be solved—health safety. You know, eat more salads, exercise, and go for a physical. This isn't tough. Wear your equipment; cancer is a big issue out there; get the soot off; wear your mask. You can start impacting the heart attack issue, the stroke issue, the cancer issue by taking care of yourself, wearing your equipment, and keeping it clean. The next solvable line-of-duty death is vehicle accidents. As a fire chief you have a great opportunity to impact this. As I mentioned, you establish the policy, train your people on the policy, and enforce the policy. The enforcement piece has to be quick. To me, if you take the citizens' $500,000 pumper and you decide to blow a red light for whatever reason without stopping to make sure it's clear, you need to accept the fact you may very well lose your job. We need to take a very, very hard stand on this.

Let's go to structural firefighting. We need to understand what we are getting into, and this ties into the size-up, preplanning, and ensuring we have appropriate staffing on first alarms so we can send enough people to do the job initially. We need to include the size-up, 360 of the building, to understand what we are getting into, to understand what the construction is, and in most cases to understand this stuff is already destroyed before we even get to the scene.[10]

DAVE DANIELS, FIRE CHIEF

I think the culture of the fire service has changed in the last five years; there is no doubt in my mind that it has. Is there one way to go? Absolutely, but it is not now what it was six or seven or eight years ago. Many of the safety discussions that people consider routine now were not so in years past. Prior to 2002, the American fire service averaged 120 plus firefighter fatalities a year since 1977. We now have been averaging only about 98. Therefore, we are actually averaging fewer than 100 fatalities, a significant drop in a short period. Though the goal is not accomplished, it is certainly progress in the right direction. Considering the facts and the culture of the fire service to date, a major step forward is the fact that the industry is making a difference by taking responsibility for its own actions. Neither the 16 initiatives of the Everyone Goes Home campaign nor even the IAFC Safety, Health and Survival Section in and of itself can change everything about the safety culture of the fire service, but combined along with other efforts, progress is being made. One sign of the movement to date is the fact that every time people turn around in the fire service, somebody is talking to them about safety. Change is really occurring, though there is still a lot of work to do.

In 1952, over a thousand air force pilots were killed during flight mishaps; and in 2002, there were only nine such occurrences. It is because the air force started to focus on those minor errors that often began a chain of events that caused the plane to crash. The fire service is trying to learn from that type of "culture shift." The National Fire Fighter Near-Miss Reporting System was formed with help from a number of experts from outside the fire service

(continued)

including the airline industry, NASA, and the military. The non–fire service partners were surprised by how quickly the fire service reporting system was embraced by the industry itself. This is just one example that we, when we decide we want to do it, can change if we want to.[11]

RON SIARNICKI, NATIONAL FALLEN FIREFIGHTERS FOUNDATION

(RB) What do you believe has been the impact of two, I think, very critical initiatives: Everyone Goes Home and Firefighter Near Miss projects.

(RS) I really believe what both of these programs have done is raise the awareness of the true consequences of firefighter fatalities in our country. Over the years, we have always said that 100 line-of-duty deaths occur on average every year, but what we haven't said is that at least half of them were preventable. This is tough because that sends a clear signal throughout our profession and the fire service community, as a whole, that we are doing things that are leading to and in some cases contributing to the loss of our own. This is a very difficult situation for us, as coworkers of the fallen members, and their families, who often ask if their loved one really had to die. I often think about a speech that Vena Drennan gave after her husband, John, died in a brownstone fire in New York City. She stated that, had the occupants of that apartment practiced some simple fire prevention efforts, most likely her husband would not have died as a result of his extensive burns. And when her story is put into the context of the foundation's mission, I realize that our primary objective must be to support the families of the fallen and meet their basic needs. These two items have to come above everything else we consider doing, but in reality the true needs of that family would have been keeping their loved one alive. The prevention of a line-of-duty death has to be what we work on so that these families do not have to do through the traumatic occurrences of losing their firefighter. Both of these programs have raised the awareness of this problem by stating the facts of the occurrence while trying not to cause any more harm to the individuals or family members who were involved in these incidents.

For me the toughest part of my job, believe it or not, is reviewing the case files of each reported LODD. Ultimately, staff brings to me the packages every couple of days, and I have to read each and decide whether or not it meets the criteria to go on the memorial in Emmitsburg. I can honestly say that if I took a stack of those file folders and whited out the firefighter names, the incident dates, and the fire departments involved, they would read almost identical. The true definition of "insanity" is doing the same thing over and over while expecting to get a different result. Unfortunately, that is what we are doing in some cases within the fire service. These two valuable programs are getting the information about firefighter near misses and line-of-duty deaths out to the masses so that we can learn from these tragic occurrences and create a positive safety-based effort so that they are not repeated.[12]

ALAN BRUNACINI, RETIRED FIRE CHIEF

(RB) What do you believe has been the impact of the initiatives Everyone Goes Home and the Firefighter Near Miss?

(AB) I think from an academic standpoint they are probably sound. I think from an applied standpoint they have not been particularly effective. Particularly if you look at the numbers; maybe it just takes a certain amount of lead time to get those numbers to respond. I don't know if getting a bunch of safety guys together going through an administrative process really matches the place where that outcome is actually going to be tested or applied in the field. I think this is a huge challenge in connectivity. I think it is easy to say what we have to do is change the culture. I think we need to look at the way we learn things in the fire service. In some cases and in some places, the problem is not that we have a set of habits; it's that we have a set of addictions. We are trying to apply habitual management to something that is indeed an addiction. What's frustrating in the service is we are in a 12-step program, which is not going to work if you really need to be an inpatient.

(RB) That's a really good observation.

(AB) We got together in Tampa, Florida, and put little dots on a flip chart; and I'm not being critical of that because it's one of those standard ways that we gather information, prioritize it, and so on, but it isn't the way Engine 1 and Ladder 4 in the Fresno Fire Department are developing opinions, judgments, reactions, behaviors, and so on. It's almost as if we have, I don't mean to make it so truncated, but it's almost like there are two separate disconnected parts to it. One is the people who are frustrated by it and are trying to fix it, and the other is the people who are involved directly in the problem; they both have two completely different sets of routines. If you can't connect those two separate pieces, then you keep being frustrated. I notice when you are frustrated you often end up doing the wrong thing harder.

The Near Miss is a perfect program, it's well done, and the people who are doing it are terrific. I'm on the advisory committee so I'm part of it. I think that, and let me not overstate it, for scholars it's terrific material to study and to apply and try to make sense out of and relay back to the field. The problem is near miss on a fire company is a way of life almost. If a near miss were a profound, life-shattering experience, you wouldn't have stayed in the fire service for 90 days. Some people don't; some people will not recover from their first near miss. They say, "Is this what I signed up to do?" Yeah, that's what you signed up to do. "You guys are nuts; I'm out of here." But not many people leave the service because they are scared; you don't want people to do that. If we were attracting really careful people who could reflect and then adjust their behavior on near misses that would be one thing, but we don't recruit people who want to play it safe.

(RB) No, that's really not the personality type that comes into our profession, is it. Yet, there has to be a nexus with Everyone Goes Home and the Near-Miss Reporting System into the firehouse. Because if you can't connect them into the firehouse, you never impact or infect the culture to change.

(AB) There is a guy who is the operations chief for the District of Columbia, Larry Schultz. He's an interesting guy who is a terrific character. He's a real smart, very big city guy, who is real passionate about operations. He has a safety program called "nobody goes home" until we do a critique out in the street immediately after the incident to see if the standard operating procedures were followed. Do you think it's an enormously powerful tool inside the system? You know that when the event is over your boss is going to stand out in the street, and he's going to say either you did or you didn't do what we agreed to do in the SOP. You're the second engine and you're supposed to go to the back; you're the fourth engine and you're supposed to lay to the second in. It's in the SOP. Now there may be a good reason you didn't do something, and you have the opportunity to explain to your boss why you did or didn't do it. He does it every time and that's pretty influential with firefighters because they tend to evaluate things a lot more with their eyes than they do with their ears.

(RB) So you need to do it right the first time.

(AB) Pretty simple. I don't know why you couldn't apply this to any piece or part of the way we operate. If you have a call, I'm going to have the conversation, and I'm going to ask, Did you wear your seat belt? Did all of you have your gear on when we went in? When did you put it on? You are the driver, you went through ten signal light intersections, and three of them were red. What did you do? I stopped at each one. It is little things like this that can make a huge impact.

(RB) That even goes down to the company officer level when we don't have the command officer there making sure that they are doing that exact same thing.

(continued)

(AB) Absolutely! We live or die with company officers, literally.

(RB) Very true. You have been around and seen a lot of different organizations. How does culture play into this whole risk-taking concept that we continue to read about where we see firefighters who die in a vacant building or get killed driving their personal vehicles to a call? How does the culture impact that whole thing?

(AB) I see culture as a line. There are some things you do where you use seniority; a lot of what we do is based on seniority. When you are a brand-new firefighter, you get at the end of the line. The way you are successful is you stay in that line and you behave. There's a set of norms, rules, and values that regulate that line. The line will protect and adjust itself. If you get out of line it will smack you to get you back in line. It will only smack you a certain number of times, and if you conform, "the line" says we define you as not adaptive, and it selects you out of the system. It will kick you out of the line. The most powerful piece, the most powerful development, is changing fire departments within the peer process because the thing that we want more than anything as far as that line is concerned, for most of us, is to be accepted by our peers. Inside the line there is a set of very powerful dynamics. When you are a kid at the end of the line, you listen to those in front of you, which is how you get an organizational perception and a fix on the inside values. The insiders are not talking about going out and doing thrilling fire inspections. They're talking about the daredevil stuff (close calls) because those are what define our culture. I can think back when I was at the end of that line, I would hear about 10 or 12 fires over and over again. This became the cultural cues and targets, which are important in the line. As an example, cops tell us there is an interesting dynamic in domestic violence with men that if their close colleagues disapprove of it, sometimes to the extent of at times being physical with the person, it will stop the domestic violence. If they do nothing (i.e., condone it), it continues. There is a little bit of that in the firefighter risk-taking line. Someone has to break the cycle . . . from the inside.

I think there is a direct connection between the time you spend with the officers as this is an opportunity to make the culture respond, strengthen it, do a functional set of things, become more resilient. The vast majority of our culture is terrific. Yet, we have some behaviors that are addictions. An addiction is an irrational behavior that ignores consequences. We need to talk about getting beat up, getting burned, getting killed. We operate in hazard zones because there's something wrong. They don't call us because everything is OK. We have to be able to fix that and not make it worse, do no harm. I know that we put our bodies in between the problem and the customer, but we have to engage a smarter set of cultural aspects that will protect our people. The more responsive that we can be to what's actually going on, the more we will help our firefighters, the more we put it in their terms. The more real change we can create.[13]

Crew Resource Management

On December 28, 1978, United Airways Flight 173 crashed on approach to Portland International Airport. Despite the fact Flight 173 was one of the most technologically advanced aircraft of its time, the pilot flying the plane ignored his crew's warnings, ran out of fuel, and crashed the plane. Ten people were killed and 23 injured. Postcrash investigations into this crash and subsequent air disasters gave rise to the concept that human error is the overriding factor in these catastrophes. The development of the error management/crew enhancement program known as Crew Resource Management (CRM) was born from these conclusions.

CRM has been standard training for all members of civilian aviation flight crews for over 20 years. Since CRM was introduced to the commercial aviation industry, air disasters have fallen from an average of 10 to 15 to two to three per year; and for several years, no commercial air disasters were experienced. Averted errors credited to CRM rank in the hundreds per year. The United States Coast Guard reports a 74 percent reduction in injuries since it began using CRM. One of the pilots assigned to work on CRM for U.S. Airways was Captain Sully Sullenberger, who landed U.S. Airways Flight 1549 into the Hudson River on January 15, 2009, and noted in his book, *Highest Duty*, the importance of CRM in the successful outcome of this airliner emergency.

The U.S. military, medical industry, and shipping industry have also embraced CRM to reduce deaths, injuries, and errors. Each industry credits CRM with capturing errors and promoting teamwork, which prevented injury or death.

Similarities between fire service decision making and air crew decision making are striking. Effective communication, situational awareness, decision making, teamwork, and combating barriers comprise the basis for CRM training.[14]

Unfortunately, the fire service in the United States has been lulled into a weary acceptance of an "average 100 line-of-duty deaths and 100,000 lost-time injuries per year" mentality. None of these deaths or injuries is considered an intentional act. Firefighters certainly do not report to the station for duty and state, "Today I will take actions that will intentionally kill and/or injure myself and my colleagues." Instead, certain voices insist that all has been done for firefighter safety, and we are living with the best possible circumstances. They say that any further change in firefighting tactics will essentially put us outside the building on all fires and essentially out of business. However, firefighters are not being killed and injured solely by flames, smoke, and heat. Reading between the lines of the line-of-duty death reports reveals the significant effects of adrenaline and machismo. Communication failures, lack of situational awareness, faulty decision making, poor task allocation, and leadership failures are listed as the contributing factors in far too many NIOSH firefighter line-of-duty death reports. Because the factors are the same as those cited in aviation disaster reports, it logically follows that CRM would benefit the fire service. What is needed is a new way to break the chain of complacency.

A comparison of the interactions among and behaviors of emergency service crews and those of flight crews reveals a number of similarities. Both crews are structured with a leader and one or more crew members. Each group functions best when it works as a cohesive team. The team can spend hours of time performing mundane activities and then be called upon to act swiftly under very stressful conditions. Some crews work together frequently and others are assembled on short notice.

CRM can be taught to the fire service using a variety of methods. The airline industry uses a three-step process to teach the five factors (communication, situational awareness, decision making, teamwork, barriers) that comprise CRM.

1. Awareness introduces the concept.
2. Reinforcement underpins the awareness level by having attendees participate in simulated activities that require action to overcome problems in the five factors that lead to disaster.
3. Refresh is a session reminding participants of the basic concepts and reinforces the five factors through lecture and role play.

The second and third steps provide for repetitive (or in-service) training to reinforce the five factors. This training is based on the concept of *recognition primed decision making*. Although the fire service certainly could benefit from a program as intensive as the airline industry's, the current airline training has evolved over a 25-year period. Incorporating the five factors into the fire service's training methodology is an essential first step to begin to change the profession's culture. Here follow some ideas on how to do this.

Communication

Communication is the key to success in any endeavor. Everyone has experienced misunderstandings that led to errors and mistakes. CRM teaches people to focus on the communication model (sender-message-medium-receiver-feedback) and speaking directly and respectfully.

Situational Awareness

situational awareness ■ A concept that discusses the need to maintain attentiveness to one's surroundings.

Situational awareness is a concept that discusses the need to maintain attentiveness to one's surroundings. It examines the effects of perception, observation, and stress on personnel. The emphasis is on the need to recognize that situations in the emergency services are particularly dynamic and require full attention.

Decision Making

Decision making can be divided into two general categories—life threatening and non-life-threatening. Non-life-threatening decisions are typically made when a decision maker has time to evaluate options in an unhurried manner and chooses the best option. Life-threatening decisions do not offer such leisurely reflection. Making decisions, regardless of threat, depends on four factors: information, experience, knowledge, and urgency. Making correct decisions on the fireground requires that the information avalanche and information chasm situations faced by fireground officers be rapidly processed and formulated into an action plan.

Teamwork

Any group that fails to perform as a team is eventually doomed to fail. Failure in the emergency service field results in excessive damage, poor crew performance, injury, and death. CRM training emphasizes team performance through exercises in the awareness tier and crew performance during the reinforcement tier. The training also focuses on "leadership–followership" so all members understand their place on the team and the need for mutual respect.

Barriers

The final factor addressed in CRM training is recognizing the effect of barriers on the other four factors. Barriers are any factors that inhibit communication, situational awareness, decision making, and teamwork. They can be external (physical) or internal (prejudice, opinions, attitudes, or stress). The CRM segment on barriers focuses on recognizing that barriers exist and taking steps to neutralize and overcome their negative effects.

Crew Resource Management requires a commitment to change leadership and operating cultures that have evolved over generations. The similarities between the flight deck of an airliner and the cab of an emergency vehicle suggest CRM also has application to the emergency services. CRM's goals are to minimize the effect human error has on operations and maximize human performance. As has been discussed, crews trained in CRM learn skills that enhance communication, maintain situational awareness, strengthen decision making, and improve teamwork. With the U.S. military, medical industry, and shipping industry having already adopted the concept, developing and adopting CRM for the nation's emergency services is the next logical step toward a safer, more effective service.

Over the next decade, if the past trend in the fire service is not halted, approximately 1,100 firefighters will die, and 1 million will be injured. If the fire service does not take some action to arrest the effects of adrenaline and machismo, it is destined to continue a history of grand funerals and mourned losses. One powerful tool in this culture change is the use of Crew Resource Management. When the fire service adopts the proven concepts of CRM, if history holds true, it will realize the same benefits that other industries have achieved in arresting catastrophic events.[15]

Firefighter Health and Wellness

The prevalence of cardiovascular illness and deaths and work-inhibiting strains and sprains among firefighters illustrates the need for a comprehensive health and wellness program in every department. The fire service realizes health and wellness programs benefit not only individual firefighters but also the fire and emergency services as a whole; such programs can yield safer and more effective action by first responders to emergencies.

Many factors influence the occurrence of an injury, its severity, and its outcome. Without a doubt, the health of the individuals sustaining the injury is one of the more important factors. Firefighting consists of periods of low activity punctuated by periods of intense, strenuous activity. Good physical condition is a critical component in the body's ability to successfully transition, without injury, between these two activity levels.

Undoubtedly, preexisting medical conditions affect the health and safety of firefighters, including underlying medical diseases as well as physical fitness. Over 80,000 firefighters are injured in the line of duty on average every year. On-duty deaths and injuries may have been avoided, or the injuries less severe, under the same conditions if there was no preexisting condition.

Despite the known risks, thousands of firefighters and emergency medical personnel lack rudimentary medical evaluation and overall wellness that can ameliorate the physical stress of emergency response. According to the USFA's 2002 publication, *A Needs Assessment of the U.S. Fire Service: A Cooperative Study Authorized by U.S. Public Law 106-398*, only one-fifth of the surveyed departments nationwide have a program to maintain basic firefighter fitness and health, such as is encouraged by NFPA 1500, Standard on Fire Department Occupational Safety and Health Program.

The 2002 study goes on to report a large share of firefighters serve as volunteers in smaller communities, where most fire departments do not have programs to maintain basic firefighter fitness and health. There may be as many as 792,000 firefighters, or roughly three-fourths of the estimated total number of U.S. firefighters, without such programs.

Implementing health and wellness programs in the fire service will likely prevent firefighter injuries and deaths. A body of evidence suggests improved lifestyles reduce the risk of injury and death. Therefore, many injuries and deaths could be reduced by implementation of a health and wellness program in the department.[16]

THE IMPORTANCE OF HEALTH AND WELLNESS

Here are 10 reasons supporting the importance of health and wellness and the need for developing and implementing health and wellness programs:

1. **Improves heart health.** The importance of aerobic exercise cannot be understated. As has been covered in this chapter, heart attacks cause the majority of deaths among firefighters. Regular aerobic exercise helps prevent heart disease, strengthens heart muscle, decreases clotting, and stabilizes the electrical activity of the heart. Aerobic exercise slows plaque buildup in the arteries and also helps to normalize blood pressure, especially in people whose blood pressure is somewhat elevated.

2. **Improves heart tolerance.** Exercise increases blood volume, which improves heat tolerance. Improved heat tolerance will help firefighters battle more intense fires.

3. **Helps prevent Type II diabetes.** Exercise improves the body's ability to regulate blood sugar, thereby preventing Type II diabetes.

4. **Reduces risk of strains and sprains.** Physical activity strengthens the muscles and joints and other structures such as tendons and ligaments that hold the body together. This strengthening decreases the risk of strains and sprains, the leading cause of injury for firefighters.

5. **May improve emotional state.** Firefighters often deal with life-and-death situations when they respond to an emergency. Taking part in health and wellness programs improves their psychological and emotional states, which will improve emotional reactions during life-and-death situations. An improved emotional state also improves self-esteem, self-efficacy, and sleep patterns, thereby reducing depression, anxiety, and stress.

6. **Maintains weight loss.** Exercise helps control body weight and is essential in any weight loss program. Weight loss is more likely to be maintained if a person continues to exercise. Weight loss increases stamina, as well as aerobic abilities, both of which are needed for firefighting.

7. **Maintains metabolic rate.** By preventing the loss of metabolically active muscle tissue, exercise helps prevent the drop in metabolic rate that sometimes accompanies weight loss and the gradual decline in metabolic rate that occurs with aging.

8. **Enhances ability to fight fires.** Exercise can slow the loss of stamina, strength, flexibility, bone density, and metabolic rate, which all affect an individual's ability to fight a fire.

9. **Prevents development of back problems.** Maintaining flexibility in the muscles of the legs and lower back and increasing strength in the abdominal and back muscles can help prevent back problems from developing. Firefighter back problems often develop from lifting hoses and equipment and moving apparatus.

10. **Encourages overall healthy lifestyle.** As fitness improves, activity becomes easier. Exercise increases stress resistance and improves sleep. An active lifestyle also encourages other health-promoting habits, such as avoiding tobacco and alcohol and developing healthy eating habits. Besides feeling better, firefighters lower their risk for injury or even death with more and consistent exercise.

WHY HAVE FITNESS PROGRAMS FAILED TO WORK?

Despite the importance of health and wellness programs for the fire services, many obstacles exist to their inception and implementation. The fire service must address these obstacles if firefighter health is to be improved. Based on health advocate and former firefighter Michael Stefano's experiences in administering a number of programs, the following are the five leading reasons for failure:

1. *Lack of information on risk to self.* Because many firefighters are not aware of the health risks of firefighting, they are therefore uninterested in changing their condition. With many preventable injuries and deaths occurring annually, pertinent health information must be disseminated to the firefighters to motivate them to change their lifestyles.

2. *Lack of individual goals.* Programs that have failed to outline reasonable and specific individual goals are less likely to succeed. Program participants who feel they do not accomplish anything drop out. In developing any fitness program, the needs and wishes of the participants must be taken into account, and the participants must be able to track their progress.

3. *Lack of appropriate training.* Fitness programs generally are not designed by professionals and thus lack the elements necessary for an effective program. Although hiring a personal trainer may be too expensive for some fire departments, they should seek professional consultation to ensure the efficacy and safety of the program.

4. *Lack of time to devote to the program.* For career firefighters scheduling time for building and maintaining fitness must be incorporated into their daily routine. The challenge for volunteer firefighters is they already donate many hours to the fire service, and few believe they have excess time to devote to health and wellness. However, firefighter health is too important to ignore. Instead, fitness programs for volunteers should be designed around the members' personal and family time.

5. *Lack of motivation.* Even firefighters aware of their elevated health risk choose not to participate in fitness programs. Lack of motivation is a serious challenge that must be addressed by each fire department.[17]

OVERVIEW OF MAJOR FIRE SERVICE WELLNESS INITIATIVES

The fire service's greatest asset is not equipment, apparatus, or stations, but rather its personnel. Through its personnel, the fire department serves the public, accomplishes its missions, and is able to make a difference in the community. By committing to a wellness program, the fire department often increases the members' trust, which enhances every program and each call answered by the fire department. Placing a high priority on wellness makes sense for everyone, including fire service personnel, the taxpayers, and the public served.

WHAT IS WELLNESS?

Wellness is a term that refers to an individual's state of mind as well as his or her physical state, balancing between health and physical, mental, emotional, and spiritual fitness. The concept of wellness also entails having access to rehabilitation, when indicated. Moreover, wellness should be an interactive process whereby an individual becomes aware of and practices healthy choices to establish a balanced lifestyle.

In fire departments, wellness programs are intended to strengthen uniformed personnel so that their mental, physical, and emotional capabilities are resilient enough to withstand the stresses and strains of life and the workplace.

A wellness program should not be perceived as just another program, but rather as a complete commitment to the health, safety, and longevity of all uniformed personnel; productivity and performance of all fire crews; and cost effectiveness and welfare of all fire departments. *The Fire Service Joint Labor Management Wellness-Fitness Initiative* is considered a total program, whereby all components must be implemented for the benefit of both the individual and the department.[18]

NFPA STANDARDS

In August 2000, NFPA released NFPA 1583, Standard on Health-Related Fitness Programs for Fire Department Members. As the NFPA states,

> The purpose of this standard is to provide the minimum requirements for a health-related fitness program for fire department members who are involved in rescue, fire suppression, emergency medical services, hazardous materials operations, and related activities. Implementation of this document shall promote the members' ability to perform occupational activities with vigor and to demonstrate the traits and capacities normally associated with a low risk of premature development of injury, morbidity, and mortality.
>
> The health-related fitness program shall include the following components:
>
> 1. The assignment of a qualified health and fitness coordinator
> 2. A periodic fitness assessment for all members
> 3. An exercise training program that is available to all members
> 4. Education and counseling regarding health promotion for all members
> 5. A process for collecting and maintaining health-related fitness profile (HRFP) data

When the wellness-fitness initiative is adopted as a methodology within an organization, it becomes a key component of any fire department's occupational safety and health program; and it is a companion to NFPA 1582, Standard on Comprehensive Occupational Medical Program for Fire Departments, which was released in February 2000.[19]

THE WELLNESS-FITNESS INITIATIVE

Labor Management Wellness-Fitness Initiative
■ This program is designed for incumbent fire service personnel. It requires a commitment by labor and management to a positive, individualized wellness-fitness program, including information on these topics: medical evaluation, fitness evaluation, injury, medical rehabilitation, behavioral health, cost justification, data collection, and implementation.

The Fire Service Joint **Labor Management Wellness-Fitness Initiative** is a historic partnership between the IAFF and the IAFC as a way to improve the wellness of fire department uniformed personnel. Ten public professional fire departments from the United States and Canada participated in the forming of this partnership in 1997. Each of these departments committed itself to this wellness-fitness initiative by requiring mandatory participation of all of its uniformed personnel in this program. This move to commit labor and management to the wellness of all its uniformed personnel will help to establish the base of fitness/wellness in the fire service into the 21st century.

The Fire Service Joint Labor Management Wellness-Fitness Initiative is a program designed to promote the health of firefighting personnel through fair and just principles, with an emphasis on nonpunitive response to personnel involved in the program.

In 2007, a new brand and logo were established for the Fire Service Joint Labor Management Wellness-Fitness Initiative. It is now known as the WFI (Wellness-Fitness Initiative).

The WFI's intention is that its implementation should be a positive individualized program that is nonpunitive. All component results are measured against the individual's previous examinations and assessments and not against any standard or norm. However, medical practice standards may be used when results indicate that lifesaving intervention is required.

Confidentiality of medical information is the most critical aspect of the WFI. The unauthorized release of personal details that may be recorded as part of a medical evaluation causes legal, ethical, and personal problems for the employee, employer, and examining physician. All information obtained from medical and physical evaluations should be considered confidential, and the employer will only have access to information regarding fitness for duty, necessary work restrictions, and, if needed, appropriate accommodations. Also, all medical information must be maintained in separate files from all other personnel information.

Components of the Initiative

As seen in Table 9.4 (page 334), the Wellness-Fitness Initiative has multiple components: medical, fitness, rehabilitation, and behavioral health, which are designed to be implemented as a whole. In the case of the volunteer service, it would be quite challenging to put all of these into practice at once.

PEER FITNESS TRAINING

The WFI also provides for a Peer Fitness Training (PFT) Certification Program. This course helps to identify firefighters who have demonstrated the knowledge and skills required to do the following:

- Design and implement fitness programs.
- Improve the wellness and fitness of their departments.
- Assist in the physical training of recruits.
- Assist the broader community in achieving wellness and fitness (e.g., fitness programs in schools).

Home study materials for the PFT Certification Program are now available. This certification—developed together by the IAFF, the IAFC, and the American Council on Exercise (ACE)—enables fire department members who have earned it to learn wellness-fitness initiative fitness testing protocols, and it will be helpful in promoting wellness and fitness throughout their departments. For more information on the PFT course, visit http://www.iaff.org/HS/PFT/peerindex.htm.[20]

Conclusion

In the last decade we have seen greater efforts toward creating a new dynamic related to firefighter health and safety in the American fire service. Yet, as a profession, it still has long way to go.

Although part of the issue has to deal with the cultural aspects of the profession, fire service leaders must continue to push for more complete research on the health effects of this profession. They need to identify methods and means by which they can make their environments safer, their firefighters healthier, and ultimately their services more effective.

TABLE 9.4   Components of the IAFC-IAFF Initiative

CATEGORY	COMPONENTS
Medical	• Physical evaluation • Body composition evaluation • Laboratory tests • Vision tests • Hearing evaluations • Spirometry • EKG • Cancer screening • Immunizations and infectious disease testing • Referrals • Data collection
Fitness	• Medical clearance • On-duty time for exercise • Equipment and facilities • Exercise specialists and peer trainers • Fitness incorporated into philosophy • Fitness evaluations (aerobic capabilities, flexibility, muscular strength, muscular endurance) • Fitness self-assessments • Exercise prescriptions
Rehabilitation	• Need for rehabilitation • Rehabilitation as a priority • Establishment of a medical liaison • Physical therapy services • Clinical pathways • Alternate duty • Injury prevention program
Behavioral health	• Professional assistance • Nutrition • Tobacco use cessation • Employee assistance programs • Substance abuse intervention • Stress management • Critical incident stress management • Chaplain services

PEARSON
myfirekit™

Visit MyFireKit
Chapter 9 for
Perspectives of
Industry Leaders.

The challenge may be great, but the significant momentum today must be carried forward by today's leadership and those who will ascend to senior leadership positions in the future. The focus on this critical issue must be kept at the forefront in dealing with every member of the fire service. Although the profession has begun to create a great deal of energy around this topic, the fire service has yet to create the tipping point (see Chapter 3), which will create dramatic results in respect to the reduction of firefighter fatalities and injuries. However, the fire service is much closer to it today than it has ever been.

Review Questions

1. Provide an overview of the nature of firefighter fatalities as noted in the 2002 *Firefighter Fatality Retrospective Study*.
2. What future research may be needed to determine the actions to take to address firefighter fatalities and injuries?
3. Provide an overview of firefighter fatalities by department application, and correlate the types of fatalities likely to be experienced by each.
4. Explain the geographical distribution of firefighter fatalities.
5. Provide an overview of firefighter injuries by injury type, by type of duty, and by region of the country.
6. Provide an analysis of firefighter injuries and injury rates to population protected.
7. Describe the initiative Everyone Goes Home.
8. Describe the initiative Firefighter Near Miss.
9. Describe the initiative Crew Resource Management.
10. What is the Wellness-Fitness Initiative (WFI)?
11. What is the importance of health and wellness in our profession?

PEARSON

myfirekit™

For additional review and practice tests, visit www.bradybooks.com and click on MyBradyKit to access book-specific resources for this text!

Register for MyFireKit by following directions on the MyFireKit student access card provided with this text. If there is no card, go to www.bradybooks.com and follow the MyBradyKit link to buy access from there.

References

1. National Fallen Firefighters Foundation, National Fire Service Research Agenda Symposium, *Report of the National Fire Service Research Agenda Symposium*, Emmitsburg, Maryland, June 1–3, 2005.
2. Interview of Ron Siarnicki, National Fallen Firefighters Foundation, 2009.
3. U.S. Fire Administration, *Historical Overview*. Retrieved October 14, 2010, from http://www.usfa.dhs.gov/fireservice/fatalities/statistics/history.shtm
4. U.S. Fire Administration, *Historical Overview*. Retrieved October 14, 2010, from http://www.usfa.dhs.gov/fireservice/fatalities/statistics/history.shtm
5. "Landmark FEMA Study: Heart Disease Is an Epidemic for Firefighters," *Fire Engineering*, March 17, 2009.
6. Michael J. Karter and Joseph L. Molis, "Firefighter Injuries for 2007," *NFPA Journal*, November–December 2008, p. 19.
7. Ibid., p. 3.
8. Ibid., p. 1.
9. Bill Manning, *Why I Support the Everyone Goes Home® Program and Firefighter Life Safety Initiatives (and Why You Should, Too)*.

Retrieved October 20, 2010, from www.
everyonegoeshome.com/partners/support.html

10. Interview of Bill Goldfeder, Deputy Chief,
Loveland-Symmes Fire Department, Ohio,
2009.

11. Interview of David Daniels, Fire Chief,
Woodinville, WA, 2009.

12. Interview of Ron Siarnicki, National Fallen
Firefighters Foundation, 2009.

13. Interview of Alan Brunacini, Retired Fire
Chief, Phoenix, AZ, 2009.

14. International Association of Fire Chiefs, *Crew
Resource Management: A Positive Change for
the Fire Service* (Fairfax, VA: Author, 2002).

15. Ibid.

16. FEMA, *Health and Wellness Guide for
the Volunteer Fire and Emergency Services,*
FA-267 (Emmitsburg, MD: Author, October
2008). Retrieved October 21, 2010, from
http://www.nvfc.org/files/documents/
HealthWellness_guide.pdf

17. IAFC and IAFF, *Fire Service Joint Labor
Management Wellness-Fitness Initiative,*
3rd ed. (Fairfax, VA: IAFC, 2008).

18. Ibid.

19. FEMA, *Health and Wellness Guide for the
Volunteer Fire and Emergency Services,* FA-
267 (Emmitsburg, MD: Author, October
2008). Retrieved October 21, 2010, from
http://www.nvfc.org/files/documents/Health-
Wellness_guide.pdf

20. IAFF, *Health, Safety, & Medicine.* Retrieved
October 22, 2010, from http://www.iaff.org/
HS/PFT/peerindex.htm

CHAPTER

10

National and International Trends in the Fire Service

(Courtesy of Randy R. Bruegman)

KEY TERMS

BFRL, *p. 338*

Centers for Disease Control and Prevention (CDC), *p. 343*

Firefighter Safety and Deployment Study, *p. 367*

Global Concepts in Residential Fire Safety, *p. 343*

National Construction Safety Team Act (NCST), *p. 339*

National Fire Service Research Agenda, *p. 339*

NFPA Fire Research Agenda, *p. 341*

NIST, *p. 338*

Vision 20/20, *p. 365*

OBJECTIVES

After completing this chapter, you should be able to:

- Explain how the anticipated demographic trends in the United States may impact future fire service strategies.
- Describe how technology may impact the profession and its approach to the fire problem in the United States.
- Describe the elements identified in the National Fire Service Research Agenda, which focuses on improving firefighter life safety.
- Explain the international best practices related to fire prevention.
- Describe the strategic objectives and action items outlined in Vision 20/20.
- Define the research currently underway that may impact the future of the U.S. fire service.

PEARSON

For additional review and practice tests, visit www.bradybooks.com and click on MyBradyKit to access book-specific resources for this text!

Introduction

The fire service has been accused of being a profession that has been slow to change, and the facts of its history would tend to support this. Yet, as the fire service evaluates current trends, it is obvious that the profession is becoming as focused on preventing incidents from occurring as it has been on responding to them. A driving force of this shift has been the expansion of research in a variety of fire service–related topics and the globalization (the sharing) of this information among the world's fire service. Chapter 10 explores the research and the focused efforts underway that will have significant impact on the course of the fire service over the next 25 years.

The Importance of Research

During the last decade, research relating to the fire service has increased significantly. Fire service research with its attendant laboratory studies will have a significant impact on the direction and actions of the fire service profession over the course of the next century.

NIST
- National Institute of Standards and Technology; formerly the National Bureau of Standards. Founded in 1901 as the nation's first federal physical science lab.

BFRL
- The Building and Fire Research Laboratory is a NIST laboratory that studies building materials; computer-integrated construction practices; fire science and fire safety engineering; and structural, mechanical, and environmental engineering. Products of the laboratory's research include measurements and test methods, performance criteria, and technical data supporting innovations by industry and are incorporated into building and fire standards and codes and often impact industry practices.

The National Institute of Standards and Technology (**NIST**) (formerly the National Bureau of Standards) was founded in 1901 as the nation's first federal physical science lab. Its mission is to promote U.S. innovative industrial competitiveness by advancing the measurement science of standards and technologies to enhance economics, security, and the quality of life for Americans. Today, NIST operates in two main locations; Gaithersburg, Maryland (the headquarters on a 78-acre campus), and Boulder, Colorado (a 208-acre campus). NIST conducts research on a wide variety of physical and engineering sciences. One of the laboratories is the Building and Fire Research Laboratory (**BFRL**), now part of the Engineering Laboratory.

BFRL works to improve quality and productivity in U.S. construction and to reduce human and economic loss due to fire, earthquakes, wind, and other hazards. The BFRL studies building materials; computer-integrated construction practices; fire science; fire safety engineering; and structural, mechanical, and environmental engineering. Products of the research conducted at the BFRL include measurements, test methods, performance criteria, and technical data that support innovation by industry and are incorporated into building and fire safety standards and codes. The primary research areas include high-performance construction materials and systems, which enable scientific and technology-based innovation to modernize and enhance the performance of these materials and systems. Fire loss reduction facilitates engineered fire safety for people, products, and facilities as well as enhanced firefighter effectiveness with a goal to reduce firefighter fatalities by 50 percent. Enhanced building performance provides a means to help predict that buildings work better through their useful lives by providing knowledge and measurement to optimize building life cycle performance. The homeland security component uses the lessons learned from the World Trade Center disaster to protect people and property better, to enhance the safety of fire and emergency responders, and to restore public confidence in the safety of high-rise buildings nationwide.

The BFRL is at the center of corporation research as well as it provides outreach to many other private and public organizations on projects of mutual interest through cooperative research and development agreements. These agreements with other well-known fire research facilities, such as Worcester Polytechnic Institute and Maryland Fire and Rescue Institute, include the hosting of guest researchers, supporting a competitive postdoctoral program in conjunction with the National Academy of Sciences/National Research Council, and collaborating with those doing fire-based research worldwide.

A new task BFRL has undertaken resulted from the events of September 11. The **National Construction Safety Team Act (NCST)**, signed into law in October 2002, authorized NIST to investigate major building failures in the United States. The BFRL has been charged with this effort. The NIST investigations will not only establish the technical causes of building failures but also evaluate the technical aspects of emergency response and evacuation procedures in the wake of such failures. The goal is to recommend improvements to the ways in which buildings are designed, constructed, maintained, and used. Two major safety investigations of note are the September 11, 2001, fire and collapse of the World Trade Center, and the February 20, 2003, fire at the Station Nightclub in West Warwick, Rhode Island. Such inquiries will have direct impact on the fire service, as lessons learned from them can be immediately applied to codes and standards in respect to building construction and response deployment.

It is evident since September 11 that the synergy of research created through NIST and BFRL has increased and will continue to promote a greater collaborative worldwide research effort into fire service–related issues. What will come of such research is yet to be known, but the fire service can anticipate the impacts will positively affect its ability as a profession to reduce the severity of property loss and life loss in this country.

National Construction Safety Team Act (NCST)
■ Signed into law October 1, 2002, by President Bush, it authorizes the National Institute of Standards and Technology (NIST) to establish teams to investigate building failures. These authorities are modeled after those of the National Transportation Safety Board (NTSB) for investigating transportation accidents. NIST is a nonregulatory agency of the U.S. Department of Commerce.

National Fire Service Research Agenda

On June 1–3, 2005, the **National Fire Service Research Agenda** Symposium was held at the National Emergency Training Center in Emmitsburg, Maryland. The NFFF conducted the symposium; funding was through a grant from the National Institute for Standards and Technology (NIST).

The symposium's purpose was to produce a document to identify and prioritize the areas where research efforts should be directed to improve firefighter life safety. The intended use of the document is as a guide for both research organizations and allied fire service organizations to support reducing firefighter fatalities.

The focus of the symposium included the following: firefighter health and wellness; structural firefighting; wildland firefighting; firefighter training; emergency vehicle design and operations; and reduction of fire risk occurrences. The symposium attendees represented several segments of the research community, including fire protection, building construction, occupational medicine and behavioral science, fire service organizations, individual fire departments, and allied professionals. Attendees identified opportunities for sharing efforts and results, and organizations that currently or could potentially sponsor, conduct, and participate in various projects aimed at improving firefighter life safety.

National Fire Service Research Agenda
■ Symposium to produce a document to identify and prioritize the areas where research efforts should be directed to support improvements in firefighter life safety.

THE PROBLEM STATEMENT

Each year, more than 100 firefighters are killed in the line of duty in the United States, and approximately 100,000 firefighter injuries are reported. Many in the fire service believe these statistics must change. The origin of this symposium was the work done at a National Summit on Firefighter Life Safety convened by the National Fallen Firefighters Foundation (NFFF) in March 2004 to identify 16 firefighter life safety initiatives. These initiatives established the key strategies to be implemented to meet the U.S. Fire Administration's goal of reducing firefighter fatalities. More than 230 participants were involved in developing the initiatives to form the basis of a major effort to reduce the risks of firefighting and emergency service delivery. The NFFF accepted the responsibility to lead the effort to implement all of the initiatives, in cooperation with all of the organizations concerned with firefighter health and safety.

One of these initiatives called for the development of a national research agenda to support the implementation of advances in firefighter health and safety; such a focused approach targeting a specific problem area in the fire service had not been done previously.

Participants in the 2005 symposium concentrated their efforts on five different areas:

- Structural fire suppression
- Wildland fire suppression
- Research and training
- Vehicles and equipment
- Health, wellness, and physical fitness and incident reduction

Within each of these areas, symposium attendees identified research issues. In many cases, the research priorities identified by different groups were interrelated and suggested areas of study that could support multiple objectives. In other cases, existing research efforts were recognized that should be expanded and coordinated with other efforts. It was acknowledged that certain research projects being conducted for other purposes could be adapted to meet the needs of the fire service.

Four separate discussion groups developed the proposed research issues and then presented their recommendations to a meeting of all the symposium participants. The individual groups were asked to focus on issues relating to the following:

- Firefighter health, wellness, and fitness
- Training and incident management
- Technology applications, fire apparatus and equipment, and transportation
- Fire prevention, public education, and data

This national research agenda will help to focus research institutions on the topics most critical to the fire service and should facilitate the prioritization of where federal money is directed for the funding of such activity.

Visit MyFireKit Chapter 10 for more in-depth information on the National Fire Service Research Agenda.

NFPA Fire Research Agenda

The next 25 years will bring many challenges to fire safety in the protection of buildings and structures. In 2008, in celebration of its 25th anniversary, the Fire Protection Research Foundation convened a series of meetings to learn from members of NFPA technical committees, staff, and leadership about the strategic issues likely to affect future fire safety. The result is the **NFPA Fire Research Agenda**, designed to focus and set priorities for the Foundation's research program in the near term and to ensure it meets the emerging research needs of NFPA technical committees and other constituents.[1]

THE IMPACT OF DEMOGRAPHIC CHANGES

According to Dr. Kevin McCarthy, senior social scientist at the RAND Corporation, the U.S. population will grow steadily over the next 25 years by about 1 percent, or 3 million people, per year. Immigration will account for 40 percent of this growth, with the Asian and Hispanic portion of the overall population growing from 22 to 34 percent. Population increases will primarily occur in southern and western states, most notably Arizona, Florida, Nevada, and North Carolina, and will be increasingly concentrated in urban areas. Some of these urban zones are subject to large-scale, human-made, and natural disaster, including wildland fires, hurricanes, and earthquakes. The composition of the population is also changing with the median age growing from 35 to 38 years. The percent of the working population will shrink, while the senior population will increase from 12.5 to 20.0 percent of the population. These factors mean more single-person households and fewer households with children, resulting in shifts in housing types.[2]

So, how do these factors affect fire safety? The panelists involved in the development of the research agenda concluded that the population that is aging and consequently increasingly demonstrating disabilities will impact needed building fire protection features. Concentration of population will result in demands for both fire protection services and the infrastructure (for example, water, roads) needed to support them. Public fire safety education programs will need to adjust to the changes in demographics, including cultural and language changes. Finally, the decline in the labor force as a percentage of the overall population will require the fire service to use its human resources effectively for fire safety; for example, through targeted allocation and collaboration.[3] This will require the following:

1. Public fire safety education programs must respond to the demographic and cultural shifts in a variety of ways, if they are to be effective.
2. Urban growth patterns show an increasing trend toward higher-density populations, and this will require a focus on all the codes and standards addressing the protection of these type of buildings.
3. As our population becomes more concentrated in urban or suburban areas, our emergency response procedures, including evacuation routes and firefighting tactics, must adapt. The fire service and other emergency responders will need to keep pace to better address large-scale, human-made, and natural disasters.
4. One particular facet of shifting urban growth patterns is the fire problem affecting the urban–wildland interface. Recent decades have already seen more of these disasters, and shifting populations continue to magnify the problem.
5. As the population ages, we must consider effective means to address those with disabilities in our fire safety strategies.
6. The decline in the workforce will affect all fire safety disciplines.[4]

NEW HOPES, NEW CONCERNS: MATERIALS AND TECHNOLOGY

The coming decades will be marked by major advances in materials and technology. Areas including information availability and utility, biotechnology, smart materials, and nanotechnologies are developing at a rapid rate, as are their applications. For example, biotechnology applications include personalized medicine based on databases of patients' information. Nanotechnology applications, meanwhile, range from new families of chemical and biological sensors and battery-capacity improvements to wearable personal medical monitoring—all with obvious and far-reaching implications for the fire service and fire safety efforts. The changing context of fire safety will focus on biotechnology, smart materials, advanced communication, sensing devices, and faster and more robust information sharing. As such:

1. Advances in electronics technology, for example, hold enormous potential to help in detecting fires.
2. The impact of nanomaterials on fire safety has yet to be fully researched and understood. These and other new materials used in building furnishings and construction—new lightweight roofing construction materials, for example—may change many of the basic fire safety design approaches.
3. Technology advances with direct application to the fire service are experiencing an unprecedented boom. The need exists to continue to stimulate this advancement and promote fire service improvements such as integrated sensors and controls and locator systems, which will greatly help firefighters.
4. Technology applied to safety and health issues is also under intensive study. Conference participant Ellen Sogolow of the Assistance to Firefighters Grant Program noted the "application to clinical techniques for real-time monitoring of firefighter safety during fire events," including improved breathing environments and enhanced monitoring of body temperature and heart rate data.
5. The trend in customization of technologies to meet specific performance objectives is directly applicable to the fire suppression systems, which will be used in future building construction to help detect and suppress fire more effectively.
6. Although many of these new technologies offer great promise for the future of fire safety, Dr. Philip Anton, director of the Acquisition and Technology Policy Center at the RAND Corporation, points out that innovative technologies bring new concerns. Safety and health technologies could raise issues related to privacy and ethics, whereas new materials could present new health hazards or new sets of fire hazards.[5]

ENVIRONMENT, ENERGY, AND SUSTAINABILITY

According to Shere Abbott, director of the Center for the Science and Practice of Sustainability at the University of Texas, more than 800 major floods have occurred worldwide since 2000, and wildfires in the western United States have increased fourfold in the last 30 years. Natural resources are declining as a result of increased demand and the degradation of natural ecosystems. Water demand worldwide has tripled in the last 50 years; thirty-six U.S. states face immediate water shortages. More than half the world's population currently lives in cities, with this figure poised to increase sharply. By 2050, the number of people worldwide living in cities is projected to double or triple to around 7 billion. So, what is the relationship among fire safety, sustainable development, and the overall quality of life issues? Issues for construction include the following:

1. The implications of these shifts may be profound and may reshape how we think of fire safety. Along with a possible increase in the number and severity of natural disasters, the increased demand on emergency responders calls for new tactics and will affect fire service–based standards.
2. Transportation-related energy issues will take on even greater urgency, and advancements will create new fire safety opportunities—and new hazards. Greater use of alternative vehicles and their various fuels—biodiesel, ethanol/alcohol, hydrogen, and electricity, for example—presents different hazards and will demand unique emergency response and firefighting tactics, as well as fire protection systems.
3. Sustainability will fuel continued pressure on environmental restrictions for a range of chemicals, a trend affecting virtually all national codes and standards, from the selection of fire suppressants to the hazard control of building contents and furnishings.
4. Declining water resources will continue to have a major impact on fire suppression systems and firefighting strategies, from residential firefighting and home fire sprinkler designs to requirements for high-rise buildings and other high-water-volume fire control strategies.
5. Energy conservation awareness will continue to fuel developments in green building design. The fire safety impact of new types of wall construction, increased thermal tightness, and alternative energy sources such as solar will need to be addressed.[6]

Global Concepts in Residential Fire Safety

As covered earlier in this text, the vast majority of fire deaths and a large proportion of fire injuries occur in the home in the United States and in most western industrial nations. As part of its research program to reduce such deaths and injuries, the National Center for Injury Prevention and Control at the **Centers for Disease Control and Prevention (CDC)** sought to identify the best global practices in community fire safety that might be transferable to the United States or would at least stimulate discussion and ideas to incorporate new approaches in the United States. Finding best practices elsewhere can speed up innovation in the United States. This CDC-commissioned study was conducted by TriData of Arlington, Virginia.

The **Global Concepts in Residential Fire Safety** study provides many examples of successful community fire safety programs in other nations known to have innovative community fire safety programs, which can be associated with reductions in their residential fire death rates. Prevention programs sought at both the national and local levels in each nation were studied.

Centers for Disease Control and Prevention (CDC)
■ The agency of the U.S. Department of Health and Human Services, founded in 1946 and headquartered in Atlanta, whose mission centers on preventing and controlling disease, and promoting environmental health and health education in the United States.

Global Concepts in Residential Fire Safety
■ The practices of fire service providers outside of the United States.

PEARSON
myfirekit

Visit MyFireKit Chapter 10 to find the reports.

STUDY APPROACH

The Global Concepts in Residential Fire Safety project began with the selection of nations to visit, with the first visit to the United Kingdom, with whom the United States has common language and culture that makes acceptability of its ideas easier. The United Kingdom has had a major shift in fire prevention strategy toward undertaking more community fire safety in the last decade. In addition, from surveys for the International Technical Committee for the Prevention and Extinction of Fire (CTIF) of European fire prevention programs, Sweden and Norway were known to have extensive prevention programs affecting residential fire safety.

For each nation they visited, researchers worked with several sources to identify communities conducting innovative programs to reduce fire injuries in the home. Researchers interviewed not only officials of the national agency in each nation that deals most with fire prevention but also officials from local fire brigades and national fire protection associations. Extensive Internet investigation was conducted on community safety programs in each nation selected and candidate cities to be visited.

For each fire service organization that was a candidate to visit, the study communicated with either its chief fire officer or its head of prevention as to his or her willingness to participate. In some cases, the study used a proxy in the national fire organization or chief fire officers' organization for making the initial contacts. The result was that all of the fire organizations that were approached enthusiastically agreed to participate.

RESEARCH QUESTIONS

A list of research questions was sent to each organization to be visited, with some variation tuned to each agency and nation. The questions, with some variations, appear in the box on the following page.

Providing this list of questions ahead of time assured that the agencies visited would invite people with the right knowledge to each interview, and that appropriate background materials would be readied upon the arrival of the research team. This process had been used in past international research.

Following site visits, several of the departments surveyed sent follow-up data and information, and follow-up dialog ensued by phone and e-mail. The draft report was sent to each agency visited to review the section on its nation and organization, with most submitting written edits.

The study did not present everything obtained during the visits and research, but rather focused on the programs likely to be of most interest to the United States fire service.[7]

Best Practices from the International Fire Service

The Global Concepts in Residential Fire Safety study provides examples of successful best practices in community fire safety programs from England, Scotland, Sweden, Norway, Australia, New Zealand, Japan, Canada, Puerto Rico, Mexico,

1. What noteworthy community fire safety programs (public education) are being or have been undertaken by your agency or others? What population groups do they target? How are they delivered, how often, and with what content? Is there any evaluation of their results?

2. How do you cope with the diversity of ethnic groups and languages in developing and delivering community safety programs? Do you tailor programs to a group or just translate the same materials for each?

3. We are aware of many of your past prevention programs, including some discussed in reports on International Concepts in Fire Protection and for CTIF. What programs have been discontinued because they were not working, too expensive, out of date, or for other reasons? Which approaches are still being continued because they have proved to be cost-effective? How is cost-effectiveness determined?

4. Regarding smoke alarms, how have you tried to get them installed and maintained, especially in low-income and immigrant households?

5. Besides various forms of public safety education, are there other efforts for reducing residential fire injuries, such as home inspection programs, chimney sweeps, increased code requirements for residences, better product safety, and residential sprinkler systems?

6. Cooking, heating, electrical issues, smoking, children playing, and arson usually are the leading causes of residential fires. Do you target each cause?

7. What role, if any, is given to firefighters for community safety programs?

8. Are your residential prevention programs multi-hazard? That is, do you combine fire prevention information with non–fire injury prevention, resilience for disasters and terrorism, and so on? Or do you have separate programs for each risk?

9. Overall, what do you consider the most cost-effective ways to reduce residential fire injuries?

and the Dominican Republic. The programs selected have helped reduce fire injuries and fatalities in the home.

Of all the best practices identified in this study, one in particular stands out. To reduce fire casualties in the home, the British fire service is visiting large numbers of high-risk households to do fire safety inspections and risk reductions, especially to ensure these households have a working smoke detector. This approach has required a major change in the culture and mission of the British fire service and should be adapted for use in the United States. The British fire service believes this approach is a major factor in the 40 percent drop in fire deaths in the United Kingdom over the last 15 years; and it probably could have a large impact in the United States and other nations as well if implemented there.

BEST PRACTICES FROM THE UNITED KINGDOM

In the last decade, the change in the prevention strategy used by the British fire service is just short of being revolutionary (see Figure 10.1). National legislation since 2004 has required the fire brigades to engage in strong community safety programs as part of an overall national strategy for improving fire safety. Every British firefighter now is expected to participate in prevention. A national-level

FIGURE 10.1 Flag of the United Kingdom

fire and resilience section was established in the Department of Communities and Local Government to be the focal point for developing national strategies, campaigns, and materials.

The best practices that have arisen out of the new prevention strategy fall into eight major categories:

1. Identifying and analyzing high-risk households
2. Increased staffing and training of prevention programs
3. Making home safety visits
4. Coordinating national and local fire safety campaign
5. Conducting extensive school and youth programs
6. Directing programs to the high-risk elderly population
7. Developing safer consumer products
8. Increasing the use of fire stations for community fire safety programs

Highlights of the best practices include risk analysis, staffing and training for prevention, home safety visits, fire safety campaigns, school and youth programs, programs for the elderly, safer products for the home, and community fire stations.

Risk Analysis

Local fire brigades use nationally developed risk analysis software that links fire data with socioeconomic data to estimate areas of high risk and to target fire safety programs to high-risk groups and households. The nationally developed risk models have been disseminated to all local fire brigades. These brigades undertake integrated risk analysis to decide on the best mix of prevention and suppression for their community.

Fire Brigade Staffing and Training for Prevention

Prevention now is considered a line service, not a support service. The prevention function, which is often supervised by the deputy fire chief for operations, makes extensive use of line firefighters. More fire department resources (person-hours) are being devoted to prevention than in past years. National standards of cover (response times) have been dropped in favor of local discretion, to allow local trade-offs between reduced fire coverage and more attention to prevention. As has been noted, all firefighters are expected to participate in prevention, with recruits being advised that prevention will be a significant part of their job. Recruit training includes practice delivering community fire safety programs.

Home Safety Visits

The British fire service is making visits to a large percentage of high-risk homes, using a combination of line firefighters and prevention specialists. The visits include installation and testing of smoke alarms, inspections for hazards, mitigation of hazards, and one-on-one education. Community safety specialists called "advocates" join firefighters in visiting ethnic or high-risk households. Their specialties include foreign languages and the specific problems of the elderly, alcoholics, and the hearing or mobility impaired. Home visits are scheduled via call centers established in each brigade.

Fire Safety Campaigns

National and local public safety campaigns use paid, prime-time television and radio spots, print media, and the Internet. They do not rely on free public service announcements, and select local radio stations and newspapers are used to reach ethnic populations who are the prime target audience for these media. Television and radio advertisements are run at times of the day that people are applying the behavior addressed, for example, cooking safety at dinnertime. Also addressed is getting coverage of the targeted issues on news and talk shows.

School and Youth Programs

School programs reach close to 100 percent of students in selected elementary school grades in many fire brigades. Significant numbers of students in secondary schools are also being contacted. The school programs are conducted mostly by firefighters but sometimes by teachers and prevention personnel. Special programs target youths who have demonstrated antisocial behavior such as fire setting, attacks on firefighters, or vandalism.

Programs for the Elderly

Social service caretakers of the elderly are trained in fire safety practices they can implement or advocate during their home visits. This includes visits to the homes of high-risk elderly by the fire service and partnerships with various social service agencies to increase resources and provide varied ways to disseminate safety information to the elderly.

Safer Products for the Home

Ten-year tamper-proof, battery-powered smoke alarms are being installed by the fire service and are available for purchase by the general public. In addition, hard-wired smoke alarms are required in all new residential premises and major residential refurbishments. National law requires flame- and cigarette-resistant upholstered furniture and bedding, and portable home sprinkler systems are used for extremely high-risk households.

Community Fire Stations

New fire stations are designed to be community fire safety centers as well as stations from which to respond to calls. These new stations offer reception areas from which to obtain safety literature; live fire demonstrations; and viewing areas to observe firefighter training and response.

The residential fire death rates in England and Scotland dropped in the past 15 years by 41 and 44 percent, respectively. The residential fire death rate dropped from 9.7 deaths per million population in 1990 to 5.7 deaths per million in 2005–2006. Even though it is difficult to attribute cause and effect to particular programs, the data suggest the new approaches are working, and the British fire service also believes that to be the case.[8]

FIGURE 10.2 Flag of Sweden

FIGURE 10.3 Flag of Norway

BEST PRACTICES FROM SWEDEN AND NORWAY

The current fire protection strategy in Sweden and Norway shares many similarities with that in the United Kingdom, though it was arrived at independently (see Figures 10.2 and 10.3). Some of the best practices found fall into nine categories: shifted responsibility for building fire safety; increased fire brigade staffing and training for prevention; expanded home safety visits; increased seasonal and year-round national and local safety campaigns; employee safety education; broadened school safety programs and safety programs for the elderly; required use of home fire extinguishers; improved consumer product safety; and inflatable cushions for jumpers.

Building Owner Responsibility

Sweden and Norway emphasize the responsibility of building owners to ensure the safety of their buildings, and not to depend on fire service inspections.

Fire Brigade Staffing and Training for Prevention

Sweden and Norway are developing highly educated "fire engineers" to form the framework for risk management and resource planning in local fire brigades. The proportion of fire department staff to prevention is much higher than in the United States. In Oslo, a city of 540,000, 40 fire prevention personnel and another 50 firefighters are in stations on a typical weekday. Norway requires one prevention full-time employee for every 10,000 population. In Sweden fire recruit training lasts two years, with 25 percent of the time spent on prevention and risk management. This commitment continues once on shift. For example, Umea, Sweden, takes a group of firefighters off shift duty for 3.5 weeks each year to deliver school programs.

Home Safety Visits

In Sweden and Norway, homes with chimneys must be inspected by licensed chimney sweeps/fire inspectors from four times a year to once every four years, depending on fuel and frequency of use. The home inspectors check heating systems and also do broader home safety inspections. The Oslo Fire Brigade annually visits all of its old, high-risk apartment buildings (on 3,000 blocks) to meet with occupants to discuss fire safety. Posters in the buildings advertise when the fire service is coming, and condominium associations are given safety checklists to pass on to unit owners.

Fire Safety Campaigns

The Swedish fire service gives a fire safety calendar to schoolchildren and households. The calendar shows two days per month on which every household should take specific safety actions, such as testing smoke alarms, checking fire extinguishers, and practicing escape plans. The fire service trains children in schools and then designates the children as the fire marshals for their homes, with specific responsibilities. For example, winter safety advice is tied to the Advent holiday in December; children participate in safety events scheduled for Advent and other winter activities. Because supermarkets are visited by almost every household, the Umea Fire Brigade stations firefighters in them to show shoppers a short safety film, discuss safety issues, and hand out safety literature; and in movie theaters in Sweden, a one-minute fire safety spot is shown to address winter safety hazards. A "Safe Home" campaign is also aimed at builders; when they comply, they can advertise "we build homes that are fire safe."

Employee Safety Education

Some Swedish and Norwegian fire brigades provide instruction to municipal workers and certain private industry workers on fire safety at work and at home. This practice has had a multiplier effect as this instruction is often shared with the families in the households of the workers.

School and Elderly Programs

The Swedish and Norwegian fire service reaches most schoolchildren twice during their school years. Programs for the elderly include the fire service training caretakers in fire safety and providing fire-resistant "smokers' aprons" to elderly who insist on smoking. Also, emergency egress features are promoted for homes of the elderly.

Home Fire Extinguishers

Norway requires extinguishers or hose lines attached to faucets in every home, in addition to smoke alarms. Home occupants are trained to extinguish small fires because the fire service cannot arrive within the 2 to 4 minutes it takes for many fires to reach flashover. Sweden estimates that 35 percent of homes are equipped with extinguishers.

Safer Products for the Home

To reduce fires from unattended cooking, timers are being built into stoves in Norway or, less expensively, stoves are plugged into timers. The timers shut the stoves off if the person cooking forgets to do so or falls asleep. The use of timers is advocated especially for households of elderly people. Electrical equipment is recommended to be plugged into "power strips," and the strips turned off at night for whatever electrical equipment does not have to operate all night long. Safety candle use is promoted; the wicks in these candles do not burn down to the bottom, and so the candles "self-extinguish" before reaching a flammable surface.

Inflatable Cushions for Jumpers

The Oslo Fire Brigade has deployed large, rapidly inflatable cushions in all fire units to rescue people trapped in residences (or other buildings) up to the fourth floor. A trapped victim can jump onto the cushion. In the first year they were deployed, the use of these cushions saved 13 people. Scandinavia, which has slightly lower residential fire death rates than does the United States, has achieved this with fewer firefighters per capita than in the United States. It is realizing better results with smaller fire departments by emphasizing prevention.

The United Kingdom has succeeded in changing its fire service culture over the past decade and transferring practices from a few innovative brigades to many. The Scandinavian fire service also is changing its culture to focus on risk management. Although it is sometimes difficult to transfer good practices from one culture to another or one country to another, chief officers, leaders in fire and life safety, and prevention leaders must begin to develop strategies on how to apply these best practices in their own communities. Adapting these best practices may help continue reduction in the fire injury and death rates in American homes, especially those at highest risk, and in time will reduce the risk to firefighting personnel.

The nations of Australia, New Zealand, and Japan have significantly lower accidental fire death rates in residences compared to the United States. Visits to their fire services found many best practices that the United States might use for reducing residential fire deaths and injuries. Fire protection in Australia is provided largely by state fire services, which allow for consistent statewide prevention practices that reach a large percentage of the population. New Zealand's national fire service has the same advantage. Japan's fire service is entirely local. Although many of these concepts are being used sporadically in the United States, they are often not as widespread, or they utilize different implementation approaches.

BEST PRACTICES FROM AUSTRALIA AND NEW ZEALAND

Prevention programs are developed and evaluated at the state or national levels in Australia and New Zealand (see Figures 10.4 and 10.5). Line firefighters trained for specific programs largely deliver them, and both volunteer and career firefighters participate. More firefighter time is spent on prevention than in the United States.

Prevention programs are targeted to high-risk groups that are identified from geographic information systems (GIS) analysis of fire experience and socioeconomic data. Prevention programs, evaluated annually, are dropped or reshaped

FIGURE 10.4 Flag of Australia

FIGURE 10.5 Flag of New Zealand

if not working. Because these prevention programs reach large percentages of these targeted populations, the result has been fewer deaths, injuries, and lower dollar loss.

Reaching Schoolchildren

Fire safety lesson programs are tailored to existing curriculum and are aligned with state or national education curricula to fit better with teaching needs and hence increase their acceptance by the schools and teachers. These programs explicitly encourage students to pass on what they learn to their parents, which is especially important when the parents are not literate in English. To facilitate these programs, prevention units appoint a coordinator as the liaison with the school system. Local fire stations solicit and maintain contacts with local schools. Coupled with tailoring fire safety lessons to the curriculum, the result has been that the majority of schools accept the fire safety programs. Almost all children in kindergarten through grade 1 benefit from school fire safety programs that cover simple basics; programs tailored for most children in grades 5 and 6 contain more advanced information. Some brigades revisit schools two to six weeks after their initial visit to reinforce the messages.

A mobile home fire safety van called a "mobile education unit" has compartments simulating rooms in a home. Fire brigades bring it to schools to vividly demonstrate the fire safety issues in each room of the home.

For children with disabilities, fire safety strategies and materials are available in Braille, simplified English, and large print. Programs are delivered to special schools and to individuals with disabilities attending regular schools.

For preschool children, it is difficult for fire safety to be taught effectively because of their limited intellectual development. Instead, fire brigades address the programs to their parents and professionals in nursery schools, child health care centers, and other venues.

The educational effort conducts a periodic evaluation and redesign with focus groups to assess all prevention materials. Surveys of parents and teachers are conducted to assess whether the school lessons changed the behavior of the child or the household members.

Reaching Juvenile Fire Setters

Firefighters who volunteer to deliver programs to juvenile fire setters receive psychological screening. Those who pass are trained and then conduct an intervention under supervision before being allowed to carry out sessions on their own. Psychologists periodically meet with the program deliverers to mentor and monitor them. Pairs of firefighters conduct a four-visit program in homes of children who set fires. Parents must be present or within earshot.

Longer, more intensive programs lasting from four sessions to 12 weeks are used for juveniles brought to the attention of the juvenile justice system. These programs can be used as an alternative to jail time. The programs strive to replace destructive behavior with positive fire safety practices.

Reaching the Elderly and People with Disabilities

Case workers and professionals who provide in-home services to the elderly and people with disabilities are trained to look for fire safety problems when in the homes, especially the lack of working smoke alarms. This training is now part of the nationally accredited certification process for community care workers in Australia, which means they all receive it.

Caregivers in agencies whose clients have an intellectual disability are trained on not only hazards to look for but also how to teach fire safety basics to a clientele group with learning difficulties.

Reaching Immigrants and Ethnic Groups

Fire safety ads are run in ethnic newspapers and radio stations or on programs favored by target ethnic groups. Fire safety spokespersons with the right language skills work with ethnic radio talk shows. Key fire safety materials are translated into the major languages spoken in the jurisdiction. Web sites list materials and programs available in each language, using the alphabet and script of the language. To assist in this effort, fire brigades appoint liaison officers whose sole duty is to work with minority and ethnic groups in the community on improving their fire safety.

State and National Campaigns

By developing programs at the state or national level, costs are reduced by not having each community develop its own program. This also provides consistency of messages and makes affordable the use of experts to develop educational materials. This approach facilitates evaluation of programs by having fewer to evaluate and larger numbers of participants in each.

Campaigns are evaluated and then reshaped, if needed. New Zealand undertakes two 1,000-person surveys a year. One tests recall of specific campaign spots; the other checks knowledge of safety awareness and practices. Australian brigades, too, use surveys and focus groups to evaluate campaigns, attitudes toward fire safety, and how best to promote awareness.

Campaigns run year-round, with seasonal themes. Television and radio ads are purchased to air during prime time or when programs are most likely to be watched by the targeted high-risk groups. Ads on sports channels reach middle-age male drinkers. Ads during soap operas reach the unemployed. A television spot innovatively used a person who cares for burn victims to deliver a fire safety message. The fire services provide spokespersons to television and radio talk shows for interviews.

Rather than tell firefighters not to deal with the media and pass all contacts to the public affairs office, firefighters are encouraged to contact local media within the constraints and advice provided in a media guidebook. The result has been expanded media coverage and dissemination of fire safety information.

The fire service also develops an annual list of prevention programs that can be sponsored by industry. McDonald's, Subaru, major banks, and other companies have provided millions of dollars for specific prevention programs.

Home Visits

Firefighters make home visits, primarily to check on the existence and maintenance of smoke alarms, especially in areas with high poverty or joblessness. Some programs go door-to-door to visit every household in a small village or target area, such as rural Maori villages in northern New Zealand; they reach about 60 to 70 percent of households in the selected areas. The fire brigade also helps the household develop an escape plan and provides safety advice. To assist local ethnic groups, unemployed people from the community are hired to help install smoke alarms. This has a triple benefit: local installation knowledge is left behind; the program is more acceptable to local ethnic groups; and some unemployed receive a step toward employment.

Homes in Urban–Wildland Interface Zone

Australia has developed an approach to dealing with wildfires that threaten homes. Radically different from the method used in the United States, the policy is to "leave early before road travel becomes dangerous, or stay and defend your home." It is based on research by CSIRO, the Australian national research laboratory. Residents can do many things to prepare their homes to make "staying to defend" a reasonably safe option. Codes governing new interface homes dictate what can be built in areas of high wildland fire danger. Potential home sites are rated on their proximity to high-risk vegetation and on the slope of the land. Depending on the assessed level of risk, requirements vary for building materials and structural features such as the materials used in the roof and floor supports, the thickness of windows, the need for screening against flying embers, or changes in landscaping.

GIS maps show which existing houses are in the highest danger areas. In some Australian states, the fire brigades inspect homes to see whether they are likely to withstand a fire if the occupants stay and fight.

Codes and standards for home fire safety are generally similar to those in the United States, other than homes built in the urban–wildland interface. Some differences do exist. All living units of high-rise apartment buildings must be sprinklered, not just the common areas. Public housing must be sprinklered and compartmented. Legislation requires working smoke alarms in every home; some states call for them on every level of a home. Some require all new residences to have hard-wired alarms. Use of high-voltage electrical circuitry has led to strong electrical safety requirements, including mandatory ground fault interrupts and circuit breakers on home electrical systems. These safety features help reduce electrical fires.

Consumer Products for Safety

Several product innovations are used to promote safety. The usage of fire blankets is actively encouraged, especially for kitchen fires such as flaming oil pans on stoves. Fire extinguisher use in the home is endorsed but not mandated. Home sprinklers in New Zealand are rare due to the fact sprinklers were built to stringent standards and were not cost-effective. As a result, a home sprinkler system has been designed and is in use at a sharply lower cost that will handle 90 percent of home fires.

Managing Fire Service Prevention Resources

Certain fire services have shifted some resources from commercial inspections to home visits. Most Australian fire services have accepted longer response times as a trade-off for doing more prevention. In New Zealand and Queensland, each fire station develops a business plan for the year, including prevention activities, and these plans are entered into a Web-based "station management system." Throughout the year, the station enters data into the system to report home visits, school visits, presentations, and inspections. Progress can be tracked and graphically displayed at any level of aggregation: shift, station, district, region, or national level. Stations can view past fire experience in their area with demographics overlaid, which helps them not only to plan their prevention activities by neighborhood or even house by house but also to track results over time. In addition, every home and other building in New Zealand is entered into a database that records all its interactions with the fire service, including calls, inspections, and education visits.

BEST PRACTICES FROM JAPAN

Japan has perhaps the most extensive public fire safety education program among developed nations (see Figure 10.6). The result is its relatively low accidental fire death rate despite having a disproportionately large elderly population, the age-group with the highest fire death rate. Japan has increasingly tied fire safety education to disaster education and training, because earthquakes lead to many fires and represent a high public concern. The best practices found in one or more Japanese cities follow.

Organization for Prevention

Typically 8 to 10 percent of fire department personnel are dedicated full time to prevention: some at the headquarters prevention division and some assigned to fire stations. This is a much greater commitment to prevention than that seen in

FIGURE 10.6 Flag of Japan

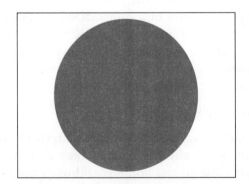

U.S. fire departments. In addition, all line firefighters are expected to participate in prevention. Major cities have large volunteer fire corps trained to provide support in large-scale disasters. These volunteers also take part in prevention efforts and carry their training to their own homes and neighborhoods.

To reach the young, almost all schoolchildren in Japan receive extensive fire and disaster safety instruction in school through a variety of programs. These varied approaches keep interest high.

To get through to the elderly and people with disabilities, the fire service visits the elderly at home to give advice on home safety, especially smoke alarms. Elderly living alone and households of people with disabilities are the priority targets for these visits. Besides conventional smoke alarms, pendant radio alarms worn around the neck, outside bells triggered by home smoke alarms, and automatic alarms linked directly to the fire department are used in selected homes of the elderly, to get them assistance more quickly. For certain elderly and people with disabilities, the fire department identifies a neighbor willing and able to assist in an emergency. These neighbors are given some training and dispatched in parallel with a fire unit or ambulance if their assigned person calls for help.

Although Japan is largely a homogenous society, increasing numbers of foreigners reside there. To reach immigrants and visitors, Tokyo offers its 370,000 foreign residents disaster training and instructions on how to make a 1-1-9 emergency call, how to use fire extinguishers, and basic first aid. English-speaking fire personnel are assigned to areas with many foreigners.

The development of both national and local prevention campaigns allows Japan to have fire safety initiatives going on throughout the calendar year. In fact, Japan probably has the most extensive system of national fire safety campaigns in the world. Weeklong fire prevention campaigns run in the spring (March 1–7) and the fall (November 9–15), with messages significantly different from what is normally emphasized in the United States. One week in January is for volunteers practicing disaster mitigation. Hazardous materials safety week takes place in June. A week in September is for practicing first aid, promoting injury prevention, and explaining proper use of emergency medical services. Another week in September is disaster preparedness week, and every September 1 is National Disaster Preparedness Day. One day each year highlights the need to preserve national heritage sites. November 9 is 1-1-9 day. Various messages and public training accompany each special week or day.

Life safety learning centers are found in the major fire departments, which run large fire and disaster training centers equipped to give practical, hands-on training in dealing with the hazards of earthquakes and the fires that may follow. Many of these lessons also apply to everyday fires and injuries. Visitors get to put out a simulated kitchen fire with a real water extinguisher and crawl low through a smoke maze. The centers also teach CPR and first aid. Some training centers give visitors a report card on how they did on each activity and show videotapes of their errors, which probably adds to the retention of the lessons. The scale of training in Japan's life safety centers is extraordinary, reaching millions of people. Tokyo's alone reached 2 million people in the last decade.

Firefighters conduct large-scale home visits to deliver safety information directly to households. Tokyo's goal is to visit all households at least every 5 years. Osaka's firefighters visited 420,000 out of 1,270,000 homes. Local fire stations decide which areas to visit. Firefighters may take along neighborhood association leaders or public welfare personnel. Visits usually stop at the door, without entering. During a visit, the firefighters talk about smoke alarms and other current safety issues, and leave literature. If requested, they test the smoke and fire alarms.

In the area of codes and standards, the largest recent change in fire prevention in Japan is its emphasis on requiring home smoke alarms, and not just one per house but rather one in almost every room. The 2006 National Fire Defense Law requires smoke alarms in all new residences; they must be installed in each bedroom and the stairway to the bedrooms. Starting in 2010, existing as well as new homes must meet the standard.

The extent of these best practices is a major reason why Australia, New Zealand, and Japan have much lower residential fire casualty rates than the United States. Their ideas should be considered as part of thinking about revising U.S. strategy for fire prevention.[9]

BEST PRACTICES FROM CANADA

Fire protection in Canada is provided by local city or regional fire services (see Figure 10.7). Some provinces supply strong prevention leadership and have innovative codes and laws requiring fire safety education province-wide.

Cultural Change

A sea of change is underway in the fire service culture of Canada to emphasize prevention and to establish a public mind-set of taking collective and personal responsibility for safety. This can be seen in regard to personal accountability, in the use of firefighting personnel, in prevention efforts, and in the design of fire prevention on a national level.

Certain provinces are making greater personal responsibility a focus as the fire service works to get citizens to recognize fire risks, realize that many of these risks are preventable, and assume personal responsibility to take recommended preventive measures. For example, Ontario now has a "zero tolerance" philosophy toward property owners and tenants who do not install or maintain smoke alarms on each story of their buildings. Not complying with smoke alarm requirements can result in a fine of $235. In instances involving death or significant injury, fines go up to $50,000 for individuals and $100,000 for corporations, and/or a jail sentence of up to one year. The Ontario Fire Protection and Prevention

FIGURE 10.7 Flag of Canada

Act of 1997 mandates that prevention activities be part of the everyday functioning of the fire service. As a result, more firefighter time is spent on residential fire safety and school programs than in the past.

In addition to changing the culture of safety for the public and firefighters, a shift is occurring of Canadian strategic concepts in developing and implementing prevention programs. A common theme in successful efforts is taking a comprehensive approach that includes good data for targeting and evaluation; partnerships to leverage resources; and multiple coordinated programs. The aim is to reach the population at large, but with a focus first on high-risk subgroups.

To target nontraditional issues, many Canadian fire departments identify problems in their incident data analysis even if the problem is outside the scope of their traditional activities, so long as the programs help prevent fires or injuries. Examples are overuse of home electrical systems for growing marijuana (Surrey, BC); bike and pedestrian safety (Waterloo, ON); and earthquake preparedness (Vancouver, BC).

The use of prevention programs that use multiple coordinated approaches and capitalize on partnerships with other government agencies and the private sector are more likely to be successful on a large scale in a city or province. They include more outreach, more repetition, and more consistent messages; and they reduce confusion across messages. In many cases, costly prevention programs are being done without tapping fire department budgets by soliciting funding from private sponsors. These initiatives measure success in terms of outreach; change in knowledge; change in behaviors; and change in rates of fires, deaths, and injuries. Some examples of their metrics: "percent of households visited"; "percent of population reached with cooking safety messages"; "average score of school-children on safety exams"; and "percent of households with working smoke alarms and completed escape plans." In the United States many programs are measured in terms of the *number* of people reached, which is less informative than the *percent* of people or households reached.

Some of the best specific practices found in one or more Canadian fire departments are summarized here and detailed in the Global Concepts in Residential Fire Safety report. Most successful prevention programs use several of these approaches.

Home Visits

Door-to-door canvassing by fire departments has become as crucial a strategy for reducing residential fire deaths in Canada as it is in the United Kingdom. Many Canadian fire departments now do home visits on a large scale. Typically the plan is to visit all households over a specified period of time (e.g., every five years), with visits to the highest risk areas first. During home visits, pairs of firefighters inform residents of the legal requirements for smoke alarms, the need to test and maintain them, ways to prevent common fires, and the importance of having an escape plan. Some departments install alarms or batteries, if needed. Most leave safety literature if no one is home and invite the occupants to request another visit when convenient. They collect data on each household visited as to its level of risk and mitigation measures taken. In many cities, these programs were followed by a 40 to 50 percent reduction in residential fire deaths.

Partnerships

Canadian fire services make excellent use of partnerships with businesses, school systems, and health care organizations to increase outreach of messages, identify high-risk homes, and obtain funds for campaigns and safety villages. These collaborations leverage the limited resources available for prevention.

City Departments

Many cities train local social services such as visiting nurses, those delivering meals to the homebound, and social workers on how to test smoke alarms, look for hazards, and provide fire safety information when they visit homes. The agency personnel contact the fire department to follow up on fire safety problems found if not readily solved. This gains fire service entry into some of the highest risk households that would otherwise be difficult to get into.

Private Industry and Wealthy Donors

Major companies provide goods, services, or money for prevention. For example, Kidde Canada discounts smoke alarms; Co-operators Insurance provides money for smoke alarms; Canada Tire allows use of store parking lots in 310 locations for fire safety demonstrations; Wal-Mart supplies money to buy Risk Watch; Duracell makes batteries available for home visits; and CTV and Global TV donate spot time and produce spots. Shell Oil and an insurance company worked together to fund 100,000 fire safety coloring books tailored to deliver Ottawa's messages to schoolchildren. Royal Bank of Canada, McDonald's, The Home Depot, Terasen Gas, and others donated money to construct a Fire Safety House in Vancouver. The Knowledge Network provided $50,000 in production services for a fire safety video.

Reaching All Schoolchildren

School programs in Canada tend to reach a larger percentage of schoolchildren than those in the United States. Fire departments work hard to reduce barriers and provide incentives to facilitate school participation. One fire department provides refreshments for teachers during their lunch periods while a fire prevention officer explains the program. Teachers and school administrators are solicited to be on fire safety curriculum development committees to help blend programs with the main school curriculum.

Safety Centers

A strategy to increase outreach and improve quality of fire safety programs is to bring classes to safety centers instead of doing the training in classrooms. Waterloo region's fire and police services teamed to develop and staff a children's safety village that includes realistic-looking, but scaled-down, buildings; streets with intersections, parking, and sidewalks; and a railroad crossing. The village has an educational center where classes in grades 2, 4, and 6 attend fire safety lessons. The children first watch and discuss a fire safety film. They then have safety lessons further explained using a "hazard house" (like a large doll house). Finally the children go to full-size mock-ups of rooms to identify and discuss safety issues and practice safety skills.

The ability to bring classes to public educators dramatically increases the number of students who can be taught by a small prevention staff. It also provides a much better teaching environment than does a typical classroom.

High School Interns

Brampton and Waterloo expand their public education resources with high school interns to provide instruction at the safety centers or during school visits. Interns receive community service credits needed for graduation.

Safety Homework

Following class sessions, many school fire safety programs reinforce messages by giving students exercises to do with their families, such as testing smoke alarms and making an escape plan. Some programs survey parents asking whether their children brought home safety ideas and the parents' perception of the program. Teachers who bring classes to the safety village also are asked to evaluate the experience.

Reaching the Elderly and People with Disabilities

Most Canadian cities have safety programs for the elderly, often a Canadianized version of the NFPA's "Remembering When" program or the Ontario Fire Marshal "Older and Wiser" program. Toronto's program uses three key approaches:

- Training care providers to check on fire safety when they visit homes of the elderly
- Transporting homebound elderly to safety presentations
- Distributing a safety calendar to seniors, to help remind them of basic safety practices throughout the year

Reaching Immigrants and Ethnic Groups

Most Canadian cities have tailored fire safety programs for their growing immigrant and ethnic populations. Some innovative strategies include the following.

Community "Ambassadors" Ottawa uses volunteers active in their ethnic community to go door-to-door with firefighters during home visits. They supplement firefighter staffing for the home visits, help firefighters gain access into households by showing a familiar face, and translate, if necessary.

Native Councils In some Native Canadian cultures, the tribal council plays a major role in safety issues, so the fire department provides safety information for the council to pass on, instead of approaching individual tribal members directly.

Inner-City Bus Ads Because high-risk populations often use buses, Ottawa purchases smoke alarm ads on inner-city bus routes.

Ethnic Media Many fire departments put fire safety messages in local ethnic newspapers and on radio and television stations to reach specific groups.

English Classes Fire departments in Ontario partner with English as a Second Language (ESL) schools to provide basic home fire safety information to new Canadians taking these classes.

Firefighters with Language Skills Vancouver maintains a list of firefighters who speak a language other than English. They assist with interpretation of educational materials and giving presentations.

Translated, Simplified Public Education Materials Brampton, Ottawa, Toronto, and Windsor have simplified public education materials translated into the major second languages prevalent in their regions.

Partnerships with Immigrant Associations   The Office of the Fire Marshal in Ontario gets assistance from the Ontario Council of Agencies Serving Immigrants to present prevention materials to frontline workers through their settlement services.

Reaching the General Population

Most cities and some provinces have fire safety campaigns throughout the year. Some of their innovative approaches follow.

Budget for TV and Radio Ads   Many Canadian provinces and cities have budgets for buying airtime for fire safety messages, typically 15- to 30-second advertisements during popular programs. They do not depend on free, late-night public service announcements, which often miss the majority of their target audiences. The airtime budgets are augmented by industry donations and TV station discounts.

Subsidized Fire Safety Literature   The Public Fire Safety Council, established under Ontario law, is empowered to raise funds from private organizations to develop fire safety campaign materials. The Council subsidizes sale of NFPA and other materials to the fire service, giving a 75 percent discount, which results in much wider use of the materials.

Media Events and News Conferences   Provincial fire marshals and fire chiefs hold media events and news conferences to demonstrate safety issues such as unattended cooking, drinking and smoking, alarm maintenance, and to announce the latest safety statistics. Some of these media events are held in fire stations.

Innovative Dissemination Venues   Ottawa firefighters hand out home safety brochures in supermarkets and grocery stores, and use parking lots of businesses that sell smoke alarms for safety demonstrations. Ottawa makes presentations in on-campus residences.

Featured Fire Station   Although citizens can visit any fire station, Brampton steers them to a particular fire station whose firefighters are handpicked for their ability to provide safety education. This enhances the likelihood that information will be properly communicated and memorable.

Cooking Safety Trailer   Unattended cooking is the leading cause of residential fires. In partnership with Coldwell Banker Realty and the Kitchener Fire Department, Waterloo built a trailer for dramatic demonstrations with live fire about how cooking with oil can be dangerous if left unattended, and what happens if one attempts to put out the fire with water. The trailer is taken to many venues and provides an unforgettable demonstration.

Firefighter Union-Sponsored Campaigns   The British Columbia Professional Fire Fighters' Burn Fund has developed large-scale burn awareness campaigns including poster contests that involve two-thirds of the schools in the province.

"Superboarding"   Vacant homes in Surrey, BC, are boarded up extra securely to reduce arson. Ontario Fire Code requires that vacant homes must be secured against entry and enforces it.

Fire Crew Messages   Ontario firefighters are expected to be the first point of contact with the public. Fire trucks are equipped with a set of 24 message cards that address the most common fire safety issues, which helps crews give impromptu presentations in the field with consistent and correct messages.

Required Residential Sprinkler Systems

The ultimate long-term solution to most of the residential fire problem is sprinklering all residences, not only high rises but also single-family dwellings. A 1990 Vancouver bylaw required sprinklering of all new residential units and all residences remodeled at 50 percent or more of their cost. By 2009 almost half of all Vancouver residences were sprinklered, and residential fire deaths decreased dramatically. Damage to non-sprinklered homes was 13 times higher than for equivalent sprinklered homes over a 10-year period. The Vancouver sprinkler ordinance also facilitates denser, greener development, which helps to reduce traffic and vehicle pollution. Over 20 other jurisdictions in British Columbia, including some with volunteer and combination fire departments, also require home sprinklering.

BEST PRACTICES FROM PUERTO RICO

Puerto Rico has a national fire service serving the entire island (see Figure 10.8). After the DuPont Plaza Hotel fire in San Juan in 1986, which killed 100 people and injured 150 others, a commission on fire safety recommended changes in public safety education, codes and laws, building retrofits for fire safety, and an overhaul of the national fire service. As in many nations visited, the safety culture of the fire service and society has been changing. Following the Dupont Plaza Hotel fire, the fire service placed greater emphasis on fire prevention and public education. Fire headquarters created a division of residential fire education that studies risk factors and plans programs to mitigate those risks. Prevention programs are funded by the Puerto Rican Emergency Management Agency, the U.S. Department of Homeland Security, and occasionally the private sector. Line firefighters are trained to assist in classroom presentations on fire and burn dangers and how to escape from a fire.

Correcting Misconceptions About Safety of Homes   Many residents had mistaken ideas that because homes in Puerto Rico generally are built of concrete, they are not vulnerable to fire. A high priority has been to help people understand that a home's contents and interior finishes are what burn and that anyone could suffer a serious fire by not being careful. An assumption of residents in high

FIGURE 10.8 Flag of Puerto Rico

rises was that if a fire were to occur, they could evacuate by getting to the roof where a helicopter could rescue them. The public had to be made aware of the need to plan escape in other ways and to practice what is planned. Also, the combination of alcohol, smoking, and living alone factors in many fatal fires, so messages to older residents targeted this risk syndrome.

Safer Consumer Products

A common cause of home fires in Puerto Rico is electrical arcing from frayed or broken electrical appliance cords, especially when plugged into overloaded outlets with combustibles nearby. These fires increase during the holiday season as residents decorate homes with lights, electric candles, and displays. The fire service now urges residents to install arc fault circuit interrupter breakers, which stop electrical current before conditions reach the point where they produce sparks and high heat. New homes must have these devices.

Reaching Schoolchildren

Preschool and early elementary school children are considered the highest priority target audience for residential fire education in Puerto Rico. This is because the fire service had seen an increase in the number of children who died in fires, many started by the victims. Much hope for the future lies with educating the current generation.

Adapted NFPA Programs   The fire service uses a widely accepted NFPA school fire safety curriculum, *Mis Primeros Pasos*, with a few changes made to reflect specific risks in Puerto Rico and to be more culturally relevant to Puerto Rican children. This included developing their own mascot, a little fire hose named *Manguerita*, and adding a puppet theater as another vehicle for teaching the lessons.

Use of Music   Music is an integral part of Puerto Rican culture, and songs have been incorporated into the school fire safety program. The fire service created its own music for the NFPA curriculum, with lyrics set to rap and hip hop styles to increase children's interest. *"El Bombero Rapero"* (rapping firefighter) raps about fire safety and how to escape. Youngsters love it and sing along. The music has been recorded and made available to United States fire departments. This is an excellent example of customizing programs in respect to local culture.

Partnerships with Broadcast Media

The Puerto Rican fire service developed a home fire safety campaign supported by major radio stations and several television stations. Messages are carried during prime time rather than off-hours. The director of residential fire safety for the Puerto Rican fire service is also its public information officer. His experience gets more attention for prevention in the media spotlight.

BEST PRACTICES FROM MEXICO

In Mexico, this study focused on residential fire safety in the State of Guanajuato, which is in the center of the nation (see Figure 10.9). Compared to the United States fire departments, Mexico has an extremely small number of firefighters per 1,000 population in most cities. As an example, the city of Silao (population

FIGURE 10.9 Flag of Mexico

170,000) has a single 4-bay station, 7 paid firefighters, and 38 volunteers. Because of the shortage of firefighters, prevention is viewed as especially critical. The head of safety for the State of Guanajuato says that *"El major departamento de bomberos no es el que mas incendios atiende; sino que el que mas incendios previene."* Translation: "The best fire department isn't the one that responds to the most calls but rather the one that prevents the most fires." To accomplish control of the fire problem, educating fire suppression personnel on the importance prevention plays, and their role in it, is as important as educating the public.

Cultural Change

Officials from the Guanajuato state government have set as their mission to change the way future generations will regard safety—their personal responsibility for being aware of risks and preventing safety problems. This strategy includes two key elements:

1. *Educate* the younger generation on all aspects of safety, especially fire safety, road safety, burn prevention, and accident safety.
2. *Provide* safer, more decent houses for the lowest income families and persons with disabilities, many of whom live in makeshift wooden shacks or broken-down vehicles. The state plans to provide the building materials and know-how to construct 36,000 basic concrete dwellings for high-risk households, with help in their construction from family and friends.

Reaching Schoolchildren

Like Puerto Rico, Guanajuato's primary target is preschool and early elementary school children. The bulk of fire prevention education in schools is delivered by teachers on their own or sometimes working alongside firefighters in the classroom.

State Education Office   The state's Department of Education, not the fire service, is the principal organizer and deliverer of fire safety education through schools, especially preschools. The government's goal is to instill a sense of personal responsibility for safety from fire at an early age. It hopes that, by creating a safety-focused mind-set, the younger generation will carry forward safety awareness and caution in the years to come.

Training Teachers   Guanajuato conducts an intensive three-day fire safety training program for teachers of young children. The teachers learn how to instruct students about preventing and escaping from fires, how to avoid burns, and assessing risks in the home. The teachers also study how to be first responders in their schools, including use of fire extinguishers, first aid, and some aspects of confined space rescue in earthquakes. The trained teachers became part of an auxiliary corps of helpers to the fire and rescue service.

Public Campaigns

Besides the school program and the program to provide safer homes are the following public campaigns.

Safer Cooking An effort is made to encourage people to make more use of indoor ranges and stoves, and less use of outside *"comales"*—cooking over open fires, often in areas surrounded by combustible materials. The tradition of cooking with *comales* is well established, and it will take time to alter cooking preferences.

Milk Cartons Fire officials got a milk producer to put fire and life safety messages on the sides of milk cartons. The messages, which vary, are expressed in bright pictures and limited numbers of words to be attractive and easily read.

Alternatives to Candles Candles are widely used in Guanajuato homes, especially on home altars. They may be left to burn for long periods of time. When combined with unattended children or combustibles near the altar, they are a major fire risk. Officials promote the use of battery-powered candles instead of open flame candles to reduce fires and burn injuries.

Alternatives to Fireworks Each year, local and state officials see many injuries from fireworks, which are used at home to celebrate several religious and civic holidays. Guanajuato implemented a campaign encouraging families to use alternatives such as confetti, noisemakers, and streamers, and to leave fireworks to professionals at public shows.

BEST PRACTICES FROM DOMINICAN REPUBLIC

The visit to the Dominican Republic was limited to the Santo Domingo Fire Department and actually was a by-product of a vacation trip as opposed to planned research (see Figure 10.10). Nevertheless, it yielded several interesting practices:

Neighborhood Visits and Talks The Santo Domingo Fire Department conducts public education in neighborhoods primarily by going to schools, industries, churches, and neighborhood associations to deliver fire safety education face-to-face. It uses a team of seven public educators, including several volunteers although it is a career department.

Fire Safety Brochure An excellent illustrated fire safety brochure is widely distributed to adults, schoolchildren, and businesses. This brochure features the

FIGURE 10.10 Flag of the Dominican Republic

fire department logo and the name of the insurance company that sponsors it. The relatively unusual feature of the brochure is its level of specificity on what to do for common local problems.

Department Newsletter and Magazine   The Santo Domingo Fire Department publishes a newspaper-type periodical every three months with safety information. It is distributed by the public education program and other means to the public, not just the fire department. This guarantees that messages are repeated and exposure increased.

Fire Station Visits   Groups of schoolchildren are brought to their neighborhood fire stations for discussions of fire prevention. The visits are documented by photos with the fire chief or other officers that are printed in the fire department magazine.[10]

The global report on the best practices in community fire protection provides a wealth of proven ideas, techniques, and programs as found in many of the countries studied, which could be utilized in the United States. To do so will require a cultural shift of the profession and a refocusing of time, energy, and organization commitment to prevention.

PEARSON
myfirekit

Visit MyFireKit Chapter 10 for the complete reports on Global Concepts in Residential Fire Safety, Parts 1–3.

Vision 20/20

Despite significant progress in the last 30 years, the United States still has one of the worst fire loss records of the industrialized world. Fire loss includes not only fire deaths and injuries but also the subsequent social, environmental, and economic impacts. For example, in 2006 the United States had 1.6 million fires attended by fire departments, and no one disputes the actual number is higher due to unreported fires. The number of deaths is in the thousands, the number of injuries in the tens of thousands, and the economic losses in the billions.

An ad hoc group formed around the concept that it was time to move forward with efforts to reduce fire losses in the United States further. Other industrialized nations have a better safety record than does the United States, and those members who ultimately ended up forming the steering committee for **Vision 20/20** felt the United States could and should collectively do better, acknowledging that significant progress in fire losses has already been made.

The group came together under the auspices of the Institution of Fire Engineers (US Branch) and applied for grant funds from the U.S. Department of Homeland Security to conduct a national strategic planning process, which would point to "gaps" in the existing service delivery systems that need emphasis. Doing so would potentially improve fire safety efforts nationally without taking away from existing programs that are already working well. A strategic planning process, complete with an updated look at national fire loss data and information about the current fire prevention "environment" nationally (current prevention efforts, futures research, etc.), was prepared so that funds for collaborative planning on a national level could be solicited.

The successful grant application allowed the formation of a steering committee for what ultimately became Vision 20/20. Earnest preparation for a national strategic planning process began in August 2007. A Web forum was developed and provided to a variety of stakeholders to place the background fire loss infor-

Vision 20/20
■ The Institution of Fire Engineers US Branch was awarded a Fire Prevention and Safety Grant by the U.S. Department of Homeland Security to develop a comprehensive national strategy for fire prevention. The goal is to help bring together and focus fire prevention efforts collectively to effectively address the fire problem in the United States.

mation before as wide an audience as possible. This established a foundation of shared reference material in preparation for the hosting of an on-site physical planning forum like no other in recent history. More than 500 people ended up participating in the Web forum at 13 different satellite locations around the nation sponsored by various fire prevention associations, and hundreds of others participated individually. Given the scope of the nation's fire problem, they were charged with developing strategies to reduce fire losses. The results of the Web forum were then refined and taken to the physical forum held in Washington, DC, on March 31 and April 1, 2008.

This Vision 20/20 report represents the refined forum results from the meeting where more than 170 fire and other agencies with a stake in the nation's fire problem met to outline the next steps that would lead toward a more fire-safe nation. It signifies the first steps in filling perceived gaps in the nation's fire prevention efforts. Ultimately the fire service must move the following strategies and actions steps forward collaboratively and in conjunction with existing efforts.

Participants identified five main strategy areas, with numerous action items listed for each that can help move prevention efforts forward. These strategies and many of the action items have been mentioned in previous reports. They represent the general consensus on issues that need more emphasis in the current environment. Data collection and analysis, and its importance to fire prevention efforts, was captured as a priority that spreads throughout the five major strategic areas identified for follow-up action. Rather than report it as a separate strategy, it is given emphasis as relevant to the success of each of the five strategies presented in this report.

Facilitators for each strategy work group have been identified. Funding to support continued activity that will keep each strategy working in conjunction with the others and existing programs is being sought. Each work group will be responsible for incorporating the volunteer assistance that was offered at the Forum while it completes the more detailed action plans associated with each strategy. Each work group is tasked with maintaining a collaborative environment where partners are welcome, and consensus is attempted for major decisions.

To some extent, each strategy relies on accurate data and analysis of the data, so that the case can be made for increased emphasis on prevention programs. That particular aspect will be part of each working group's deliberations about action steps. It is evident that some modification of the action steps will occur based on the planning efforts of the work groups that form around each strategy. The major strategies include the following:

1. Increasing advocacy for fire prevention
2. Developing and implementing a national fire safety educational/social marketing campaign
3. Raising the level of importance for prevention within the fire service
4. Promoting technology to enhance fire and life safety
5. Reducing fire loss by the refinement and application of codes and standards, which enhance public and firefighter safety and preserve community assets

PEARSON
myfirekit

Visit MyFireKit Chapter 10 for the major strategies and action plans as well as an interview with Chief Jim Crawford.

Firefighter Safety and Deployment Study

In 2000, at the NFPA meeting in Anaheim, California, NFPA 1710, Standard for the Organization and Deployment of Fire Suppression Operations, Emergency Medical Operations, and Special Operations to the Public by Career Fire Departments, was passed. For those of you who were there, it was a very contentious meeting, with a significant amount of rhetoric from both the firefighting and city management communities. The fear of mandated staffing, lawsuits, and loss of management rights were the main focus of the debate. Now, over a decade later, much of what was feared has subsided, yet a great deal of animosity continues to exist between the various factions.

However, as I witnessed the debate, it became clear that if we did not put science behind this and other such consensus standards, we would continue to create such animosity between all the different groups of local government; the result would be very unhealthy for the fire profession. At the time, as the second vice president of the IAFC, I submitted to the board of directors a proposal to seek a research study on firefighter safety deployment and community risk. My objective was to gain support to approach NIST and others to perform such a study. Approval was gained; and a process began to obtain approval of the other major fire service players, such as the IAFF, ICMA, NVFC, and CPSE. Although I thought this was to be the easy part, it took almost four years to get everyone to support it and NIST to agree to do the study once funded.

CPSE created a nonprofit CFAI-Risk and was awarded an AFG grant for $1 million to begin what I believe will be the beginning of a series of very important research studies: the **Firefighter Safety and Deployment Study**.

STUDY'S BACKGROUND

Although the nature of the fire service has changed dramatically in sophistication, techniques, and scope over the years, it has always remained committed to a central mission of protecting lives and property from the effects of fire. Every statistical analysis of the fire problem in the United States identifies residential structure fires as a key component in firefighter and non-firefighter deaths, as well as direct property loss.

The scope of fire service activity has increased in recent years. Notably, more fire department resources are committed to emergency medical service (EMS) calls every year; and the fire service is often called to lead or contribute to the mitigation of natural disasters, hazardous materials incidents, and terrorism. Whether current resource allocations match existing service commitments remains unclear, due primarily to a lack of scientifically based analytical tools.

Despite the magnitude of the fire problem in the United States, no scientifically based tools are available to community and fire service leaders to assess the effects of prevention, deployment, and staffing decisions. Presently, community and fire service leaders do have a quantitative understanding of the effect of certain resource allocation decisions. For example, a decision to double the number of firehouses, apparatus, and firefighters would likely result in a decrease in community fire losses, whereas reducing the number of firehouses, apparatus, and firefighters would likely yield an increase in community fire losses. However, decision makers lack a sound basis for quantifying the total economic benefit of more fire resources or the number of firefighter and non-firefighter lives saved or injuries prevented.

Firefighter Safety and Deployment Study
■ A CFAI-Risk study, funded by an AFG grant, is underway and is intended to evaluate firefighter safety in various deployment models versus the risk protected.

Studies on adequate deployment of resources are needed to enable fire departments, cities, counties, and fire districts to design an acceptable level of resource deployment based upon community risks and service provision commitment. These studies will assist with strategic planning and municipal and state budget processes. Additionally, as resource studies refine data collection methods and measures, subsequent research and improvements to resource deployment models will be based upon a sound scientific foundation.

PURPOSE OF THE STUDY

This project in firefighter safety and deployment of resources seeks to enable fire departments and city/county managers to make sound decisions regarding optimal resource allocation and service based upon the following: scientifically based community risk assessment; safe, efficient, and effective emergency response system design; and the local government's service commitment to the community.

SCOPE OF RESEARCH

The scope of year 1 research on this project was focused on producing a scientifically based resource allocation model based upon an international search of the published literature and expert elicitation. The model is useful for analysis of the impact of existing and alternative prevention and mitigation strategies at the station or fire department level. Model development was followed by the production and testing of Web-based fire department surveys designed to collect data at the department level, the station level, and the incident level. Participating departments were selected using a two-stage stratified probability sampling methodology. Selected departments have signed onto the study and completed a department-level survey. The focus of year 2 is to generate survey data to provide the technical basis for modeling equations and to conduct field experiments to develop an in-depth understanding of deployment analysis. Year 3 will analyze the data collected from the field experiments and reports entered from actual incident by selected departments. Year 4 will deliver the results to the fire service through software development, model verification and testing, documentation, and dissemination.

METHODS FOR YEAR 1 RESEARCH
OVERVIEW OF METHODS

The Center for Public Safety Excellence subcontracted with three research organizations to conduct the study: the National Institute of Standards and Technology (NIST), Worcester Polytechnic Institute (WPI), and the International Association of Fire Fighters (IAFF). Principal investigators, one for each of these vendors, conducted the fundamental study activities. In addition to the vendors, the International Association of Fire Chiefs (IAFC) was brought in as a fire service partner. These five partner organizations worked collaboratively to establish the technical basis for risk assessment and deployment of resources by local fire departments. Here is a list of the partner organizations and individual council representatives.

PARTNER ORGANIZATION COUNCIL REPRESENTATIVES

Kathy A. Notarianni, Ph.D., P.E.
Department of Fire Protection Engineering
Worcester Polytechnic Institute

Jason D. Averill
Group Leader, Engineered Fire Safety Group
National Institute of Standards and Technology

Bill Grosshandler, Deputy Director
Building and Fire Research Laboratory
National Institute of Standards and Technology

Lori Moore-Merrell, Dr.PH., M.P.H.
Assistant to the General President
International Association of Fire Fighters

Randy Bruegman, Fire Chief
President, Center for Public Safety Excellence, Inc.

Paul Brooks, Executive Director
Center for Public Safety Excellence, Inc.

Ed Plaugher
Director of National Programs
International Association of Fire Chiefs

This project's ultimate goal is to create tools the departments can use to do the following:

- Better assess the risks and hazards in their communities.
- Plan adequate resource deployment to assure safe, efficient, and effective mitigation of emergency events.
- Measure effectiveness in responding to and handling events.

In addition to the partners, the fire service has identified project technical experts in each major area of the project, expanding upon the expertise of the principal investigators. Fourteen experts have provided invaluable time and talent to the study in fields including fireground operations, fireground strategy and tactics, consensus standards and codes, fire engineering, emergency medical services, economics, decision analysis, risk assessment, statistics, and geographic information system mapping. Each individual's name, affiliation, and area of technical expertise are shown here.

In addition to expert elicitation, a comprehensive literature review was performed with the objective of developing a state-of-the-art understanding of the correlations and causal links between factors that correspond to variation in community outcomes. Historic and contemporary literature sources were compiled on a diverse group of subject matters related to the study. This search included articles in refereed journals in a variety of disciplines, fire service publications, textbooks, and relevant Web sites. Researchers reviewed several

PROJECT TECHNICAL EXPERTS

William "Shorty" Bryson, Chief
Miami Fire Department
Expertise: Fireground Operations

Bob Chapman, Ph.D.
NIST
Expertise: Economics

Ronny J. Coleman
Fire Service Consultant
Expertise: Standards of Cover and Fire
Department Risk Assessment

Dennis Compton, Chief
City of Mesa, Arizona
Expertise: Fire Department/Fireground
Procedures/Tactics

James J. Corbett, Jr., Ph.D.
University of Delaware
Expertise: Technology Policy; Decision
Analysis

John A. Granito
Fire Consultant
Expertise: Operational Performance;
Deployment and Staffing

William Guthrie
NIST
Expertise: Regression, Experiment
Design, Robust Statistics, Data Analysis

Russ Johnson
ESRI/GIS
Expertise: GIS

Sanjay Kalasa
Vice President
ACS—Firehouse Software
Expertise: Software Programming and
Web Development

Dan Madrzkowski
NIST
Expertise: Field Experimentation

Mike McAdams
Montgomery County Fire Department
Expertise: EMS and Quality Measurement

Jonathan Moore
International Association of Fire Fighters
Expertise: Fire and EMS Operations; GIS

Phil Pommerening
Fairfax County Fire Department
Expertise: EMS

Russ Sanders
National Fire Protection Association
Expertise: Consensus Standards and
Codes

hundred related studies, articles, expert opinions, standards, codes, and academic papers and compiled them into an executive summary, which synthesizes the findings (including consistent findings and conflicting findings) as appropriate to the study, and an annotated bibliography, which includes all resources. The final product will provide a technical basis for screening and eliminating non-predictive concepts, as well as assessments of the statistical significance/power of the final product.

Based on the literature review and input from technical advisers, a formal model structure was developed containing decision variables, objective variables, and boundary variables.

Decision Variables

Decision variables within the model are variables with levels that can be determined by the fire service and include prevention programs, apparatus and staffing levels, first alarm (initial) response levels, dispatch protocol, training, and mutual aid, among others.

Objective Variables

Objective variables were selected to reflect the primary measurable concepts that correspond to the actual risks observed in the community under consideration including the number of firefighter injuries and deaths per year; the number of civilian injuries and deaths per year; total annual direct economic losses; firefighter work hours lost per year; and total firefighter work hours spent on work restrictions per year.

Boundary Variables

Boundary variables were identified as predictors of community risk that are outside the routine decision set of the fire department. Examples include characteristics of the built environment (number and type of buildings), characteristics of the population (population, socioeconomic factors, etc.), community loss history, and elements of critical infrastructure that require unique consideration.

Data Collection

A representative sample of approximately 400 fire departments was asked to provide incident response data using a secure, Web-based data entry portal.

The Web-based surveys were compiled using technical experts to formulate questions based on the study model. Surveys were then assessed for correlation with data typically available in most fire departments. Data specialists from three geographically diverse metropolitan departments assisted in the development of the assessment.

Data Analysis

Once data are collected and compiled, a regression analysis will be performed to determine which variables significantly impact the dependent variables (community outcomes). The regression analysis will form the basis for development and testing of a predictive computer model for risk assessment and deployment of resources.[11]

Preliminary Conclusion

A scientific foundation for the assessment of community risk by fire and city officials will ensure efficient expenditure of public resources and raise the bar for technical discussion of the community impact of changes to resource levels.

What exists today is a lack of accepted research that validates risk versus deployment. In many jurisdictions, the discussion revolves around how much the community can afford to allocate to fire protection. The influence of the fire chief, the political power and influence of the local union, and the economic ability to fund of the jurisdiction frequently drive this.

Often when new development occurs, money gets tight, and new priorities are set. The fire service has little influence in part because it has lacked the solid science behind what it intuitively knows from field experience.

This not only affects the ability to articulate risk versus deployment but also impacts related discussions such as those about the use of residential sprinklers. The ability to quantify community risk, deployment, and strategies to lessen the risk scientifically will be a tremendous leap forward for the fire service.

PEARSON

Visit MyFireKit Chapter 10 to read Phase I and II of the Firefighter Safety and Deployment reports and the Final Report on Fire in Residential Dwelling Field Experiments and EMS Field Experiments.

Firefighting Robotics

One aspect of technology that has received minimal attention in the United States is robotics. A significant amount of research has been and continues to be conducted in this field that is directly related to fire service application. In fact, many robots are in use today internationally. Yet, they are not widely used in the United States.

Robotics has been developed to go into the heart of fires and remove flammable or dangerous chemicals. Fire Spy's developers—the United Kingdom's West Yorkshire Fire & Rescue Service and JCB, a manufacturer of construction equipment—believe this device could save firefighters' lives. Fire Spy is based on a vehicle developed to withstand temperatures of up to 800 degrees Centigrade that is controlled remotely. The driver can see what is happening in the fire through two cameras, one infrared and one standard, which beam back video pictures. At the front is a powerful grabbing arm.[12]

In service for several years now, Fire Spy can recover gas cylinders from the scene of a fire, locate and shut off gas valves in a blazing building, and cool chemical drums with its high-pressure jets of water.

There is a robot truck that shoots 4 tons (3.6 metric tons) of water and 1 ton (0.9 metric ton) of foam. It is used at the fire department in Shenyang, Liaoning Province, China. Also driverless, this robot vehicle can start, stop, speed up, or slow down—all by remote control. Its long-range water cannons and option to use a driver make it very versatile.

The Tokyo Fire Department, which has been developing firefighting robots for more than 30 years, followed its early tank-like, monitor-nozzle vehicle with smaller and larger devices that can squeeze into cramped areas and douse them with water, open doors and valves, walk up walls, and heft Haz-Mat drums.

"Theodor" is a human-sized robot working for a German BASF chemical plant's fire department. It can ride its track-like platform up stairs, pulling its own trailer behind; can perform tasks with a telescopic arm that lifts weights of up to 132 pounds (60 kilograms); and can be used to seal off leaks, detect explosives, and perform active firefighting.

Smaller robots were used after September 11, 2001, to find bodies at the site of the World Trade Center disaster. Despite their diminutive size, such devices can provide a range of services, including communication, light, audio, color and infrared camera input, global positioning, mapping and sonar, and biological and chemical detection.

The shape-shifting marsupial robots hold the most promise for firefighting professionals and urban search and rescue (USAR) workers, according to Dr. Robin Murphy, director of the Center for Robot-Assisted Search and Rescue (CRASAR) and professor at the University of South Florida in Tampa. Murphy describes her team of robot designers, builders, and handlers—who come from robot manufacturers, think tanks, and the U.S. Navy's robot lab at the Space and Naval Warfare Systems Command—as a "strike force" that continues its research work while pre-deploying these devices regularly at such high-target events as the Super Bowl.

According to Murphy, the shape-shifting robots can insert sensors into rubble and position them to assess structural damage by collecting visual and seismic data. They can also carry radio transmitters or even small amounts of food or medication to trapped survivors, guide jaws-of-life devices, and pinpoint the location of a person's limbs to keep rescue workers from injuring them during extraction. Their ability to crouch down and rear up in spaces of varying sizes makes them much more versatile than traditional robots.

Dr. Howie Choset, associate professor of mechanical engineering and robotics at Carnegie Mellon University in Pittsburgh, Pennsylvania, believes that the most efficient robots under development for use by firefighters and rescue workers will be so-called snake robots. About the size and shape of a human arm, they can "thread through tightly-packed volumes, accessing locations that people and machine otherwise cannot," says Choset.

"Actually, they ride on a mobile robot that will carry the snake to the insertion point and go from there," he says. "They extend the reach of the rescue worker, and they do this in a minimally invasive fashion. They don't disturb the surrounding rubble, which is important when you have a fragile, collapsed structure."[13]

It will be interesting to see whether the American fire service will begin to utilize this technology—such as robotics or other spin-off technology from the DOD and NASA—more in the future. The fire service must take full advantage of innovation in the protection of the citizens and firefighter personnel.

INFORMATION AND TECHNOLOGY

A colleague sent me a YouTube clip on the progression of technologies. A few excerpts give one an idea of just how fast the world has changed and is expected to change in the future.

We Are Living in Exponential Times

- It was estimated a week's worth of the *New York Times* contains more information than a person was likely to have been exposed to in a lifetime in the 18th century. It is estimated four exabytes of unique information will be generated this year, which is more than in the previous 5,000 years.
- The amount of technical information is doubling every two years.
- Years it took to reach a market audience of 50 million:
  - Radio—38 years
  - Television—13 years
  - Internet—4 years
  - iPod—3 years
  - Facebook—2 years
- The number of Internet devices:
  - 1984—1,000
  - 1992—1,000,000
  - 2008—1,000,000,000
- The first commercial text message was sent in 1992. Today, the number of text messages sent every day exceeds the population of the planet.
- Thirty-one billion Google searches occur every month; in 2006, it was 2.7 billion every month.
- Predictions are that, by 2049, a $1,000 computer will exceed the computing capabilities of the entire human species.[14]

It is hard to imagine what the impact of this explosive informational exchange and technology revolution will have on the fire service in the future. Yet, it is a safe assumption that this revolution in technology application in the fire profession has just begun.

Conclusion

PEARSON
myfirekit
Visit MyFireKit Chapter 10 for Perspectives of Industry Leaders.

If this book had been written 10 years ago, the amount of research conducted at that point was fairly localized at NIST and other academic institutions. This research was to a great degree narrowly focused on the true physics and science of fire behavior. The last decade has seen a broadening of research encompassing not only science and physics but also the applications and related analysis of what is and is not working. The blending of the academic and field experiences is helping to shape new perspectives on the strategies that should be pursued in the future. It is clear that in the future the fire service not only must continue to use the latest technologies but also must concentrate more time and resources in preventing emergencies from occurring in the first place.

Review Questions

1. What is the National Institute of Standards and Technology, and what role does it have in fire service research?
2. Explain the National Fire Service Research Agenda. Select one of the issues identified and explain how research may help to impact the issue.
3. Explain the the role of the National Construction Safety Team Act and its potential impact on future code and standard development.
4. Provide a detailed analysis of the global concepts in fire safety.
5. How could these global concepts be implemented in the United States?
6. What is Vision 20/20?
7. Explain the five strategies identified in Vision 20/20 and their potential impact on fire safety.
8. Provide an overview of the Firefighter Safety and Deployment Study. Explain the potential future impact on the fire service profession.

PEARSON
myfirekit™

For additional review and practice tests, visit www.bradybooks.com and click on MyBradyKit to access book-specific resources for this text!
Register for MyFireKit by following directions on the MyFireKit student access card provided with this text. If there is no card, go to www.bradybooks.com and follow the MyBradyKit link to buy access from there.

References

1. Kathleen H. Almand and Casey C. Grant, "The Future of Fire: What's Next," *NFPA Journal,* March–April 2009.
2. Ibid.
3. Kathleen Almand, *2008 Symposium.* Retrieved October 30, 2010, from www.nfpa.org/displayContent.asp?categoryID=1285
4. Kathleen H. Almand and Casey C. Grant, "The Future of Fire: What's Next," *NFPA Journal,* March–April 2009.
5. Ibid.
6. Ibid.
7. Philip Schaenman, *Global Concepts in Residential Fire Safety; Part 1—Best Practices from England, Scotland, Sweden, and Norway,* TriData Division, Arlington, VA, for Centers for Disease Control and Prevention and Assistance to Firefighters Grant Program, October 2007, pp. vi–xi and 1–4.
8. Ibid.
9. Philip Schaenman, *Global Concepts in Residential Fire Safety: Part 2—Best Practices from Australia, New Zealand, and Japan,* TriData Division, Arlington, VA, for Centers for Disease Control and Prevention and Assistance to Firefighters Grant Program, August 2008, pp. 5–16.
10. Philip Schaenman, *Global Concepts in Residential Fire Safety: Part 3—Best Practices from Canada, Puerto Rico, Mexico, and Dominican Republic,* TriData Division,

Arlington, VA, for Centers for Disease Control and Prevention and Assistance to Firefighters Grant Program, July 2009, pp. vi–xvii.

11. Jason D. Averill, Lori Moore-Merrell, Kathy A. Notarianni, Robert Santos, and Adam Barowy, *Multi-Phase Study on Firefighter Safety and Deployment of Resources*, Year 1 Final Report, September 19, 2008. Retrieved October 15, 2010, from www.firereporting.org

12. BBC News, *Fire Spy, the Robot Fire Fighter*. Retrieved October 19, 2010, from www.robotbooks.com/fire-fighting-robot.htm

13. Latayne Scott, "Robots to the Rescue," *NFPA Journal*, March–April 2004.

14. Fantastic video on the progression of information technology, researched by Karl Fisch, Scott McLeod, and Jeff Brenman, retrieved October 17, 2010, from www.youtube.com/watch?v=cL9Wu2kWwSY

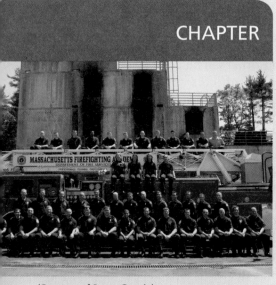

CHAPTER **11**

The Future: The Legacy of Leadership

(Courtesy of Bruce Gauvin)

KEY TERMS

attitude, *p. 378*
creativity, *p. 396*
direction, *p. 380*
integrity, *p. 378*

legacy, *p. 378*
make a difference, *p. 389*
priorities, *p. 395*
purpose, *p. 391*

resilience, *p. 385*
significant, *p. 378*
success, *p. 378*
vision, *p. 380*

OBJECTIVES

After completing this chapter, you should be able to:

- Describe the leadership characteristics needed to overcome adversity.
- Define the term *success*.
- Define the term *significance*.

PEARSON
myfirekit™

For additional review and practice tests, visit www.bradybooks.com and click on MyBradyKit to access book-specific resources for this text!

Introduction

In leadership roles at certain times, what seems to have been a minor gesture on one's part has left a lasting imprint on others. The conversations one has, the notes shared, and the calls made as chief officers can have a deep impact on others. On many occasions, people have related to me about how my conversations or notes helped them in choosing a path or making a difficult decision, even though I often do not remember the level of detail that these people do.

The point here is that leaders touch people with their words, actions, thoughts, and attitudes. Every day as leaders move through their personal lives and professional careers, they are building a **legacy** of leadership. This legacy can be either positive or negative, and lasts long after the leader has left his or her post. Creating a positive legacy of leadership requires a focused effort on others, as much as on oneself as a leader. It requires a mind-set that can face adversity and turn it into an opportunity. It entails an **attitude** that failure is not an option nor is it ever final. It requires leadership that strives for **success** yet understands the importance that significance has on those one leads, manages, and interacts with. This chapter looks inward toward the qualities that leaders possess that make a difference in the lives of the people they lead. So often the rank and file witness people who have achieved great success, lose everything due to the way they achieved it in the first place (**integrity**), because of their perceptions of being better than everyone else (character) and of believing their success granted them special privilege (attitude). Many in leadership positions today may be viewed as successful; yet because they lack integrity, character, and attitude, they will never be **significant**.

As senior officials in the fire service reflect, dream, and envision what the future of the fire service will be, they are the leaders who will be creating a positive legacy. Those leaders of significance will not only influence the future of the service but also spawn the leaders to take their place in the future.

Adversity Often Introduces Us to Our True Leadership Characteristics

Whereas all leaders are no strangers to challenges, exemplary leaders seem to thrive on them. There are a few tips for leaders to do to enable others to learn to thrive as well. Many can be found in the research by Salvatore R. Maddi and Deborah M. Khoshaba, who wrote *Resilience at Work, How to Succeed No Matter What Life Throws at You*. Others are from their own experiences or those that have been witnessed in others as they have been faced with extreme challenges. They include the following:

1. Clarify what you value most.
2. Fully commit to what is important.
3. Embrace the challenge.
4. Paint the big picture.

legacy
- The knowledge and wisdom that are transmitted by or received from another.

attitude
- A complex mental state involving beliefs, feelings, values, and dispositions to act in certain ways.

success
- The achievement of a favorable or desired outcome.

integrity
- The personal inner sense of "wholeness" deriving from honesty and consistent uprightness of character. The etymology of the word relates it to the Latin adjective *integer* (whole, complete). Evaluators, of course, usually assess integrity from some point of view, such as that of a given ethical tradition or in the context of an ethical relationship. Moral soundness: "he expects to find in us the common honesty and integrity of men of business"; "they admired his scrupulous professional integrity."

significant
- Having meaning, influence, or effect.

5. Engage others.
6. Control what you can.
7. Take charge.
8. Tell positive stories.

1. *Clarify what you value most.*   A great example of one who clarified what he valued most was the late Randolph Frederick "Randy" Pausch (1960–2008), an American professor of computer science and human–computer interaction and design at Carnegie Mellon University (CMU) in Pittsburgh, Pennsylvania. Pausch learned he had pancreatic cancer, a terminal illness, in September 2006. Yet he turned his challenge into a leadership opportunity when he gave an upbeat and adage-laden lecture entitled "The Last Lecture: Really Achieving Your Childhood Dreams" on September 18, 2007, at Carnegie Mellon, which became a popular YouTube video and led to several other media appearances. Professor Pausch spoke about the importance of understanding what is important in life: to understand the mountains we must climb to be successful, to have a happy home life, to be financially stable, and to strive to leave a positive legacy. All are intricate parts of what makes life's journey so challenging and yet so inspiring. Professor Pausch coauthored a book titled *The Last Lecture* on the same theme, which became a *New York Times* bestseller. Pausch died of complications from pancreatic cancer on July 25, 2008. "The brick walls are there for a reason," he says. "They're not there to keep us out . . . [but] to show how badly we want something." Leaders must always be clear about what drives them, but challenging times require being especially clear, which means reassessing what is important and where they are headed.

It all boils down to keeping hope alive, which is much more than a slogan. Keeping hope alive is essential to not only energetically attain the highest levels of performance but also lead an active and healthy life. People with high hope, compared to people with low hope, have a greater number of goals across various arenas of life, select more difficult goals, see their goals in a more challenging and positive manner, and attain higher grades in school.

Hope is an attitude in action that enables people to mobilize their healing powers and their achieving powers. With hope they are able to transcend the difficulties of today and envision the potentialities of tomorrow. Hope helps people to bounce back even after being bent, stretched, and depressed by finding the will and the way to aspire to greatness. Hope is testimony to the power of the human spirit.[1] .

2. *Fully commit to what is important.*   This change demands one's full attention, imagination, and effort. Leaders must keep reminding people about the meaning and significance of the work.

3. *Embrace the challenge.*   When everything is going well—revenue is increasing, the community is growing, and the department is along for the ride—any leader or organization can easily slip into a comfort zone. Leaders get lulled into applying the same models, and they measure performance the same way, using an almost effortless managerial approach with a little growth thrown in. The true test of any leader is that, when the tide turns and the organization faces a significant challenge, he or she uses it as an opportunity for change and not as an excuse to do what has always been done, but with fewer resources. I have found that adversity always opens people up to new schools of thought, to a

willingness to explore new ideas and concepts, and to being more creative. However, it requires a leader to lead them there.

4. *Paint the big picture.* Most difficult situations take place in a much larger context. It is vital for leaders to determine what is the bigger picture so they can help others understand what is happening and how it affects the organization. A good example of this is the effects of the recent economic downturn in our nation and around the world. It is important for fire service leaders to explain to their firefighters what is occurring with the budget. For example, sales tax is down by X percent, property tax is down by Y percent, and other revenues may be less as well. Explaining the relationship between the budget and services paints a bigger picture of what is occurring and its impact.

5. *Engage others.* It is crucial to look beyond ourselves or our department to realize that others are also being affected by a particular hardship. Who are they, and how can you and others engage with them? Strengthening relationships and developing wise counsel are two very important elements for fire chiefs when moving through hardship. I have found one of the best ways to strengthen relationships is to be open, honest, and as transparent as possible. Most people appreciate this; even if they may not agree, they respect the delivery. Seeking wise counsel is also very important for every chief officer. These may be peers in the profession or other department directors, but some of the best advice comes from subordinates whom you trust. Their opinions and willingness to give you the answer as they see it, not the answer you want to hear, are of great benefit.

6. *Control what you can.* Even though leaders do not control the broader environment, they can still take charge of their own lives. They must determine the extent of their control and recognize that of others in order to positively influence the outcome. It is important to stay focused on what can be controlled and let go of what cannot be.

7. *Take charge.* It is a healthier to be proactive. Leaders with high hopes acknowledge reality, but move quickly to mobilize personal and group resources to deal with the problems. It is crucial for leaders to analyze and strategize, but they must also make something happen. They must determine even what little things can be done to get moving in the right **direction.**

direction
■ A course of action.

8. *Tell positive stories.* Exemplary leaders characteristically look at the future through a positive lens. Optimism is not only essential in tough times; in some cases, it may be the only thing that people hang on to! Being upbeat and positive about an outcome is key to believing that the future will work out for the best. Great leaders can always find reasons for hope. They are always self-assessing to determine the following: what positive steps the organization has taken and to relay that information to others; how to keep a positive and hopeful outlook; how to recognize others who are making a positive contribution; and how to keep themselves and others enthused about the work. For chief officers it is vital to keep personnel focused on the big picture, the vision for the organization. It is so easy for many people to get bogged down into issues that will last only a short period of time, but can create lasting negativity for the organization to overcome.

vision
■ The manner in which one sees or conceives of something.

Great leaders always inspire and have a unique talent to frame the future in a **vision,** which others can move toward.

Failure Is Not an Option, nor Is It Final!

The way in which leaders handle adversity will determine how they recover from the challenges they face and often dictates their future success in the years ahead. How they handle adversity always sets the course: one course leads to recovery; the other, to desperation. When the path is to recovery, leaders are already laying the foundation to move the organization forward. They are beginning to envision the new possibilities and to plant the seeds (ideas and concepts) on how they intend to reach recovery.

The other course, desperation, is the antithesis of the organization's victim cycle. On this course, all the leaders' emotion and energy are focused toward feeling sorry for themselves. There is often public display of negative emotion and unprofessional behavior, as if it were going to change anything. These two very distinct paths dictate what the organization is to become in the future. The choice is always that of the leader. How leaders handle adversity and lead during difficult times is the true measure of their leadership and value to the organization and the community. Two stories, one which happened in space and the other on the high seas, depict such adversity and focus on two leaders. People who lead by their actions have shown failure is not an option nor is it final!

Do You Have the Right Stuff?

The movie *The Right Stuff,* which focused on the Mercury and Gemini astronauts of the 1960s and 1970s, promoted the popular image of these early astronauts. These men liked to push the envelope—driving fast, flying the highest, and living life to the hilt. All were test pilots who had immersed themselves in the technology of the day and were willing to sacrifice their own lives to be the first to challenge the outer atmosphere. These same traits can be found in many firefighters today. It is through the Mercury, Gemini, and Apollo missions that Americans were able to witness what leadership and grace under pressure were all about.

The confidence of these astronauts was due, in part, to their knowing they had a great team around them—flight directors, mission control experts, and technicians who watched every piece of data at mission control. They were led by mission commanders, not unlike the fire chief, whose job is to look out for the well-being of the crew, the ship, and the entire mission control team.

One of those mission commanders, Gene Kranz, stood out. His reflections on his days at the National Aeronautics and Space Administration (NASA) involve a great deal of history as lived by those involved in mission control. The title of his book, *Failure Is Not an Option,* was his mantra during the *Apollo 13* flight crisis.

Many from my generation still remember hearing the words of Captain James Lovell, commander of the ill-fated *Apollo 13* mission: "Houston, we have a problem." These words marked the start of a crisis that nearly killed three astronauts in outer space. In the four days that followed, the world was transfixed as the crew of *Apollo 13*—Jim Lovell, Fred Haise, and Jack Swigert—fought cold, fatigue, and uncertainty to bring their crippled spacecraft home. Those four days also provided lessons in leadership, in overcoming adversity, in innovation, and in faith—all characteristics needed to become a successful chief officer.

To the outsider, it looked as though a stream of engineering miracles was being pulled out of a magician's hat as mission control identified, diagnosed, and worked around one life-threatening problem after another on the long road back to Earth. From the navigation of a badly damaged spacecraft to impending carbon dioxide poisoning, NASA's ground team worked around the clock to give the *Apollo 13* astronauts a fighting chance. But what was going on behind the doors of the Manned Spacecraft Center in Houston—now Lyndon B. Johnson Space Center—was not a trick, or even a case of engineers on an incredible lucky streak. Instead, it was the manifestation of years of training, teamwork, discipline, and foresight that to this day serves as a perfect example of how to do high-risk endeavors right. It is very similar to what the fire service does every day, 24/7.

For most of the way to the moon, the command/service module (CM) and the lunar module (LM)—dubbed the *Odyssey* and *Aquarius,* respectively, on the *Apollo 13* mission—were docked nose to nose. But the astronauts generally remained in the command module, because the lunar module was turned off to preserve power. Most of that power came from a cluster of three fuel cells in the service module. The fuel cells were fed by hydrogen and oxygen from two pairs of cryogenic tanks, combining them to produce electricity and water.[2]

The crew ran into a couple of minor surprises during the first two days of the flight, but generally *Apollo 13* was looking like the smoothest flight of the program.

As the crew finished a 49-minute TV broadcast showing how comfortably they lived and worked in weightlessness, Lovell stated: "This is the crew of *Apollo 13* wishing everybody there a nice evening, and we're just about ready to close out our inspection of *Aquarius* (the LM) and get back for a pleasant evening in *Odyssey* (the CM). Good night."

Nine minutes later, oxygen tank number-two blew up, causing the number-one tank also to fail. The *Apollo 13* command module's normal supply of electricity, light, and water was lost; and it was approximately 200,000 miles from Earth.

The big question was, How to get back safely to Earth? Before the explosion, *Apollo 13* had made the normal midcourse correction, which would take it out of a free-return-to-Earth trajectory and put it on a lunar landing course; now the mission had changed to returning the astronauts back to Earth safely.

The trip home was uncomfortable, lacking food and water; sleep was almost impossible because of the cold; and, when the electrical systems were turned off, the spacecraft lost an important source of heat. The temperature dropped to 38°F and condensation formed on all the walls.

One of the most remarkable achievements of mission control was quickly developing procedures for powering up the CM prior to reentry. Flight controllers wrote the documents for this innovation in three days, instead of the usual three months. The walls, ceiling, floor, wire harnesses, and panels were all covered with droplets of water. It was suspected conditions were the same behind the panels. The chances of short circuits caused apprehension, but thanks to the safeguards built into the command module after the disastrous *Apollo 1* fire in January 1967, no arcing took place. As the spacecraft decelerated in the atmosphere, it rained inside the CM.

Four hours before landing, the crew shed the service module; mission control had insisted on retaining it until then because everyone feared what the cold of space might do to the unsheltered CM heat shield. Photos of the service module showed one whole panel missing and wreckage hanging out. Three hours later the

FIGURE 11.1 *Apollo 13* Crew Arrive on the Prime Recovery Ship USS *Iwo Jima* *(Scan by Kipp Teague © Kennedy Space Center/NASA)*

crew left the lunar module *Aquarius* and then splashed down gently in the Pacific Ocean near Samoa and were taken to a recovery ship (see Figure 11.1).

After an intensive investigation, the *Apollo 13* Accident Review Board identified the cause of the explosion. In 1965, the CM had undergone many improvements, which included raising the permissible voltage to the heaters in the oxygen tanks from 28 to 65 volts DC. Unfortunately, the thermostatic switches on these heaters were not modified to suit the change. During the final test on the launch pad, the heaters were on for a long period of time. "This subjected the wiring in the vicinity of the heaters to very high temperatures (1,000°F), which have been subsequently shown to severely degrade Teflon insulation. The thermostatic switches started to open while powered by 65 volts DC and were probably welded shut." Furthermore, other warning signs during testing went unheeded; and the tank, damaged from 8 hours of overheating, was a potential bomb the next time it was filled with oxygen. That bomb exploded on April 13, 1970— 200,000 miles from Earth.[3]

What transpired on this mission has an important message for those of us in the fire service, which is that leadership and teamwork are vital in overcoming adversity. For the *Apollo 13* mission command team, "Failure is not an option" became a shared value.

Although the story of NASA mission control teams as they guided the *Apollo* spacecraft through successful lunar landings and helped to save the lives of the *Apollo 13* crew is a never-to-be-forgotten chapter in the history of space flight, it also offers an insight into the world of teamwork and leadership. It has been said that there are three kinds of people in the world: those who make things happen, those who watch things happen, and those who wonder what happened. Gene

Krantz and the *Apollo 13* team were people who made things happen. They did so through visions and declarations, and by having the ability to create a shared value system with a core group to get things done. As those in the fire service aspire to move up in the organization, they have to be willing to ask themselves whether they have what it takes to be a good leader. Do they have the right stuff?

Most fire chiefs today can tell the story of an intelligent, skilled fire officer who is promoted to the position of fire chief, only to fail. They can also tell the story of someone with solid but not extraordinary intellectual ability or technical skill who, on becoming fire chief, achieved great success. These stories support the belief that identifying individuals with the right stuff may be more of an art than a science. Max DePree, chairman Emeritus of Herman Millen, Inc., and author of several leadership books, observed that "Leadership is much more an art, a belief, a condition of the heart, than a set of things to do. The visible signs of artful leadership are expressed ultimately in its practice."[4] The most successful leaders are those who can build lasting relationships rather than cite the management theories of different leadership styles. After all, personal styles of leaders can vary greatly. But a common thread exists among those who succeed. Successful leaders made their declaration early in their careers, understood the enormity of the position, and made the commitment to prepare themselves, from both a relational and an educational perspective, to succeed. Part of having the right stuff is knowing the right questions to ask and the right direction in which to move.

Failure Is Not Final

When Chuck Burkell of the NFA introduced a speaker for an Executive Fire Officer (EFO) graduation at the academy, he stated,

> What is the true measure of a person? Is it the number of wins against losses, triumphs over failures? How many or how few mistakes are made? The truth is we have all made mistakes, most of us have learned from them and then we moved on with a more educated and enlightened perspective. But what does one do when a mistake is made that has life-altering consequences, the kind of mistake that sends out ripples that extend far beyond the shore; ripples that turn into huge swells, and come crashing back to engulf your daily life. How one stands against that tide is the true measure of a person. The story of Retired Commander Scott D. Waddle, former Captain and Commander of the USS *Greenville* and the eight minutes that changed his life forever, is a testimony to leadership, accountability, and facing adversity.[5]

On February 9, 2001, the USS *Greenville*, commanded by Commander Scott Waddle, USN, collided with a Japanese fishing training ship, the *Ehime Maru*, about 9 miles off the south coast of Oahu, Hawaii, USA. In a demonstration for some civilian visitors onboard, the *Greenville* performed an emergency surfacing maneuver. As the submarine surfaced, it struck the *Ehime Maru*, and within minutes of the collision, the *Ehime Maru* sank, killing nine of its crew members, including four high school students.

You may remember watching the news reports of the tragic collision between the USS *Greenville* and the Japanese fishing vessel, *Ehime Maru*, in 2001. Everyone felt great compassion for the victims and their families, but many in leadership positions could also sympathize with the submarine commander who, despite his own grief over the nine deaths and the ending of his career, found the strength to try to do what was right. Commander Waddle's book, *The Right Thing*, tells the behind-the-scenes story of how he felt left adrift at sea by the Navy in the aftermath of this tragedy with little direction, support, and inadequate legal representation. He and his family were feeling intense pressure from the guilt over the loss of life, the pain of being relieved of command, the expense of a mounting legal campaign to salvage his own life, being in the center of an international incident and media firestorm, and experiencing a degree of disappointment over the actions or inactions of the Navy he had loved and served for 24 years. Through it all, Commander Waddle took full responsibility for the accident and the performance of his crew. His request to apologize officially after the accident was unheeded, and yet he felt compelled to apologize personally to the Japanese government and to the families of the victims. This story shows how a leader can take the witness stand at great peril, against the strong advice of his lawyers. The paradox is that his actions after the event upheld the honor of the Navy and remind us that, even in the face of adversity, leaders must do the right thing.

The following are excerpts from a speech given by Commander Waddle (Ret) at the National Fire Academy (NFA), National Emergency Training Center, on June 1, 2005.

> The topic is "Leadership is a Choice, not a Position," and failure is not final. In our lives we have setbacks, we have disappointments. Those disappointments, although they are defining moments in our lives don't necessarily have to define who we are as people. What occurred on February 9, 2001, was an incident that will forever remain not only with me but it altered and shaped and changed the lives of nine families. Four teenagers, four 17 year old students, two of their instructors, ages 33 and 35, and three crew members aboard a Japanese fishery training vessel called the *Ehime Maru* . . . 150 feet in length, 500 tons displacement, about half the length of the USS *Greenville*, got underway on a Friday afternoon from Honolulu harbor heading out to sea returning back to Japan and without knowing it, was in a perilous position, especially when a submarine came up underneath in a emergency surfaced maneuver, collided with the vessel and a rudder cut a hole 30 feet in length, three feet diagonally to the underbelly of the vessel causing it to sink in three minutes. Four teenagers, two instructors, three crew members perished in that accident.[6]

Waddle's willingness to share his experience and the leadership lessons learned from this tragedy are reflective of many of the challenges facing those in leadership positions today.

It is a very interesting time in the fire service. Although leadership has never been easy, the complexities faced today require a leadership skill set that is quite unique in the history of the fire profession. The generational issues, financial challenges, increasing demand for services, and increased focus on government transparency leave very little latitude for mistakes. Many chiefs today, especially in time of resource reductions, find themselves in the middle of competing interests, between elected officials and labor groups. The elected officials had to make tough decisions to balance the budget, whereas labor groups attempt to protect what they have. In the center of the dialogue is the chief fire officer who must make the recommendation to the governing body and live with the resulting fallout from the various labor groups; at times such as these, a leader's integrity, character, and **resilience** will be fully tested. As Waddle states, "Those leaders

resilience
■ The ability of a system or organization to adapt itself to a changing environment.

who survive and are successful admit when they are wrong and have the ability to overcome adversity." Successful leaders instill a viewpoint into the culture of their organizations, which is one of their personal core beliefs and that is, Failure is not an option, nor is it final!

From Success to Significance

As noted in Scott Waddle's story, in life, it often takes experience and going through personal and professional challenges to make us humans reflect on and realize what is important. The fire service is a unique group because the public and the elected officials see those in the fire service as very special, successful individuals. Sometimes many in the profession let that perspective go to their heads. Let me share an experience that helped me reflect on what is important. At the time, I did not fully appreciate the message, but, over time, I came to realize its importance.

It was shortly after September 11, 2010, toward the end of that month; I was the incoming president of the IAFC. On September 11 and for the weeks that followed, the fire service was engaged with the rescue and recovery efforts in New York and at the Pentagon, with what was going on nationally, and having conference calls with the White House, with FEMA, and with many others. The fire service was front and center of the dialogue, and these phone calls were pretty important stuff; at least I thought so.

At the end of that month, my secretary came in and said, "Chief, there's someone in the lobby who wants to talk to you." I walked out and there was an 8- or 9-year-old boy with one of those giant pickle jars, probably a 10- or 15-quart pickle jar, stuffed with money. It was all he could do to hold onto it; even though his mother was with him, he would not give it to her to hold. When I walked up and introduced myself, I noticed he had the "We Will Never Forget" symbol pinned on his shirt. He asked, "Are you in charge?" I said, "Yes, I am." He said, "I want to give you this money. Will you make sure the firefighters in New York City, their families, and their children get this money?" I said, "We will absolutely do that." Then he just turned around and began to walk out. I said, "Thank you." I asked his mother, "Where did he get this money?" She said, "He had watched what happened on September 11, and he asked if he could take his money out of his piggy bank to help. He went to the store, bought a bunch of candy bars, and began selling them to raise money." There was $187.50 was in the pickle jar. Why did he do this? Why did he take it upon himself to get involved, make sure the money was brought to us, and ask that we ensure it got to the right place? I think it is because that boy had a servant's heart.

The fact is that almost everyone in the fire service I have met has a servant's heart. When I got back to my office, I reflected that it had been a very emotional event to have experienced. However, I really did not understand the significance of that event—and what that young man's giving of himself meant to me—for almost a year, until I had actually assumed the presidency of the IAFC. I kept reflecting on that 8-year-old boy, the challenge he undertook, and the reason he did it. Was he a successful person or a significant person? At age 8 most are not considered successful, but he sure had significance. Then I started to look at myself, as I was going through the process of being the IAFC president, being able to do some pretty unique things, and reviewing my own career; yes, I would be perceived as being successful.

The reality is we sometimes let the rank or the positions we attain define our success. How many of you have had this thought? When I reached the rank of fire chief at an early age of 34, I considered it as a success. I now have been the chief of five departments, president of the IAFC, president of the CPSE board of directors, and have been involved with a lot of other stuff. At the end of the day, what does it all mean?

The successes we have are moments in our lifetime of noted accomplishment. Yet, what makes the difference is if what one has contributed lasts beyond a lifetime. It is with this thought I want to challenge you as we complete this book. Sometimes when we put on these uniforms we wear every day—and I love mine, I love being fire chief, and I love putting on my uniform and being recognized as a member of our profession—it lends itself to helping each of us forget what is really important. People really love the fire service. I can tell you the respect goes all the way to the White House and Capitol Hill. But is this really what our success is about? Think about it in the context of your own career. Do you view yourself as a fire officer or fire chief in your own community? Do you put yourself on a pedestal as do other people? Is this the success you are trying to gain in your life?

Dennis Waitly has written numerous books and lectured internationally. As a very inspirational and articulate speaker on the importance of values and leadership, he talks about the real leaders in business, in the professional community, in education, in government, and in the home. They all seem to draw upon a special cutting edge, separating them from the rest of society. The winner's edge is not in a gifted birth, in a high IQ, or in talent. The winner's edge is in the attitude, not aptitude. I do not know about you, but I would fall into that category. I have never assumed I was the sharpest tack in the drawer nor do I have the most talent. You may be a rocket scientist; I'm not. What I am is driven, and I have a really good attitude. Think about the people you work with every day. The fact is, when you walk into a fire station, you can often predict whether it is going to be a good visit or bad visit before you walk through the door. What drives this? It is not because the firefighters you are meeting have a low IQ; they are often the best of the best. It is not because they do not have a lot of talent; in most cases they are our best people, the most talented in society. What drives the difference? It is the attitude. It impacts how we live our lives, our organization, and our influence on others.

Like many of you, I have traveled around and visited a lot of fire departments. In doing so, I can just see how people react—from the secretary to the chief, the deputy chief, the firefighters on the line, the mechanics—and within a very short period of time, I can pick up the organizational attitude within the group. All I have to be able to do is step back and observe. Honestly, there is no way to hide it for very long. Attitude is absolutely one of the most important tools we have to propel our careers, and it dictates where we will be in the future. If our attitude dictates our altitude, then our *significance dictates our success*.

What is significance? Let us begin by asking, what is significant to you? Each of us will have a different definition. So, what is the significance of your life? Does putting on a uniform every day make you significant? Does it make you a good person or a person of success? Although you have worked hard to have the right to wear this uniform or hold your rank to be seen as successful, are you having significance on others? This is quite a different question, is it not? At the end of the day when we all die (by the way, we are all going die), our success will not be measured by the uniform we wore but by the significance we have had on other people. Jackie Robinson once said, "A life isn't significant except for the impact on others' lives." This is what I learned the year of my term as the IAFC president. The position afforded me the opportunity to do some cool stuff: I met with Secretary Tom Ridge three times. I went to a White House Christmas party and testified to congressional committees—pretty amazing. Not everybody gets to do these kinds of things. I met a lot of important folks, went to places most people do not ever go to, but at the end of the day does it mean anything? The accolades, the position, and the access that comes with it do not mean I am significant; they just indicate I had some momentary successes.

When I am introduced to speak today, most in the fire service probably do not know I was president of the IAFC and most probably do not care; this is OK. Yet, there is usually someone in the room I have had significance on as I have helped his or her career, just like others have helped me. John Maxwell calls it applying the law of legacy to one's life. So, what do you wish your legacy to be? When you reflect on your career, you may realize it may not have been the most successful person in your community who has had the biggest impact on you. It may have been the basketball coach who took the time to say you are not the greatest player in the world, but I like your attitude. Or the English teacher who said you really are not the best writer I have had, but if you come in after class, I will help you. For me, it was my Geometry teacher, who said I do not know whether you are ever going to get this, but I am not going to fail you. I got a D, but she did not fail me because she said I was working hard and trying my best. These people can be significant in our lives and their influence can make a difference in us that lasts a lifetime. They leave a lasting imprint on us that can never be erased!

For many, success is a diploma on the wall or the collar brass we wear; however, significance is the phone call from someone you invested your time in and supported who says thank you for helping me because now I have the job I always wanted. It is the colleague who says thanks for helping me because if you had not given me a piece of advice or shared some of your wisdom with me, I would have made a really bad mistake in my organization, which would have cost us a lot of money and perhaps my job.

As mentioned previously, it always starts with attitude. John Maxwell lists several extremely insightful factors about attitude. Attitude is the advance man of our true shelves. Every time you meet somebody, you can tell fairly quickly what kind of attitude he or she will bring to the table. Have you found this to be true? Certain people walk into a room and suck the life out of the room. Other people walk into a room, and it lights up; they have a big smile on their faces, even if they do not have all the answers. You are happy to see them. They are just effervescent, and it bubbles over and has a positive impact on individuals, on the crew, and on your organization. You want to be next to those people because they are fun to be with and have a constructive influence on others and their environment.

Attitude's roots are inward but its fruit is always outward. Whether a person has a bad or a good attitude, it will ultimately become evident. Whatever is inside is going to come out; it cannot be kept hidden. Attitude is our best friend and our worst enemy. How many times have you dealt with people whose attitude took them down a path they should not have been on? They just could not help themselves. Most of it can be attributed to their attitude. When you are dealing with personnel issues, step back, look at the situation, and think about the why. What is the motivation for what you are dealing with? Many times it starts with the attitude of the individual. It is more honest and more consistent than our words. It is an outward look based upon our past experience.[7]

One of the things I find interesting in life is people who have poor attitudes; nothing really makes them happy. Unfortunately, some in the fire service fall into this category. This is true even though firefighters have a pretty good job with a pretty good wage and benefit package; most would say firefighters have the best job in the world. We have the highest respect rating of any profession in the country. What else do we need? Yet, everyone runs across people in the fire service who say the organization or the city is a mess; or they were not paid enough, did not have enough time off, were not respected enough, or were just getting the bad end of the stick here. I often wonder whether those who are not happy doing this job could possibly be happy doing anything. What are these people like on the home front? What is their impact and significance on their families? What types of seeds are they sowing for their own future and for those around them?

Our attitude is the one thing that draws people to us or repels them, and it is never content until it is expressed. When I go into firehouses and do station visits, it is not uncommon for at least one person not to say anything for a long time and then all of a sudden launch off. After everybody has had a lot of constructive dialogue, he or she just cannot hold it in. The bad attitude will always come out. It is the librarian of our past; it is the speaker of our present; but, more importantly, it is the prophet of our future. The attitude we bring with us—not once a month, twice a month, but each and every day—dictates what our future is going to be.

A good example of how attitude is infectious can be found in the following story of a young employee at a grocery store. While leading a customer service program for a grocery store chain, Barbara Glanz touched on the concept how everyone can **make a difference** for the customers and make them feel special. When an employee can do this, customers will want to come back. Shortly after her presentation, she received a phone call from a 19-year-old bagger by the name of Johnny. Johnny, who has Down syndrome, had an idea of how he could make a difference for his customers—by sharing a thought for the day with his customers.

make a difference
▪ To cause a change in effect or change the nature of something in a positive way.

Each night with his father's help, Johnny would choose or make up a positive thought for the day. His father would print multiple copies on their home computer, and Johnny would cut out each quote. The next day, when he was done bagging each customer's groceries, he would stick his thought for the day in one of the bags. It did not take long for Johnny's customers to become infected by his attitude.

His checkout lines were often three times longer than anyone else's; and even when other lines were open, customers would wait for Johnny and his thought for the day. Johnny, a person who may have been perceived by some as having minimal influence, had transformed the entire store. His story, now on video, which I would encourage you to see, has influenced the millions who have watched and been inspired by his heartfelt commitment to service.[8]

It comes down to the fact that it is the little things that make a difference, and every little thing counts. Just look at this story of Johnny, a person who was not at the highest level of the organization, did not really have a lot of responsibility, and yet he changed the entire focus of the store and the organization that he worked for. He took it from being just another place to buy groceries to being an experience. Look around and notice the people within your organizations who are doing this today. They are there, and we each need to ask whether we are doing what is needed in our organizations to have this same level of influence. In many cases we are not because we have lost sight of the fact that these little things do add up. In Johnny's case, we can see how he actually transformed the entire store and the way it thought about customer service. Does attitude really make a difference? Absolutely!

Good leaders realize that life is a team sport. Typically, one does not get into positions of influence without the help of our parents and then, as we move through life, our spouse, significant other, kids, friends, and all the people who have helped us along the way. So often, as we ascend in our careers, we totally forget those personal investments that have helped shape what we have become.

Attitude is not only about how we approach our work but also about how we approach our family and our life, which are both very critical. It is not just about organizational importance; it is about the influence we have on others as we attempt to fulfill our life's dreams. Many of you are likely sports enthusiasts. Some teams have great talent on paper but when they get on the field perform poorly. Why is this? Is it because they could not get along? Did everybody want to be the star or be out in front and not play as a team? Even teams with great talent are not going to perform well if they have a rotten attitude. Teams with great talent and a bad attitude will be an average team; teams with great talent and an average attitude will be good team; and teams with great talent and a good attitude will be great teams. Last, but not least, average talent combined with a great attitude will produce a great team (see Table 11.1).

In 2008, the Fresno State Bulldogs men's baseball team exemplified this. It was one of the lowest ranked baseball teams in the West Coast league entering the playoffs. I don't know whether they had expected they would have the success they achieved; but, as they began their journey, they started to believe. Their attitude was very inspiring because it was not about them; it was about the team doing whatever it took from a team perspective to win. When I had the chance to talk to the team after they had returned home as champions, the attitude was the same.

TABLE 11.1		Abilities, Attitudes, Result		
Abilities	+	Attitudes	=	Result
Great talent	+	Rotten attitudes	=	Bad team
Great talent	+	Bad attitudes	=	Average team
Great talent	+	Average attitudes	=	Good team
Great talent	+	Good attitudes	=	Great team
Average talent	+	Great attitudes	=	Great team

Source: John C. Maxwell, *Attitude 101: What Every Leader Needs to Know*, Maxwell Motivation, Inc. (Nashville, TN: Thomas Nelson Publishers, 2002).

It was absolutely incredible. They got it. It was about the adventure. It was all about living the experience as a *team*. It was about winning as a group, not achieving individual accolades. They went from underdogs to wonderdogs. Pretty amazing! Compared to other teams, they had average talent; only three or four players had professional baseball potential and the rest will be or are doing other things in life after college. Many knew they were not going to play baseball after the next couple of years. But they had a great team, and the reason was their attitude drove them to be their best and win the College World Series.

Success is often about knowing the critical elements that drive each of us, knowing our **purpose** in life. When did you know the fire service was your calling? Everybody is different. Did you really know it was your calling, or did you just happen to fall into a good deal and learn to love it? Some people do this. Sometimes life just happens.

Everyone needs a purpose to which one can give the energies of his or her mind and the enthusiasm of his or her heart. Stop, sit back, and ask yourself the following questions: Is this your purpose or is it your job? What actually drives you? What really turns you on? What is the thing you really want to put your energy into? Is it part of your inner being that truly floats your boat? What is it? Is it being a fire chief? Many fire chiefs and mentors of young officers coming up share a long-term concern that many do not see being in senior leadership in the fire service as part of their future. We have to change this, or the survivability of the fire profession is going to be based on the leadership of the past or those who take the job but are not inspired by it. Not a lot of people seem to have the big desire to be the chief executive. We have to look at the fire profession and ask why. What really motivates you to put a lot of energy, attention, and heart and soul into something? What is it? Has this changed from the time you were age 18, 25, 35, 45, or 50? Regardless of the generation you come from, has it changed for you? Are you still plugging along at the same thing you were doing 15 or 20 years ago because it is what you have always done? However, perhaps in your heart you know that maybe it is not what you should be doing. Each of us has struggled with this question during our careers. The answer may be as simple as success opens up more opportunity. Sometimes you need to ask yourself, Is what I am doing today what I really want to do? We have all had this discussion with ourselves, to explore other opportunities whether it is in the fire department, in the private sector, or in federal government. These can be very difficult decisions to make, and it all comes back to having a discussion with oneself about what is right step. The basic question: Is your soul in it or not? I do not care what job you have once you get there; it is hard to do it if it is not in your heart.

Doing what you love to do is part of what drives having a great attitude. Another critical question to ask is, What am I searching for? Many people seem to be on a constant search for the perfect job, perfect boss, perfect __! All of us have a strong desire deep within us, something that speaks to our innermost thoughts and feelings, that really motivates each of us. What is in it for you? We all live under the same sky; however, we do not all see the same horizons. Think about it. We can see that with the people whom we work with in dealing with the same issue but from vastly different perspectives. There is an old saying: you cannot mistake the edge of the rut for the horizon. As leaders, our role often is to lift people up so they may see over the ledge, so they can understand there is a much bigger horizon out there, which sometimes they do not allow themselves to see.

purpose
■ Expected achievements if the organization is successful in completing its mission. It can be expressed in either qualitative or quantitative terms, within the parameters to be able to objectively verify them. That which we hope to create, accomplish, or change with a view toward influencing the solution to a problem.

Have you ever allowed your perspective to taint the vision of your horizon? Have you ever been in a situation—whether a relationship at home or at work, a relationship with your kids, spouse, girlfriend, boyfriend, or boss—where your perspective was like looking with two eyes through a keyhole? This vantage point provides a pretty limited view, does it not? Well, I have been at that keyhole on many occasions. However, when I step back and finally get through with the emotional issues I was dealing with and reflect on the bigger picture, I find myself asking: How did I get so wrapped around the axle over this issue? Sometimes our emotions, lack of information, or ego can take us down a path we should not be on. Remember: the horizon is much bigger than we can see much of the time. Although individuals are all different, we each have opportunities if we allow ourselves to look at the horizon to see those opportunities instead of limiting ourselves by a fixed vantage point.

I love this quote by the late actor Danny Kaye: "Life is a great big canvas and we should throw all the paint on it we can." The question is, Are you throwing all the paint you can on your own canvas? A great book by Bob Buford is called *Halftime: Changing Your Game Plan from Success to Significance*. The book highlights the move from success to significance in the context of the four quarters of a football game. Many people live their entire life as if they were in the first half. The first half is about achieving success by paying your dues and doing all the things you need to do to be successful; but what do teams do when they go in at halftime? They regroup as they evaluate the first half, and they ask, What are we going to do in the second half, and what do we need to change in the second half to win the game? Do we need to change the defense? Do we need to change up the players? What will it take to win the game? This football analogy can be applied to the fire service. Most, not all, of us are eager to enter the second half. When you go into halftime, what are you going to think about for the second half? I would propose you think about moving from what has made you a success to how you are going to be significant in the last half. At the end of the game, the goal is to be able to say not only that you had a life of success but also that you had a life of significance.

This is a different discussion and requires a revision in our playbook for the second half. Just because we are all successful, does not mean we have all gained significance. As you enter the locker room at halftime, start to ask yourself the following: How do I gain significance in the second half of my life? Where do I go from here? What do I need to be different? Unfortunately, many people go through life and at the end of the day and on their deathbed they say, "I wish I would have done this, I wish I would have done that, I wish I would have kept in touch with my mentor, wish I would have, I wish, I wish, I wish." Have you witnessed this experience with someone close to you? I think many of us have. Our family lost a good friend about a year ago. Talk about someone who did not leave anything on the table, it was she. When they had the funeral, 4,000 people showed up from all over. She was a Christian woman who went above and beyond to help people, be a friend, and do whatever it took to make a difference in others' lives. It was the most inspiring hour and a half I think I have every sat through. Sometimes at funerals the testimonials are little bit trite; you can tell they are not really from the heart. However, the testimonials to this woman were totally from the heart. You could see she had a life of significance for 20 to 25 years. This was because she deliberately chose to be significant in how she lived.

Which path are you on? At the end of the day, what are people going to say about you? I know I think about this a lot. I look at my kids and wonder, If I get old and grumpy, they may say I am old and grumpy now; and when I get older and grumpier, what are they going to say? What have I brought into others' lives that is significant? What are my colleagues going to say about what I did in the organizations I was a part of? What are my friends going to say about what I did for them, helping them in their lives? What is my wife going to say about the type of husband I was? Often we do not ask ourselves these very, very difficult questions until it is too late. How many of you have policies written 25 years ago because somebody screwed up, and everyone who was part of the issue is gone, but the department still maintains the policy. This is a classic example of how we and our organizations carry baggage around forever. We often see it in our labor–management relations. Sometimes people cannot let go of something because the former chief or former labor president did something really stupid, inappropriate, and did not bring the right attitude to the discussion and now we are living with it 5, 10, 15 years later. Although it happened under somebody else's watch, we are still dragging it around with us.

As leaders and as people who want to move from just being successful to having significance, we have to let go of the baggage because a person cannot run a marathon with two five-pound balls strapped around his or her ankles. It can be very tough to let go because at times we are so emotionally vested in the issue that we cannot move beyond what happened in the past. This creates scar tissue, as I call it, that we look at every day and is a constant reminder of what happened to us. It can keep us from being able to focus on our future. It is just as simple as that. The past is a dead issue, and we cannot gain any momentum moving toward tomorrow if we are dragging the past behind us.

As I watched the campaigning in the 2008 presidential election, the dialogue was disturbing. Such behavior of candidates tearing down one another in the name of politics will leave scar tissue for years to come; and, unfortunately, it has fueled more of the same destructive dialogue in the next cycle of elections in 2010 and likely will become the norm. It can also be seen at the state and local levels, because there is a trickle-down effect on what is acceptable. It is not healthy for the country or its citizens because it perpetuates a negative mind-set. This is happening in fire service organizations as well. For some it is not about a position on an issue, but about tearing down the other side enough so one's side can win. Leaders can do this, too, in their organizations, whether it is the chief or the informal leader at Station 3. Although there can be no success or significance without an attitude of sacrifice, I think the sacrifice being referred to does not mean sacrificing other people's feelings and/or dignity.

Rusty Rustenbach, in his pastoral article, "Giving Yourself Away," writes, "Today, we live in an age when only a rare minority of individuals desires to spend their lives in pursuit of objectives which are bigger than they are. For many people, when they die it will be as though they never lived."

The people in our organization whom we lead are often directly reflective of our leadership. If leaders are not prepared, if they are not well rounded, and if they are not focused on those work issues for our organizations, then everybody does not go home safe and happy.

John Maxwell says it best: "Success is when I add value to myself; significance is when I add value to others." When I went in for halftime, I made a choice. I said we are going to play the second half with the same team, but we are going to change the plays. I am not going to focus so much on success as I am on significance. I made this choice because, at the end of the day, I want people to look back and say, you know what, he made a difference in my life. I am already successful, and I can check the box off, but at the end of my adventure, other people will not remember this. What people are going to remember are the times when someone made an impact of significance in their lives, and this will last forever. So where are you in the game? In the second quarter? At half-time? What are you thinking about as you come out of the locker room after halftime? Here is a great example of someone who chose to be significant over just being successful.

When Rick Hoyt was born in 1962 with the umbilical cord wrapped around his neck, cutting off the oxygen to his brain, resulting in his being born as a spastic quadriplegic, nonspeaking person. His parents, Dick and Judy, were told there would be no hope for their child's development. When he was eight months old the doctors told the Hoyts they should just put him away—he would be a vegetable all his life. However, the Hoyts were determined to raise him as "normally" as possible, and within five years, Rick had two younger brothers. While the Hoyts were convinced Rick was just as intelligent as his siblings, getting the local school authorities to agree was not as easy. Because he could not talk, they thought he would not be able to understand, but that was not true. Because the Hoyts had always wanted Rick included in everything, they wanted to get him into public school, and ultimately they were successful in doing so.

With the assistance of a group of Tufts University engineers, once they had seen some clear empirical evidence of Rick's comprehension skills, they built an interactive computer that would allow Rick to write out his thoughts using the slight head movements that he could manage. Rick came to call it "my communicator." A cursor would move across a screen filled with rows of letters, and when the cursor highlighted a letter that Rick wanted, he would click a switch with the side of his head. When the computer was originally brought home, Rick surprised his family with his first "spoken" words. They had expected perhaps "Hi, Mom" or "Hi, Dad," but on the screen Rick wrote, "Go Bruins." The Boston Bruins were in the Stanley Cup finals that season, and his family realized he had been following the hockey games along with everyone else.

In 1975, Rick was admitted into public school, and two years later, Rick told his father he wanted to participate in a five-mile benefit run for a local lacrosse player who had been paralyzed in an accident. Dick, far from being a long-distance runner, agreed to push Rick in his wheelchair. They finished next to last, but they felt they had achieved a triumph. That night, Dick remembers, Rick told his parents that he just did not feel handicapped when he and his father were competing. Rick's hope to participate provided a new set of opportunities for him and his family, as "Team Hoyt" began to compete in more and more events. It is hard to imagine now the resistance that the Hoyts encountered early on, but attitudes did begin to change when they entered the Boston Marathon in 1981, and finished in the top quarter of the field. Dick recalls the earlier, less tolerant days with more sadness than anger. Most of all, perhaps, the Hoyts can see an impact from their efforts in the area of those with disabilities, and on public attitudes toward the physically and mentally challenged. Today Team Hoyt is an inspiring story of a family that chose to be significant in the lives of their children and have become an inspiration for many throughout the world.

Rick's own accomplishments, quite apart from the duo's continuing athletic success, have included his moving on from high school to Boston University, where he graduated in 1993 with a degree in special education. Rick now works at Boston College's computer laboratory helping to develop a system code-named "Eagle Eyes," through which mechanical aids (for instance, a powered wheelchair) could be controlled by a paralyzed person's eye movements, when linked up to a computer.

Together the Hoyts do not only compete athletically; they also conduct motivational speaking tours, spreading the Hoyt brand of inspiration to all kinds of audiences, sporting and nonsporting, across the country. Their story is captured on video and is a must see. The message of Team Hoyt is that everybody should be included in everyday life.

This is a pretty amazing story. The Hoyts have certainly lived a life of significance. The question is, Where are you in your life? Are you successful, significant, or neither? We live in an age in which only a few know where they are going. Think about this. A lot of people in our organizations may fit this description. We are so focused on career success that we do not recognize the importance that significance plays in our lives. I wish somebody would have had this discussion with me when I first came into the fire service, because I would have viewed my career quite differently. I do not know whether it would have changed my path, but I think I would have appreciated my success along the way a little more and appreciated the people at the time who helped me to be successful. It would have helped me share with my employees to appreciate that all aspects of our lives are important. The truth is that we make a living by what we get; we make a life by what we give.[9]

Maybe the most important question to ask yourself about your job is not what you are getting, but what are you becoming?

As important is this question: Are you the person you want to be today? There are some days when the city manager calls with an urgent request from the mayor or a council member that needs to be addressed immediately. This may not be a crisis from your vantage point, but it is from his or her viewpoint. Sometimes you are perplexed or frustrated at the level of energy and amount of time spent on such issues, some which are mundane at best. Then about a half hour later you get a call from dispatch and are called to a three-alarm fire; and you watch the firefighters work. You quickly realize this is why you put up with all the other stuff. We get up each day, put on our uniforms, and go to work to make a difference. When the alarms go off, as these young men and women are getting on the fire trucks heading out the door, it is the work we do every day to help prepare and train them that will make a difference in how they perform and in keeping them safe. Every day we have the opportunity to give back to everybody who works for us, hopefully in a positive way, which is what is so rewarding about being a chief officer who has chosen to be significant.

What are you bringing with you that makes you different from everyone else? (See Figure 11.2.)

While pursuing an elected position in the IAFC, I would see people who would run for the IAFC presidency who really had no **priorities**. The only reason they were running was to say they had been president of the IAFC. This is the

priorities
■ An authoritative rating that sets precedence and order or a sequence.

FIGURE 11.2
Attitude

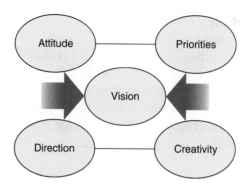

wrong reason, but we have all seen people seek and take positions for the wrong reasons. We also have people who do this in our own communities as elected officials. They could care less about the job they do once they get there. It often becomes about the power, the notoriety, the position—all are the wrong priorities. In a life of significance, you have your priorities in order. You need to have a direction where are you going in the future. Whether you are in the first quarter, second quarter, halftime, or third quarter, know where you are going. It is also about **creativity**, about looking at things a little bit differently, and about the ability to see the whole horizon and not the edge of the rut.

Janet Wilmoth of *Fire Chief* magazine wrote a great editorial regarding the Charleston, South Carolina, fire service entitled "The Naked Chief." I copied it and gave it to all my chief officers. It spoke to the blind spots that we all have in our organizations and we have as leaders. Sometimes we have been doing things the same way for so long we do not see they are not working very well anymore. We have lost sight of the whole horizon. So part of what is very important is not only to have confidence in what kind of job we are doing, but we are open to look at things differently so that we do not put ourselves, our organizations, our firefighters, and our families in positions that we should not. This is a difficult task as many of us in the fire service are Type A personalities, like to be in charge, and know what we want. We are a very unique group; we can process from 1 to 50 very quickly and we know at the end of the day what direction we want to go in. These may be good traits to be successful, but these can also limit our ability to be significant.

Separating significance from success falls into three areas. First, the motivations for success are about you, numero uno! The motivation for significance is about other people. Second, with success your influence is rather limited, but with significance it becomes unlimited. Third, your success will last no more than your lifetime. We have lost several key fire service leaders this past decade, and many in our profession probably will not even remember they are gone. Yet, the people in their lives whom they touched will know they are not here because they had significance on their lives. Success lasts only a lifetime; significance can be forever because the impact you have on others will trickle down to their children, their grandchildren, their grandchildren's children, and so on. You can leave a lasting legacy. Vince Lombardi once said, "The quality of a person's life is in direct proportion to their commitment to excellence, regardless of their chosen field of endeavor." Is it not true? In committing to excellence, we should commit not only to the success of our organizations and to our careers but also to excellence in having a life of significance.

creativity
■ A mental and social process involving the generation of new ideas or concepts or new associations of the creative mind between existing ideas or concepts. The process of either conscious or unconscious insight fuels creativity.

PEARSON
myfirekit™

Visit MyFireKit Chapter 11 for the Perspectives of Industry Leaders.

Conclusion

All those in leadership positions should strive to achieve significance in their lives as it truly provides the ability for them or those in positions of influence to have maximum impact.

One thing I started to do several years ago is thanking those who had an impact on my life and making sure I called them to do so. In doing this, I received a lot of feedback. In fact, one of these people was the city manager I had worked for in Illinois. He said he really appreciated my doing that. He said he always wonders how I am doing. I shared with him why I called, thanked him for his mentorship, and let him know that he was important to me. I still use much of what he taught me today. It was important to me to make sure he was doing well and to let him know how much I valued his help. I believe these people did not invest in you to send you off and never hear from you again. They care about you. Make the call and tell them how you are doing; share with them the reason you are doing so well is because they had significance on your life.

A life of significance does not look the same for every person. Every one of us is different. We bring different skill sets, varied experiences, and unique personality characteristics to the mix. Some of us grew up very poor, some wealthy; some of us had very stable families and some of us did not. From small to large families growing up in different areas of the country, we still have a common bond as leaders in the fire service. Yet, as we all strive to be the best leaders when we talk about excellence, we must remember it is often the little things that count the most. In our positions, when we take the time to reflect, we are often at 30,000 feet, going 600 miles per hour, and forget that if something falls off the plane it will be gone forever. Do not be so focused on being the biggest fish in the pond or be going so fast that you forget the small things that really make a difference. What is of enduring worth is caring about others, not in a superficial way, but in a real way, as proven by your investment of time in others. People will recognize when you invest time in them; they will invest time back into you, and that really is the leadership push we often miss. It is not about our rank nor about who we are and the successes we have. Instead, it is about the investment of time. It is about the person who consistently tries to improve him- or herself. This occurs as we challenge ourselves to get out of our comfort zones. How many of you are in your comfort zone right now? Have you been doing the same thing over and over again? Although each of us has been there at times, it is crucial to continually push ourselves out of it. There are chiefs, captains, and firefighters I have run across who are basically retired on the job, which is extremely unfair to the people they work with and the people they serve.

When I graduate new firefighters, one of the things I share with them is that the badge they are about to receive comes with a great amount of responsibility. In having chosen to enter the fire service, and in accepting the badge as a firefighter, they accept with this the public's respect, appreciation, and expectations of public servants who will be looked upon as heroes in their local community. With this badge and the uniform also comes an opportunity to make a significant difference in the lives of a great number of people. You can see it as firefighters drive down the street and the reaction of the general public, especially children, have as they are looking in awe at the fire truck; you can see it in the waves and smiles and the respect our profession is given. This respect opens many doors for us to make a very positive impact on people and not just from an emergency response perspective. It is about the little things we do when we are out in public wearing the trappings of our profession: stopping to say hello to a young child in the supermarket, stopping along the roadway to help someone who may be in need, and taking a little bit of time after an emergency call to straighten up the area. All make a lasting impact on the public perception of what our profession does and who we are as individuals. We have an opportunity to make a positive difference on so many through the course of our careers. We can choose the attitude we bring to the job every day. We can choose whether this attitude will be negative or positive. Know that the choice you make, especially if you are in a position of leadership, will impact and influence those who work around you and will reflect on the profession when you engage the public. If our attitude dictates our attitude, then our significance will define our true success at the end of our careers.

My challenge to these new firefighters is as it is to you. Choose to make a difference for yourself, your family, your profession; choose to make a difference in a positive way and have an impact on people you work with and those you will influence as you continue to have a successful career in the future; and choose to have a life of not only success but also significance. This is motivated by the positive impact you can have on others, influenced by the desire to leave a legacy long after you are gone. Recognize that if you are to truly grow to your potential, then you should not go to the grave with a life unused. Live each and every day to the fullest and choose to make a positive impact on others' lives, and when your journey is complete, when you have done this, you will have chosen a life of significance.

Be safe.

Chief Bruegman

Review Questions

1. Describe the leadership lessons from *Apollo 13*.
2. Describe the leadership lessons of the *Ehime Maru* and USS *Greeneville* collision.
3. How would you define success for yourself?
4. How would you define significance for yourself?
5. What are the challenges your organization is facing today? What path is your organization setting for itself for the future?
6. Write your vision of what the fire service will be in 25 years.
7. Describe an adversity or significant challenge you have faced, and explain how you handled it. In retrospect, would you have done anything different?

PEARSON
myfirekit™

For additional review and practice tests, visit www.bradybooks.com and click on MyBradyKit to access book-specific resources for this text!
Register for MyFireKit by following directions on the MyFireKit student access card provided with this text. If there is no card, go to www.bradybooks.com and follow the MyBradyKit link to buy access from there.

References

1. Jim Kouzes, *Leadership Challenge: Adversity Introduces Us to Ourselves,* May 13, 2008. Retrieved October 23, 2010, from www.leadershipchallenge.typepad.com/leadership_challenge/2008/05/adversity-intro.html
2. Stephen Cass, *Apollo 13, We Have a Solution,* IEEE Spectrum, April 13, 2005. Retrieved October 23, 2010, from http://spectrum.ieee.org/aerospace/space-flight/apollo-13-we-have-a-solution
3. NASA Apollo Mission Apollo-13. Retrieved October 23, 2010, from http://science.ksc.nasa.gov/history/apollo/apollo-13/apollo-13.html
4. HeartQuotes™: Quotes from the Heart. *Leadership Quotes and Proverbs.* Retrieved November 1, 2010, from http://www.heartquotes.net/Leadership.html
5. Chuck Burkell, Opening Remarks, EFO Class, National Fire Academy, Emmitsburg, MD, 2007.
6. Scott Waddle speaking at the National Fire Academy, Emmitsburg, MD, 2007.
7. John C. Maxwell, *Attitude 101: What Every Leader Needs to Know,* Maxwell Motivation, Inc. (Nashville, TN: Thomas Nelson Publishers, 2002).
8. *Johnny the Bagger, Service from the Heart* (Sarasota, FL: Barbara Glanz Communications, 2010).
9. *About Team Hoyt, Racing Towards Inclusion.* Retrieved November 1, 2010, from www.teamhoyt.com

INDEX